Blood Relations

Blood Relations

Transfusion and the Making
of Human Genetics

JENNY BANGHAM

The University of Chicago Press   Chicago and London

PUBLICATION OF THIS BOOK HAS BEEN AIDED BY A GRANT FROM THE BEVINGTON FUND.

The University of Chicago Press, Chicago 60637
The University of Chicago Press, Ltd., London
Published 2020
Printed in the United States of America

29 28 27 26 25 24 23 22 21 20   1 2 3 4 5

ISBN-13: 978-0-226-73997-7 (cloth)
ISBN-13: 978-0-226-74003-4 (paper)
ISBN-13: 978-0-226-74017-1 (e-book)
DOI: https://doi.org/10.7208/chicago/9780226740171.001.0001

Library of Congress Cataloging-in-Publication Data

Names: Bangham, Jenny, author.
Title: Blood relations : transfusion and the making of human
    genetics / Jenny Bangham.
Description: Chicago : University of Chicago Press, 2020. |
    Includes bibliographical references and index.
Identifiers: LCCN 2020023703 | ISBN 9780226739977 (cloth) |
    ISBN 9780226740034 (paperback) | ISBN 9780226740171 (ebook)
Subjects: LCSH: Blood groups—Great Britain—History—20th century. |
    Blood groups—Europe—History—20th century. | Blood groups—
    Research—Great Britain—History—20th century. | Human genetics—
    History—20th century. | Human genetics—Research—Great Britain—
    History—20th century. | Blood—Transfusion—Europe—History—
    20th century. | Blood—Transfusion—Great Britain—History—20th century.
Classification: LCC QP98 .B36 2020 | DDC 612.1/18250941—dc23
LC record available at https://lccn.loc.gov/2020023703

♾ This paper meets the requirements of ANSI/NISO Z39.48-1992
(Permanence of Paper).

For my father, J. Andrew Bangham

Contents

Prefatory Note

In English-speaking contexts, the terms "blood group" and "blood type" are generally used interchangeably, though with a preference in the United States for "blood type" and in the United Kingdom for "blood group." Because this book is largely about events and people in the United Kingdom, I have chosen to use "blood group" and "blood grouping" throughout.

Blood, Paper, and Genetics

In July 1939, British citizens responded for the first time to a nationwide appeal for blood. War was threatening and the Ministry of Health hoped that a nationwide transfusion service would help mitigate the bloody effects of aerial bombardment. Responding to street posters, advertisements placed in newspapers, and radio appeals, tens of thousands of people in London, Manchester, and Bristol traveled to local hospitals to have their earlobes or fingertips punctured with needles. At recruitment centers, nurses took a few drops of each volunteer's blood into a glass tube, diluted it in saline, and passed it to a trained serologist, who determined the donor's "blood group"—a crucial measure to ensure compatibility between donor and transfusion recipient (figure 0.1). While nurses and serologists handled the blood, clerks filled out forms and index cards with donors' names, addresses, and general health conditions. A few days later, each volunteer received a donor card through the mail, color-coded by blood group, readying him or her to answer the call. Blood transfusion was not new—small-scale local enterprises had been operating in several countries for nearly two decades—but this was the first time the British government had directly appealed to its citizens for their blood. In a remarkable commitment to the nascent war effort, by the end of July, the Emergency Blood Transfusion Service had enlisted 100,000 people. Being a card-carrying blood donor was a novel way in which the British people could commit to the war effort.

As donors came together in this collective act of self-defense, scientists used the mass bloodletting for a new

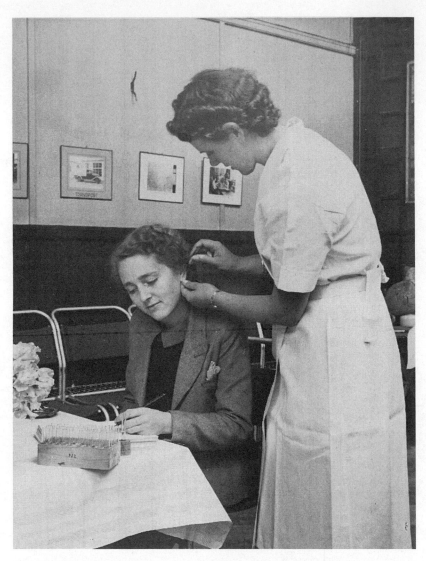

0.1 Photograph of a donor having a sample taken for a blood grouping test at the North West London Blood Supply Depot in Slough. A nurse in a white uniform stands over a potential donor to prick her earlobe and withdraw a drop of blood. On the table, next to a bunch of flowers and on a crisp tablecloth, sits a wooden block with test tubes for collecting small samples. Made as part of a series of publicity photos for the Emergency Blood Transfusion Service between 1940 and 1943, the image conveys the calm atmosphere of the depot and the serene demeanor of the donor. 21 × 16 cm.
Reproduced with the kind permission of the Bodleian Libraries, University of Oxford.

kind of genetics. A community of geneticists associated with Britain's Medical Research Council (MRC) was already engaged in a project to use blood groups to transform human heredity into a mathematically rigorous science. At the Galton Serological Laboratory at University College in Central London, one of those scientists was statistician and geneticist R. A. Fisher, who had recently been appointed professor of eugenics. Fisher believed blood groups might be used as prognostic tools for heritable diseases and as data for testing theoretical evolutionary models. When transfusion-service planners appealed to the Galton Serological Laboratory for urgent assistance in testing the July rush of volunteers, Fisher saw a magnificent opportunity to scale up his research. His serologist colleague George Taylor and other lab members began training hundreds of young women in the techniques of blood grouping; meanwhile Fisher and his secretary Barbara Simpson transcribed the blood group results from thousands of donor cards, transforming this clinical information into genetic diversity data. The London donors were unaware of it, but the scientists were turning their blood into a valuable resource for studying genetic diversity. In fact, they were taking part in one of the first large-scale surveys of human genetics ever undertaken.

This book explores how midcentury human genetics was built on the practices of extracting, moving, and transfusing blood. July 1939 was a special moment in the forging of this relationship. Since the 1920s, transfusion had gradually transformed from a perilous surgical procedure into a routine therapy. This was in part owing to the realization that the success of transfusion could be improved by paying attention to the blood groups of donor and recipient. As transfusion expanded its reach, donor registries grew and lists of blood groups swelled. Meanwhile, researchers interested in human heredity and eugenics gained an object to reckon with: in the 1930s, blood groups became widely understood as human traits inherited according to the clear-cut pattern predicted by the pioneer of genetics, Gregor Mendel. To many, the ABO groups—the first blood group system to be identified—represented the most promising path to mapping human chromosomes and understanding "race," and their study was highly prized by those who felt that human heredity needed a firmer footing. The abundant bureaucracy of the transfusion service offered the perfect material for this enterprise. Then, on the eve of war, transfusion and genetics became institutionally linked in Britain for the first time. Researchers studying blood group genetics became integrally involved in the practical work of the transfusion services, and these enterprises remained closely intertwined for the next twenty years. Wartime transfusion brought massive numbers of people into a bureaucratic system that was capable of defining and elaborating human genetic difference.

For the two subsequent decades, the transfusion services in Britain and around the world made large quantities of data available to researchers studying human heredity and diversity. Reciprocally, the study of genetic identity and inheritance contributed to significant advances in techniques for safely procuring and transfusing human blood. Fisher's lab was reincarnated postwar as two new laboratories at the Lister Institute of Preventive Medicine, in the London borough of Chelsea. The Blood Group Reference Laboratory was overseen by hematologist Arthur Mourant, whose talent for scientific management would make him a world authority on blood group population diversity. Next door, the Blood Group Research Unit was directed by Robert Race, whose warm relationships with doctors and serologists in Britain and the United States made him the leading expert on blood group genetics. Both labs carried out practical work for the transfusion services while furthering their genetic inquiries. These resulted in one of the earliest world archives of human population genetic data and the first detailed analyses of human genetic loci. As some of the first human traits that were known to be genetic, blood groups offered a vision of what human genetics could be: mathematically rigorous and drawing on large quantities of data. And all of this was created largely before human and medical genetics became highly visible fields in the late 1950s: before the structure of DNA, before chromosome changes were linked to complex bodily conditions, and before the structures of biological molecules were associated with inherited disease.[1]

Today many of us are familiar with the powerful narrative that genetics yields secrets of population identity, family relationships, and biological ancestry, and offers crucial predictions about our health.[2] This book relates how we came to understand genetics in this way. Modern genetics is not just a theoretical achievement or a triumph of experimental science: its origins lie in nationalism and midcentury politics, in the movement of materials and knowledge between the lab and clinic, and in the mundane realities of administrative work.[3] Reflecting on the early history of human heredity, which drew on the bureaucracy of the asylum, Theodore Porter reminds us to think of "the great filing cabinets of data from armies, prisoners, immigration offices, census bureaus, and insurance" that made its study possible.[4] Here, the midcentury transfusion depot plays a crucial role. This is a history of genetics in which blood, bodies, and bureaucracy take center stage.

Britain was an important site for this kind of human genetics. As routine blood transfusion got underway in the 1920s, this was a country at the center of the world's largest empire—a network of wireless and tele-graph communication, shipping lines, trade links, an administrative civil service, and colonized people. Even as that empire was in decline, its gov-ernment still had a keen belief in Britain's central status within this vast periphery, and the roles that science should play in keeping it there.[5] Postwar, through organizations like the United Nations, Britain's scien-tists expressed confidence in their ability to create a rationality that suited a postwar internationalist world order.[6] It was also a state with an es-tablished culture of technocratic voluntarism, which helped to fashion notions of blood donation as a service to humanity.[7] During the Sec-ond World War the government established an emergency system of nationally organized health care—including blood transfusion—which became the basis of the peacetime National Health Service. This institu-tional context makes Britain a tightly focused case study for depicting the relationships between blood and genetics, and for showing how sci-ence was transformed by wartime public health and tied to redemptive narratives of community and internationalism.

Materials

What were blood groups? They were not entities that could be seen and handled directly; they were immunological properties of a blood sample that could be inferred using a series of simple tests. In 1939, crime writer Dorothy L. Sayers captured both the mystery and the everyday material-ity of blood grouping in her short story "Blood Sacrifice." More psycho-logical drama than crime thriller, Sayers's story is narrated by playwright John Scales, who witnesses a life-threatening car accident outside his theater. Half hallucinating, he watches a doctor transform the empty stage of the theater into a makeshift surgical theater, readying for a life-saving blood transfusion. The doctor carries out tests on the blood of the available donors using whatever materials are at hand, including a porcelain plate glazed with pink roses. Scales follows the doctor, who carefully draws rings on the plate with grease pencil, transfers drops of blood to the plate, and adds the testing sera:

Blood and serum met and mingled. . . . Scales gazed down at the plate. Was there any difference to be seen? Was one of the little blotches . . . beginning to curdle and separate into grains as though someone had sprinkled it with cayenne pepper? He was not sure. On his own side of the plate, the drops looked exactly alike. Again he read the labels; again he noted the pink rose that had been smudged in the firing—the pink rose—funny about the pink rose—but what was funny about it? Certainly one of [the] drops was beginning to look different. A hard ring was forming about its edge, and the tiny, peppery grains were growing darker and more distinct.[8]

Scales is watching the doctor practice blood grouping, in which series of testing sera might (or might not) cause the red cells to agglutinate— that is, to "curdle" and form "peppery grains." Soon the doctor comes over, "examining the specimens closely, with the help of a pencil microscope." With a small sigh of relief he straightens up: "No sign of agglutination. . . . We're all right now."[9] To a doctor, those patterns of agglutination would indicate the "group" of the blood in question and, therefore, whether it could be used for a specific transfusion. Sayers portrayed blood grouping as simultaneously mysterious, commonplace (taking shape on a dinner plate), and technical (requiring expert interpretation by the doctor and his microscope). At the end of the procedure, Scales is told his blood group, though he remains baffled about what this means.

He was not alone. Even by the late 1930s, few people would have known their blood groups. Although the latter were by now familiar to surgeons and doctors, transfusion in Britain and elsewhere in Europe remained in most places patchy and local, and not even the names of the groups were fully standardized. As real-life donors were recruited, they shared the bewilderment felt by the fictional Scales. One volunteer from the early wartime blood drive—evidently thrilled at being part of the campaign—narrated the mystery of being tested and of receiving a card that informed him he belonged to group O. He recalled being puzzled by this information, not knowing what it meant, but he was later excited to learn that "'O' blood is the marvelous stuff that mixes with anybody's."[10] These real and fictional responses underline that blood groups were hidden; they could not be seen or felt; they were properties of blood that ordinary people could not discern for themselves but were told by a transfusion donor card.

To serologists and doctors, meanwhile, blood groups were objects that they could *make* (figure 0.2). When they were first defined at the turn of the century, blood groups had been taxonomic categories for grouping people. Viennese immunologist and serologist Karl Landsteiner had observed that mixing samples of blood on a slide drawn from colleagues often (but not always) caused red cells to clump together, or agglutinate. Landsteiner had accounted for the patterns of agglutination he observed by categorizing his donors into groups, eventually standardized to A, B, O and AB. Revealed on porcelain or white opal glass slides, or in test

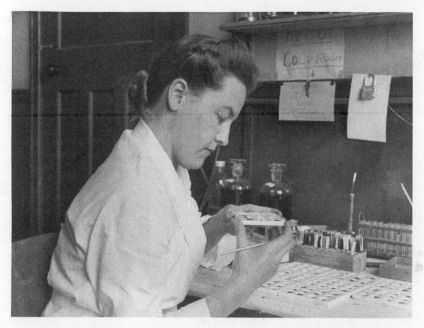

0.2   Photograph of "group determination" at the North West London Blood Supply Depot,
      taken between 1940 and 1943. A female serologist in a pristine lab coat examines
      agglutination reactions on white tiles with small depressions for mixing blood. Depot
      laboratories routinely used the "tile" technique for blood grouping. Behind the files are
      wooden blocks with labeled test tubes containing diluted samples of donors' blood.
      21 × 16 cm.
      Reproduced with the kind permission of the Bodleian Libraries, University of Oxford.

tubes, blood groups were devices for ordering patterns of serological relations. "Serum" is the term given to the fluid part of blood, which separates when blood clots—the word comes from the Latin *serum*, meaning
"whey." Serum from humans and other animals contains antibodies and
other soluble proteins; Landsteiner's practices belonged to the field of
serology. Since the 1880s, bacteriologists and immunologists had been
using sera to identify bacteria and taxonomically classify animals (and
later plants).[11] Landsteiner showed that serological techniques also offered a way to classify healthy humans—a finding that would soon seem
remarkably suggestive to those interested in race.

Meanwhile, for Landsteiner and other immunologists, *Blutgruppen*
were not just taxonomic categories but also referred to biochemical entities. Landsteiner and his colleagues understood that observed patterns
of agglutination were produced by a simple immunological reaction.

Soluble antibodies (then called "agglutinogens") could bind to antigens ("agglutinins") associated with the red cells of the blood sample, causing the red cells of that sample to clump together ("agglutinate"). Immunologists like Landsteiner understood people of group A to have "A" antigens on their red cells, people of group B to have "B" antigens, people of group AB to have both, and people of group O to have neither.[12] Immunologists understood patterns of serological agglutination on a porcelain slide to be a way of making visible specific protein antigens on the surface of red cells. Blood groups, for these scientists, were real biochemical entities that could be seen using serological practices.

What was not obvious to immunologists like Landsteiner, or to anyone else at the time, was the notion that blood groups were relevant to transfusion. During the 1910s, the movement of blood from one person to another was simply too dangerous for blood compatibility to be either important or practical. But after the First World War, surgeons increasingly adopted blood preservation techniques that could prevent blood from clotting in syringes, and possibilities for transfusion began to expand. Hospitals began compiling lists of people willing to donate—students, patients' families, and nurses. As such bureaucracies of procurement expanded during the 1920s, clinical pathologists (who had been trained to classify infectious microorganisms using serological techniques) began applying their expertise to human blood. It was now much clearer that the blood groups of donor and recipient could determine the success of a transfusion—A and B blood were incompatible, but O was generally suitable for all recipients. Hospital donor lists became longer, lists of blood groups accumulated on registries and cards, and blood flowed further and faster. The story of blood groups is about the fluidity of blood—on battlefields and on the surgical operating table—and its increasing mobility as transfusion expanded after the First World War.

By the 1930s, for those working in the transfusion services, the practice of blood grouping was fairly simple and mobile, requiring only everyday equipment. Yet it also required a highly specialized material, namely, serum. Animal sera were already central to bacteriology and public health: antibodies produced by rabbits or guinea pigs inoculated with specific microbes were used as diagnostic reagents in bacterial taxonomy and as routine treatments for the diseases those microbes caused; "serotherapy" was so-called passive immunization, used to bolster a patient's immune system.[13] By the 1920s, institutions responsible for making and distributing animal sera were an essential part of the contemporary public health apparatus, and their standards were coordinated by the League of Nations.[14] As transfusion expanded, some institutions came to specialize in making sera containing antibodies for blood group testing. Blood grouping antiserum—sometimes liquid, sometimes frozen, and later freeze-dried—was often derived from human blood itself, and it became a crucial substance circulated between transfusion centers as the practice expanded. Prepared serum would become the material with which several of the labs in this story would consolidate their authority, allowing them to ask depots around the country for specimens and data. Institutions for the circulation of serum also became centers for elaborating blood group genetics.

If these were all "wet" laboratory practices, then blood grouping also had its "dry" side, namely, paperwork. Written protocols, registries, indexes, and record cards function as tools of "scientific bookkeeping" to generate, or constrain, knowledge about the natural world.[15] Blood groups could not be visualized and handled directly. They were made from blood samples, slides, pipettes, spatial arrangements of tests—and also pen and paper.[16] Blood groups were inscribed categories that were designed to account for observed patterns of agglutination. Figure 0.3 shows a serologist interpreting the patterns of agglutination on a porcelain tile, inscribing the blood group symbols directly onto that tile.[17] As "direct" transfusion (connecting a donor's body to a recipient's body with a tube) gave way to the more straightforward "indirect" transfusion (using a bottle or syringe to contain the donated blood), disembodied blood had to be labeled, so that it could travel from and to the right people. As blood and its labels moved further and faster, the mid-1930s saw new technologies for preservation alongside international efforts to standardize blood group nomenclature.[18] By the time the Second World War was underway, blood could be stored for up to two weeks with the help of anticlotting chemicals, fridges, antibiotics, and fractionation equipment for freeze-drying sera and plasma. Moving that blood depended on bottles, iceboxes, vans, telephones, and postal networks. Holding all of these together was paper, and lots of it. Call-up letters directed specific donors to give blood at particular times and places; labels determined where blood should travel; index cards moved between transfusion centers and hospitals. This paper trail connected donor, bottle, and patient and enabled blood to move (figure 0.4). To allow the outcome of transfusion to be traced back to individual donations, the Emergency Blood Transfusion Service used labels that could be tied to and untied from bottles of blood, linking donor and recipient across space and time.

From the 1910s onward, blood groups also gradually consolidated as "genetic" objects.[19] In the 1920s, German actuary and mathematician

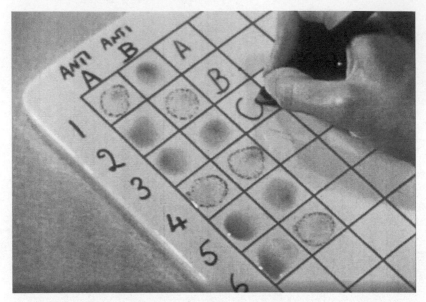

0.3 Still from the color film *Blood Grouping* (1955), the purpose of which was to show students and house officers some of the techniques used in routine blood grouping in the hospital laboratory. This section of the film explains the preparation, interpretation, and writing of blood group serological reactions carried out on a white tile. Six unknown blood samples (on separate rows) have been mixed with two kinds of antisera, anti-A and anti-B (columns). A technician studies which of these reactions have resulted in agglutination—visible as peppery, curdled grains. The technician writes the interpretation of the results on the porcelain slide with a grease crayon. Filmed at the Group Laboratories, Mile End Hospital, London. Cyril Jenkins Productions Ltd., *Blood Grouping* (Imperial Chemical Industries Limited, 1955), 20:33 min, sound, color. Still image from 00:05:07. Wellcome Collection, London, https://wellcomelibrary.org/item/b17505963.

Felix Bernstein applied novel mathematical techniques to blood group results collected from donors. Using that data, Bernstein demonstrated that the ABO blood groups—that is, the ABO antigens—were inherited via a single locus with three possible alleles: A, B, and O. Like blood groups, these alleles could not be seen directly but were inferred from agglutination patterns and calculations on paper.[20] Consensus over the genetic inheritance of the ABO groups opened up a range of new potential uses, especially for scientists eager to apply techniques of Mendelian genetics to humans.

One immediate consequence of this in Germany was in forensic science. Before the decade was out, blood groups had been presented as

10

evidence in thousands of paternity cases.[21] In Britain, they were seized upon for another purpose: blood group data were deployed in visions of new methodological standards for research on human heredity. By the 1930s, geneticists were making blood groups the basis for studying theoretical population genetics, for mapping genetic diversity, and for probing the genetics of other, more complex, human traits. At labs in Cambridge and London, the results of blood group tests were mathematically transformed and decomposed into genotypes (sets of genes that determine a characteristic; in this case, a blood group). These became the working objects for experiments on schemes of inheritance and diversity. Blood group records were clinical devices in the transfusion center and hospital but were transformed into research objects in the genetics laboratory.

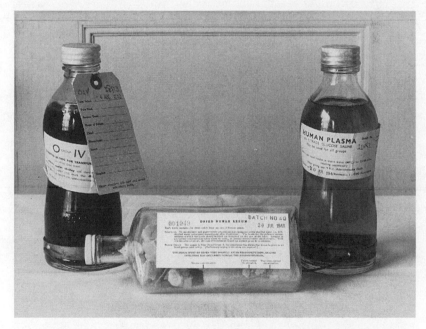

0.4   Photograph of paper labels attached to bottles of (from left to right) whole blood, dried serum, and plasma. Made as part of a series of publicity photos for the Emergency Blood Transfusion Service between 1940 and 1943. The label on the whole blood gives in large print both the official (O) and still occasionally used (IV) blood group nomenclatures. The additional tie-on form attached to the neck of that bottle would be completed after the transfusion and returned to the depot; it gives the date the blood was taken and used, the reason it was used, the name of the patient, the outcome of the transfusion, and the name of the hospital. 21 × 16 cm.
Reproduced with the kind permission of the Bodleian Libraries, University of Oxford.

Because paper was flat, cheap, and malleable, it could be moved from blood depots and reused for new purposes by other social groups.[22] In paper form, blood groups were mobile and could be repurposed in the serological lab, the bleeding center, the anthropological clearing house, and the hospital. The circulation of blood transfusion records brought doctors into new relationships with scientists: letters accompanying antisera determined the conditions of exchange; labels attached to samples brought donor and patient identities to bear on the methods and conclusions of research.[23] These discrete, sortable blood group records became ideal genetic material: fitting into a mathematically tractable science of large numbers. The wet practices of transfusion became the foundation for a dry, objective, paper-based genetics.[24] Thus, the burgeoning paper bureaucracy of transfusion medicine did not just shape the organization of research but also became the very material on which a new human genetics was based.[25]

Bodies

For all that I focus on the paperwork of the new human genetics, this was no bloodless revolution. The rise of human genetics needed not just paper and colorless reagents but also people with blood running through their veins. Bloodletting is not difficult: blood can spill from wounds; it can leave traces in inconvenient places. But nor is it easy: drawing blood is potentially dangerous; it can be messy; it is sometimes painful. Just as paper has affordances and limitations, so too does the human body. Whether for research or for therapy, blood extraction requires needles, cotton wool, bottles, sterilizing apparatus, specialist training and persuasion. And disembodied blood is always highly charged with meaning. Donna Haraway has articulated its apparently inescapable suggestiveness: "The red fluid is too potent, and blood debts are too current. Stories lie in wait even for the most carefully literal minded."[26]

Historical, literary, and anthropological studies have pointed to the varied meanings and purposes of blood in history and across cultures: as an object of religious veneration, of individual and communal identity, or of notions of racial purity.[27] These blood stories bind some people together and exclude others. Rituals of sharing blood have connoted allegiance and affiliation, and stories about its theft have expressed anxieties about and resistance to colonial domination.[28] Donors have been persuaded that their blood would fulfill obligations of citizenship, or help defend the nation.[29] Long lines of donors volunteering after terrorist atrocities articulate grief, shock, and support.[30] Protests against bans on blood donation by homosexual men have built and consolidated communities.[31] Attachments to these varied meanings powerfully affect encounters for giving and withdrawing blood.[32] In the stories of this book, giving and taking blood in different places and times have affirmed commitments to family, community, ethnicity, nation, and humanity.

Reconstructing the circumstances under which people chose to give and take blood draws attention to *whose* bodies become subjects of genetic research.[33] Relations produced by the movement of blood were deeply consequential for the kinds and quantities of data that could be collected. The blood extractions described in this book took place prior to practices of informed consent and formal bioethics.[34] But encounters around blood were strongly conditioned by the institutions in which they occurred, and by the power relationships between donors, doctors, and scientists.[35] Such encounters occurred in highly variable political circumstances, involving, for example, imperial British scientists in rural Kenyan villages, doctors in British hospitals, and nurses in wartime mobile bleeding units. The authority of collectors (usually scientists, doctors, and nurses) and the settings in which they subjected donors to bleeding (hospitals, wartime factories, schools, people's own homes) affected how often collectors could call on donors and how much blood they could take.

In turn, this meant that these places, people, and circumstances determined how much data could be aggregated, what kinds of sampling strategies were possible, who could be relied upon to give repeat donations, and whether family data was available. In other words, places and power relations strongly affected whose blood was collected and what could be done with it. In some places, collectors could go back time and time again for repeat extractions, perhaps collecting blood from members of whole families; in others, only one-off collections were possible. Depending on the outcomes of these interactions, some kinds of samples and data were suitable for studies of diversity, others for linkage mapping, and still others for elucidating new blood groups and proteins. The power relations that operated at the moment of bloodletting gave shape and meaning to the aggregation of blood and data.

This is a story of the practical links between blood and a formal science of kinship; that is, genetics. The metaphorical connection between *blood* and *kinship* is powerful. The literal meaning of the term "blood" is "the red fluid flowing in . . . arteries, capillaries and veins." But in English, for over eight hundred years, "blood" has also been used metaphorically to refer to inheritance, lineage, birth, family, and nation.[36]

"Blood" has become a synonym for the kind of relatedness that we also call "biological." Even in scholarly anthropological discourse today, the term is used remarkably often as a synonym for "biological related-ness."[37] So in telling a story of the links between genetics and blood transfusion, it is provocative to think about the ways in which these literal and metaphorical meanings did—and did not—come together.

In many places and contexts, exchanging bodily substances such as blood, organs, and sperm can result in solidarities and communities that go beyond family and race.[38] During the early years of routine transfu-sion, many noted that blood group compatibility did not follow the expected rules of kinship: family members were often unable to donate to one another.[39] Building on this, in the 1950s, films, pamphlets, and novels argued that compatibility could cut across and dispel traditional notions of family and race, and so had particular power to flatten and neutralize racial hierarchies. The 1952 film *Emergency Call* proclaimed: "White, black, brown, yellow: human blood's the same the world over."[40]

In reality, disembodied blood often continued to flow along familiar routes.[41] Sharing blood was often closely linked to affiliations and exclu-sions based on family (for example, local practices of transfusion in the 1920s), race (such as blood-segregation practices in the United States), and citizenship (during the British war effort).[42] Because blood group ge-netics depended on the infrastructures and social practices of the trans-fusion services, research was shaped by demarcations and structures of administration that often reproduced power along racial lines.

Meanwhile in laboratories, scientists created and sustained relation-ships with colleagues by exchanging blood samples. Researchers rou-tinely sent specimens to colleagues in other institutions and countries to strengthen their professional and social ties. Testifying to the power of institutions to condition collections, many laboratories used the blood of their own workers as convenient testing reagents. The war-time transfusion services regularly recruited and bled many of its do-nors in factories and offices. Such drives drew on relationships between colleagues, communities, races, and families (figure 0.5). Transfers and exchanges followed the contours of nation, class, friendship, institu-tion, and ethnicity—and these, in turn, made blood groups available for fixing pedigrees and for drawing maps of genetic diversity. Human genetics was made possible by social relationships forged and articulated through the exchange of blood.

These social exchanges also draw attention to the question of what kind of science this was, and what kinds of scientists were doing it. Many of the laboratory directors in this story were talented managers:

14

0.5   "This might happen to your husband, brother or sweetheart. You'd give your blood to help them, wouldn't you?" In an apparently posed photograph, a woman looks at a recruitment poster for the Emergency Blood Transfusion Service. The poster evoked kin relationships as a reason to give blood; the implication of this photo was that the blood of this woman could be used to save the life of a loved one. Underneath the poster an additional note informs its audience, "The Ministry of Health Transfusion Centre will be working at the F. A. P. Royal Mews, Winsor Castle on Friday, Saturday and Sunday, Jan 1st, 2nd and 3rd, and on Wednesday and Thursday, Jan 6th and 7th, 10.30am–12.30pm, 1.30–3.30pm." One of a series of EBTS publicity photos taken between 1940 and 1943.
Reproduced with the kind permission of the Bodleian Libraries, University of Oxford.

of data, of blood samples, of people, and of networks. They also exhibited forms of sociability that helped to ensure the flow of blood and paper. The liveliness and friendliness of Robert Race and Ruth Sanger enabled the passage of interesting samples through their labs. Mourant's proclivity for endless correspondence ensured the steady arrival of paper data at the Royal Anthropological Institute. Meanwhile, these materials gendered laboratory roles. To sort, order, and analyze blood groups and records, lab managers hired many women, who worked as clerks, secretaries, statisticians, and librarians—doing the "actual laboratory work" of genetics, to quote Fisher.[43] Female serologists carried out routine blood grouping tests and followed up on intriguing serological phenomena. Their laboratory directors believed that they were particularly well suited to serological research, and women apparently found this new field particularly open for them to forge productive careers in science.[44]

The movement of blood in and out of people in wartime and postwar Britain draws attention to bodies as porous entities—created, bounded, and sustained by the disciplines and processes of statecraft and medicine.[45] Bodies were permeable in another way too: just as vaccination and serotherapy left their immunological imprints on human bodies, so too did transfusion. Researchers made use of the fact that in response to a dose of donated blood, patients might produce new antibodies, sometimes to specific, as-yet-unknown blood groups. Some of those patients became highly prized research subjects: by the 1950s, a decade of nationwide therapeutic transfusion in Britain had turned some people—especially patients with chronic anemia—into veritable archives of antibodies. These individuals became precious resources to be mined by researchers in pursuit of novel blood group antigens. This was a recursive process in which the antigens of a donor could stimulate antibodies in a patient, the blood of whom might, in turn, be used to define a new blood group (often named after the initial donor). Donors and patients became indispensable parts of a technological system for circulating antisera, detecting difference, and classifying blood.

This recursive creation and labeling of serological and genetic specificity also underlines the status of human bodies in this story as *relational*. The antibodies in the blood of a transfusion patient were turned into reagents for discovering new antigens (blood groups). These groups, in turn, became categories for further specifying blood. Blood groups and antisera could *only* be defined when donor and recipient were brought into relation with each other on a porcelain slide or in a test tube. This was made possible by an administrative system that enabled doctors to retrace the provenance of blood after it had been mixed or transfused. These wet and dry serological relationships brought antibody into contact with antigen and made both visible to research scientists.

Populations

The elements of the story discussed so far—sera, paper, and human bodies—were brought together by a bureaucracy (donor registries) that made visible a specific kind of bodily similarity and difference: that is, genetic variation. The study of genetics depends on defining variation: a phenotype, or observable characteristic, has to be circumscribed before it can be followed across generations or across space. It is a concept that requires collectives of bodies and body-derived information. Like other kinds of archiving or cataloguing practice, the recruitment of British citizens to nationally standardized transfusion registries brought people into relation with one another, making visible the variation between them. Donor registries were devices for managing blood and people but were also technologies of human variation. As movement of blood in and out of bodies made these new kinds of human variation available to geneticists and serologists, researchers used that expanding apparatus for specifying blood in increasingly complex ways. Postwar, as more donors were recruited (and patients treated), researchers and doctors discovered more blood groups. These were looped back into the transfusion-service system as donors then had their blood further specified. The more donors and patients that were recruited into this system, the more finely differentiated blood became.

In this respect, the wartime and postwar blood transfusion service— its people, instruments, protocols, and documents—functioned as an infrastructure for disciplining human difference, to paraphrase sociologist Nikolas Rose.[46] Rose explains that it is when people are gathered together en masse—in hospitals, schools, factories—that their differences and similarities become visible: by bringing people together, such institutions produce a world in which people have distinctive characteristics.[47] Variation is made though efforts to record and manage attributes and deficiencies; humans are made into "individuals" with distinct traits and characters by practices that classify and calibrate those characteristics. Rose was writing about the psychological sciences, but wartime and postwar transfusion bureaucracy (and associated research programs on blood group genetics) did precisely these things. Donor index cards brought people into a standardized system of alignment that ushered into existence different types of blood. Such processes of alignment and specification happened at all scales: serologists' practices of bringing samples into contact with each other on a porcelain tile resulted in patterns of red-cell clumping; clerks' writing of donor lists meant blood groups could be compared and ordered; doctors' recording of the multiple transfusions of an anemic patient meant that their donors could be compared and noted.

This was also how differential value was made. Some blood was therapeutically safer than others; some bodies were better sources for high-titer reagents; other bodies were particularly rich archives of antibodies.[48] Individual people could be marked by their blood groups. During the Second World War, people with group O blood had been treasured as "universal" donors (those with "the marvelous stuff that mixes with anybody's," in the words of the anonymous donor quoted above). But the postwar transfusion services ordered people according to much finer blood group specifications. As more blood groups became known, the way that people were organized to give and receive blood became irretrievably bound up with patterns of immune specificity, which helped to make some kinds of blood (and some people) more scarce, and more highly valued, than others.[49]

Arthur Mourant's attempts to elucidate human difference operated on a different scale. His project brought into a single institution data from diverse parts of the world, producing genetic variation on a global scale and making some populations rarer, more valuable, more intriguing than others. Thus, the alignment, recording, and ordering of difference was the very stuff of blood group genetics, whether for studies of inheritance or (on the other side of the same coin) diversity. The wet serological and dry clerical practices of blood transfusion together produced an effective machinery for specifying and valuing human genetic difference.

Thus, the kind of genetics described in this book was dependent on state-managed bureaucracy.[50] Transfusion bureaucracies also link this story to accounts of midcentury eugenics, especially the delineation and management of populations. In the 1920s, the League of Nations had made human populations central to the geopolitical issues of migration, demography, land economies, and colonial expansion.[51] In the 1930s, German National Socialists had attempted to order and control citizens according to judgments about biological difference—blood groups included. After the devastation wrought by the Second World War, and with the emergence of the Cold War, several new institutions took up population management.[52] The United Nations, UNESCO, the

World Health Organization, and the Food and Agriculture Organization were all devoted to the international administration of public health, education, food production, and science.[53] These institutions made the study and management of populations central to the reorganization of international communities, to the politics of decolonization, and to the negotiation of global standards for public health.[54] In Britain the postwar government, faced with an economic crisis and a desire to tighten control over its empire, crafted migration policies to manage the population of the home country as well as the flow of people in and out of its colonies and dominions.[55] Attention to how such "populations" were constructed (in Britain and overseas) reveals links between human genetic research and migration, nationalism, race, and public health.[56] My focus on the local politics and classificatory practices of blood donation links broad issues of population management to human genetics.[57]

"Populations" also became central to programs of research that addressed evolutionary history.[58] In the late nineteenth and early twentieth centuries, comparative anatomy and paleontology dealt with human evolutionary history, linking its narratives to anthropological studies of race.[59] From the 1960s, terms like "molecular anthropology," "anthropological genetics," and "genetic anthropology" came to refer to the comparison of protein (and later DNA) sequences to infer evolutionary relationships.[60] Chronologically, blood group population studies fell squarely between these programs, starting in the 1920s and reaching a crescendo in the late 1950s. These studies generally focused on recovering the historical relationships between races, rather than on phylogenetic species relationships between humans and other primates.[61] With high-profile exponents such as Julian Huxley, J. B. S. Haldane, Fisher, and Lancelot Hogben, many of those deeply involved in questions about genetics and evolution were also interested the lessons that might be drawn for human society.[62] To some, blood groups seemed to offer a route to linking research on genetics in the natural world to pressing social problems.

Indeed, one powerful motivation for research on human inheritance and evolution was a desire to define human racial difference.[63] The notion that people could be classified into distinctive groupings drove the racial taxonomic blood collection projects of the 1920s and 1930s and the "anthropological" blood group diversity studies of the 1940s and 1950s.[64] In most of these surveys, collections relied on researchers' prior convictions about how people might be divided and ordered.[65] Workers in colonial hospitals or in US blood banks (in particular) often categorized patients' and donors' bodies according to local racial taxonomies—and these labels sometimes moved into research labs with blood samples and data. In the United States, hospitals and transfusion centers used the local racial taxonomies "white" and "colored," which were often imported into published papers.[66] Indeed, from the 1940s onward, differential blood group gene frequencies in "white" and "colored" populations often counted as evidence for a new blood group. Later, observations of pathologies linked to race and blood underpinned questions about the effects of selection on human populations.[67] In significant practical ways, race was made part and parcel of blood group research.

Especially in its connection to race science, blood group genetics served highly visible—and very flexible—rhetorical and political functions. In interwar Europe some anthropologists and physicians believed blood groups validated policies based on racial difference.[68] Others framed blood group diversity as evidence for the fluidity of racial boundaries. After the Second World War, blood group distributions were used to argue against notions of racial superiority. Scientists involved with UNESCO made this claim in the organization's high-profile antiracist campaign of the early 1950s (of which Fisher was a notable dissenter). If human diversity could be understood as genetic, the argument went, then race was divested of its prejudicial power.[69] Chiming with UNESCO's broader philosophy, the organization argued that genetics offered a scientifically objective basis for recognizing "unity in diversity."[70] This postwar public reframing was symbolized by the removal of the term "eugenics" from several major genetics journals.[71] At the same time, the discipline was given new urgency by questions about the hereditary effects of atomic radiation, and new links between molecular variation and inherited disease.[72] In this way, blood groups helped to usher in a new phase of human genetics, which was now seen as a means for understanding the past, present, and future of humankind.

The "relations" of this book's title, then, operates on multiple registers. It alludes to the power relations that defined who could extract blood from whose bodies. It refers to the social relationships that allowed blood samples and paper data to move between scientists, or between doctors, transfusion workers, and scientists. It suggests the in vivo immunological relations between antigen and antibody, as well as the in vitro agglutination that made antigens/antibodies legible to scientists on a porcelain tile. It hints at the carefully managed relationship created between donor and recipient through the transfusion of blood, and the ties reinforced between donor and nation. It points to the fact that human biological difference (that is, blood group difference) was brought into existence when donors were incorporated into a bureaucratic system devoted to the management of bodies and blood. And it alludes to the familial and kinship relationships that can be both defined by and disrupted by the science of genetics. All of these are relations animated by blood, its metaphorical meanings, its materiality, and its therapeutic value.

Synopsis and Sources

Blood Relations explores how the nascent field of human genetics was formed by the instruments, people, customs, materials, and networks of blood transfusion. The book's structure is part thematic, part chronological. Chapter 1 explores the valences and uses of blood groups across Europe during first three decades of the twentieth century: as immunological and biochemical curiosities, as Mendelian traits, as markers of racial difference, and as information medically relevant to transfusion. Chapter 2 focuses on 1930s Britain, when a community of influential biologists seized on blood groups in their attempts to reform the study of human heredity, in a period of intense disciplinary and political dispute. These chapters demonstrate the extraordinary flexibility of blood groups as they were deployed for medical, scientific, and political purposes.

On the eve of the Second World War, human genetics and transfusion became institutionally linked for the first time, in ways that would be enduring. Chapter 3 describes how this operated with the rapid establishment of a distributed network of wartime blood depots for extracting, preserving, and mobilizing blood. Accompanying this, the index cards and donor lists that made up the nationwide paper-based bureaucracy for keeping track of donors were reconfigured as materials for studying genetics. A variety of paper practices brought together the large-scale, modernizing project of blood transfusion with the large-scale, data-rich study of human population genetics. Chapter 4 turns to the clinically important Rh blood groups. In the 1940s a fierce, high-profile controversy over blood group nomenclatures highlighted the competing demands made on blood groups and the roles of nomenclatures as part of the practical and material apparatus for both transfusion and genetics. Following names, symbols, and other inscriptions reveals the rich array of functions of blood groups: from labels, to markers of identity, to diagnostic markers, to scientific data.

The postwar years saw dramatic changes to the organization of British medical care, including the foundation of the National Health Service, which brought blood transfusion into its administration.[73] The therapeutic potential of blood had expanded: it was now used for routine surgery, for neonatal care, and for disease treatment—innovations that went hand in hand with the standardization of transfusion bureaucracy. All of these changes altered and expanded the activities and authority of the transfusion services and blood group geneticists. Chapter 5 describes how the regional-center organization of Britain's transfusion services was linked to the discovery of new blood groups, and therefore the sharper specification of blood. Focusing in particular on the work of the Blood Group Research Unit in London, and the scientist Robert Race, it follows the recursive processes of specification, alignment, patterning, and diagnosis that happened as blood and its labels moved between donor, doctor, serologist, researcher, and patient. This interest in blood group specificities brought into focus enormous interest in "rare blood" and the precious donors who could offer it. Chapter 6 explores the cultural and medical significance of rare blood and examines other ways in which particular bodies were understood to be especially valuable for the blood that they could donate. This included patients inoculated with the blood of multiple donors, who therefore had the potential to carry a vast array of antibodies, some against wholly new blood groups. Multiply transfused individuals became exceptionally precious resources for Research Unit workers and their colleagues.

While the Research Unit became famous for its work on blood group inheritance, Arthur Mourant's Blood Group Reference Laboratory became a preeminent center for research on blood group diversity. Chapter 7 describes how this lab made and distributed reference standards for blood grouping antisera, not just to Britain's transfusion services but also for the World Health Organization. Once antiserum was routinely freeze-dried, it could circulate internationally, and Mourant took the opportunity to use those contacts to gather and map blood group frequency data from expeditions, hospitals, missions, and laboratories around the world. Mourant brought these data together in a small building at the back of the Royal Anthropological Institute—which became the site of the largest collection of human genetic data ever assembled. Chapter 8 looks at what Mourant and his colleagues did to make those data speak to human history and diversity. It examines how they ordered, racialized, calculated, tabulated, and mapped blood group frequency data, and how they used these maps to define historical patterns of human movement in a tumultuous period of postwar migration and decolonization.

Meanwhile, Mourant's collection of blood group data was being mobilized to underpin a postwar internationalist agenda. UNESCO's antiracism campaign was emblematic of the organization's early-postwar confidence in the power of "neutral," universal scientific knowledge to function as a social remedy and diplomatic tool. Chapter 9 explores the rhetorical, organizational, and strategic work through which UNESCO and the BBC presented blood group genetics as universal, explanatory, and politically neutral. These efforts continued during the 1950s and 1960s, as blood refracted into an array of new protein polymorphisms, including hemoglobins, enzymes, and white-cell antigens. Blood group data were collected as abundantly as ever, but this new array of blood-related proteins began to eclipse the status of blood groups within genetics. Chapter 10 recounts the decoupling of human genetics from the infrastructures of blood transfusion, and the consequences of this on Mourant's research.

The story told in this book is defined by the composition of archives and other sources. Those archives have themselves been shaped by the practices and politics of midcentury human genetics. My main protagonists worked in London-based, state-funded scientific and medical institutions, with resources and networks that were specific to wartime, imperial Britain and a postwar, internationalist Europe and United States. Accordingly, my principal archival sources are kept in London: three collections at the Wellcome Library (the papers of Arthur Mourant, of the Blood Group Research Unit, and of Robert Race and Ruth Sanger) and two collections at the UK National Archives (those of the Ministry of Health and the Medical Research Council). The cast of influential doctors and scientists visible in these sources used the status and funding that came from living and working in the principal metropolitan city of a European colonial country to build centers of calculation and collection. They turned networks that spanned many "regional" geographies into instruments for defining human genetic variation, and in so doing they assimilated the labor and expertise of correspondents in places remote from London. As a result, these particular sources—focused on this small cast of institutions and scientist-administrators—bring into view a broader (although still partial) array of interlocutors, subjects, observers, and collectors: a physician working in a Glasgow transfusion center, a transfusion patient in an Oxford hospital, a family of donors in rural Essex, a serologist at a private New York blood bank, a Basque lawyer in Oregon, a doctor's wife in West London, a "medical assistant" in Kenya. These rich sources spin an important international story and offer a new way of thinking about the history of genetics. They also reproduce the center-periphery perspective of Race, Sanger, and Mourant in London, which was the product of the political circumstances of state-funded, postwar British biomedicine.

Some of these archival sources have posed useful challenges during the course of this research. The Wellcome Library's Blood Group Research Unit archive and Mourant archive are extensive, carefully kept, and supported by detailed catalogs. Halfway through my book project, the library began a program to make its materials relating to human genetics freely available online.[74] This venture, which emphasized accessibility to audiences regardless of institutional affiliation, was resonant both of the Wellcome Trust's promotion of freely accessibly genomic data and of its recent promotion of "open access" publishing.[75] During the process of digitization, however, the contents of the Blood Group Research Unit and Mourant archives were reassessed for their content of "sensitive personal data," in compliance with the Wellcome Library's access policy.[76] During that labor-intensive revision process, which was carried out by two archivists, a large number of the papers that I had used during the earlier phase of my research were closed.[77] Like other kinds of medical information, blood group results attached to names and additional family and medical information were reclassified as sensitive. The archivists perceived that in some instances, such as extended correspondence, a series of letters might be capable of attaching a blood group to a disease, and then to a personal name or a pedigree, which would constitute a breach of privacy of personal data. The difficulties of assessing such large quantities of material made the task formidable for the archivists involved. To make the task manageable, in some instances, folders containing a large proportion of such data had to be closed entirely. This also made inaccessible some material relating to individuals whose names had been (for example) memorialized in the names of blood groups.

This episode highlights some of the contradictions historians face as they attempt to negotiate the visibility of the people they write about.[78] On the one hand, privacy and anonymity are central to the protection of individuals who have been the subject of medical and scientific scrutiny. On the other, rules about privacy have the potential to erase from historical records many individuals who have contributed labor and expertise to the processes of making medical and scientific knowledge (such as blood donors in a range of settings and in different eras). Such rules can be vitally important to protect privacy, but they might also make it difficult to give due credit. The weighing up of privacy and protection, visibility and credit is not easily resolvable, and it requires fine-grained attention and negotiation.[79] The history of human genetics is in part a history of visibility, value, and identity; historians must remain in active dialogue with archivists.[80]

The access decisions made by the Wellcome archivists also underscore some of the central themes of this book. The archives that came under scrutiny in 2012 are records of a set of enterprises that helped to create genetics as a field with authority over human identity, family, and history—the story that I tell here. Not only have those claims about the authority of genetics been remarkably enduring, but they were dramatically amplified in the 1990s and 2000s as money and research were committed to the Human Genome Project. The power that genetics now has to reveal aspects of human identity has itself shaped the access policies of biomedical archives, such as those of the Wellcome Library. The notion that archivists today should treat blood groups as "personal data"—in a similar way to genetic sequences—is the legacy of some of the very work that is represented in the Wellcome blood research collections, and the work that this book attempts to reveal. So while I tried to recover the story of how blood groups were collected and ordered using commonplace human identifiers, such as name, nationality, and "race", these very features of the material rendered it less accessible. Even as they shape what is possible to know about the past, present day notions of access, privacy, and protection are constantly being remade and adapted to our own understanding of history.

ONE

Transfusion and Race in Interwar Europe

Blood groups first came to matter to transfusion on the battle-fields of the First World War. With heavy artillery causing many soldiers to suffer acute blood loss and potentially fatal shock, field hospitals brought large numbers of casualties together.[1] There, experiments to prevent blood from clotting allowed doctors to move blood between bodies via bottle and syringe. Although blood groups had been known since 1900, the sheer numbers of transfusions taking place on the Western Front made them clinically meaningful for the first time. So successful was this blood grouping that by 1920, British transfusion expert Geoffrey Keynes could note in the medical journal *Lancet* that "there can seldom be any reason . . . for performing a transfusion without first testing the donor for his blood group."[2]

As transfusion services expanded, so did the bureaucracies of blood procurement. Records of donors and their blood groups began accumulating in large numbers, to be swept up in currents of interest in identity, nationalism, race, history, inheritance, and forensics. The First World War had reconfigured the peoples and territories of Europe and its empires, provoking debates about what defined a nation and spurring an obsession with the idea of national character.[3] Disputes over territory were frequently elevated to supposedly scientific questions about who belonged where—often on the basis of language, customs, and physical traits. This was a high point for eugenics movements, in which, across the world, governments and activists sought to apply theories of heredity to human reproductive policies and practice.[4] There was keen interest in how mental and physical characteristics were inherited, and fierce debates over what traits could be examined and how they should be studied.[5]

Meanwhile, geneticists, who specialized in the study of inheritance, were becoming increasingly interested in populations as units of analysis. Research programs that sought to map genes onto chromosomes diversified into studies that explored natural genetic variation, and the important question of how evolutionary change might be understood in Mendelian terms.[6] Much of this research was theoretical or based on observations of plants and animals, but suddenly blood groups looked as though they might serve as ideal traits for modeling how this could work for humans.[7] Blood groups could potentially link the science of heredity to the most urgent social problems of the era. Amid these discourses of populations and planning and the formation of new nations, blood groups were coupled to metaphors of racial identity, belonging, and kinship.

The meanings and uses of blood groups were defined by the pressures of war, nationalism, and eugenics, by the material properties of blood and changing technologies for handling it. Between 1910 and 1940, blood groups became biochemical entities, racial markers, forensic tools, and genetic traits. This chapter outlines the contexts in which blood groups were considered, the technologies through which they circulated, and the political uses to which they were put.

Taxonomy

In 1900 Karl Landsteiner was a young pathologist carrying out "serological" experiments on animal and human blood at the University of Vienna. Serology was a taxonomic practice that had been invented in the 1880s by researchers studying the properties of immune sera within the orbits of three laboratories in Europe: those of Louis Pasteur in Paris, Robert Koch in Berlin, and Max von Gruber in Vienna. They began using sera to identify and distinguish bacteria and other microorganisms. They found that inoculating a guinea pig or rabbit with a weakened form of a specific bacterium protected that animal against infection. They also reconstructed this phenomenon in test tubes: immune sera extracted from such inoculated animals responded to bacteria in specific ways, causing what immunologists called "agglutination," that is, the clumping together of bacterial cells. Using sera to identify and classify bacteria, the researchers understood agglutination to be caused by a soluble blood protein (later generally known as an "antibody") reacting in specific ways with proteins (later called "antigens") on the surface of microorganisms.[8]

While immunologists viewed serology as a fast-moving, cutting-edge biochemical science, clinical pathologists turned it into a collection of routine techniques practically applied in clinics to distinguish between morphologically similar microorganisms.[9] By the turn of the century, one of the tests most frequently undertaken by clinical pathologists was an in vitro serological test for diagnosing typhoid fever.[10] Serological testing became so crucial and so routine that by the 1920s, institutions responsible for making and distributing animal sera were an essential part of the apparatus of public health, and these procedures were standardized by the League of Nations.[11] Serology was a fast-moving field of research, but above all it was a set of routine medical diagnostic methods.[12]

Meanwhile serological reactions were being developed as methods for studying the taxonomy of larger organisms. In the 1870s, German physician Leonard Landois had observed that animal blood sometimes agglutinated in response to the blood of another species.[13] Building on that finding, Cambridge biologist George Nuttall developed systematic serological methods for studying the biochemical diversity of animal species. Nuttall's approach was to inject one animal (such as a rabbit) with the blood of another (such as a dog). He reasoned that the injected animal (rabbit) would produce antibodies against the foreign (dog) blood and that the rabbit "anti-dog" serum could then be used to test the dog's biochemical distinctiveness from another animal (such as a cat). For Nuttall, the strength of the resulting agglutination reaction indicated the degree of relatedness between the two animals, and from this data he drew phylogenetic trees. By 1904 Nuttall had carried out 16,000 tests on almost 600 different species, work that others further developed between 1920 and 1960.[14]

In these ways, the study of immune sera around the turn of the century crystallized into three distinct but overlapping domains: an academic field of immunology that dealt with immune proteins and their behavior; a practical field of clinical pathology, which used specific immune sera for diagnostic purposes; and a form of biochemical taxonomy for understanding evolutionary relationships. This constellation of agendas and practices—immunology, diagnostics, and relatedness—would define human blood grouping.

Back in Vienna, Gruber's former student Landsteiner found that serological reactions happened not just as a consequence of mixing sera and blood of different species but also sometimes from mixing the blood of different human individuals.[15] Medically qualified, Landsteiner had also studied chemistry, and he began researching the properties of "isoagglutinins," that is, antibodies that react with the sera of other individuals of the same species.[16] Testing arrays of blood samples from colleagues and patients, Landsteiner accounted for patterns of agglutination by categorizing individuals, initially into three groups (later known as A, B, and O).[17] Publishing his first paper on the topic in 1901, Landsteiner proposed that red-cell clumping was caused when antibodies in the serum of one person reacted with antigens on the blood cells of the other, with the implication that different people carried biochemically different antigens on the surfaces of their cells. His work was soon confirmed by other researchers, most famously the Czech serologist Jan Janský and the US physician William Moss, who independently defined the fourth and rarest group (later known as AB). During the next decade, serologists reached consensus that the four blood groups corresponded to the presence or absence of two antigens, A and B, on the red blood cell surface, and this resulted in the existence of blood groups A, B, AB, and O. Diagnostic serology represented cutting-edge biochemistry, and it now seemed to reveal for the first time that humans varied biochemically in consistent ways. Fundamental differences between people could apparently be revealed under a serologist's microscope, a notion that would later became deeply consequential for uses of blood groups in racial taxonomy.

These biochemical classifications came to fascinate physicians, serologists, and pathologists over the next thirty years. Researchers in Europe and the United States eagerly followed up Landsteiner's results, as they investigated the complex chemical composition of these so-called isoantibodies, their stability, and their secretion from a range of human tissues. In a similar vein as Nuttall, Landsteiner himself developed a program of research on human evolution, systematically comparing the serological reactions of human blood with those of other primates, and correlating the results with phylogenetic trees (figure 1.1).[18] Other researchers studied the blood groups of fetuses, children, adults of different ages, people with mental health problems, individuals with tumors, and patients with infectious diseases. They gradually reached consensus that blood groups were stable and constitutional; they were a fixed characteristic of a person and did not change with age.[19]

Blood grouping was classification: diagnostic methods for identifying microorganisms and techniques for detecting biochemical relatedness had yielded a method for categorizing human blood. These categories

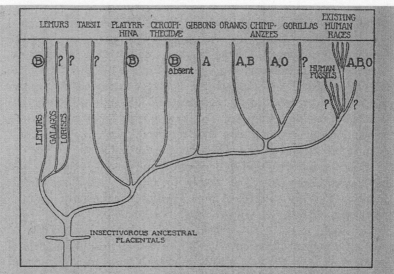

TEXT-FIG. 1. Adapted from Keith, A., The antiquity of man, London, 2nd edition, 1925, and Sonntag, Ch. F., The morphology and evolution of the apes and man, London, 1924.

A = agglutinogen of human Group II red cells.
B = agglutinogen of human Group III red cells.
Encircled B = agglutinogen similar to, but not identical with, B.
O = blood and serum corresponding to human Group I.
? = not examined.
Of the gibbons only one individual was examined.

1.1 A branching evolutionary tree of the primates, by Karl Landsteiner and colleague Philip Miller in a journal article in 1925. The authors reported that "serological studies on the bloods of thirty-six species of lower monkeys have shown that there exists a correspondence between the distribution of a certain hemagglutinogen [i.e., blood group B] and the place of the species in the zoological system." The diagram indicated four human "races" alongside eight other species of primate, suggesting that the human races were biologically distinct. The positioning of the letters "A," "B," and "O" next to those branches implied that the blood groups were linked to race, although the paper itself offered no further reflections on this. From Landsteiner and Miller, "Serological Studies on the Blood of the Primates: III" (1925), 871.
Copyright Rockefeller University Press.

also pointed to the biochemical character of human red blood cells, so that the term "blood group" referred not just to a category of people who belonged to a particular group, but also to a biochemical entity: an antigen (or two antigens, in the case of blood group AB). The notion that healthy people could be distinguished in a way only visible to scientists—that is, biochemically—was tremendously suggestive. By the 1910s, serologists were asking the crucial question: What did such differences mean?

Transfusion

Soon after Landsteiner's discovery, serologists began speculating whether blood group compatibility might be relevant to transfusion. But at that time the surgical procedures for moving blood in and out of bodies were far too hazardous for Landsteiner's grouping tests to be judged useful.[20] Complex, messy, and dangerous, blood transfusion was generally practiced only in extremis.[21] One of the most serious obstacles was clotting. Although fluid, blood is remarkably resistant to flowing out of the human body and will coagulate in tubes and syringes. Among the extraordinary solutions to this problem, in 1902 the US-based French surgeon Alexis Carrel reported delicately suturing together the blood vessels of donor and recipient, a fiendishly difficult procedure that imperiled both. Another US surgeon, George Washington Crile, modified this technique by using a silver cannula to connect the vessels, a practice that doctors in other countries adopted.[22] Although Crile himself tested blood compatibility by cross-matching (mixing together samples from donor and recipient before an operation), for many, such testing was neither possible nor necessary.[23] Most transfusion in the early 1900s was carried out without any awareness of blood grouping. After all, the surgical procedures used for transfusing human blood were precarious and intimate, and these encounters needed exceptional skill and careful monitoring, so blood group compatibility was far from being a priority.[24]

The First World War was pivotal for improving techniques for preserving and moving blood. Heavy artillery, a huge number of severely wounded soldiers, and field hospitals behind the trenches brought together large numbers of injured people suffering from shock. Doctors from the United States and Canada worked with Allied medical teams, who were inclined to experiment.[25] They found that they could routinely use sodium citrate to prevent blood clotting and could thus transfer blood between lightly injured and critically injured bodies using cannula and bottle or syringe. In that context, many people operating close to the front lines of battle witnessed and practiced blood transfusion for the first time.[26] Those workers also encountered the consequences of blood incompatibility, and although the intense pressure on surgeons often precluded such tests, they took knowledge of blood groups back to peacetime hospitals in their own countries.[27] With sodium citrate as an anticoagulant, indirect transfer made transfusion quick and safe enough that blood groups began to be relevant to the procedure. Now, transfusion textbooks could insist on the "absolute necessity of using bloods of the same group."[28]

Transfusion depended on people sharing their blood. In the 1910s and early 1920s, hospitals viewed patients' relatives as vital sources of this life-giving substance.[29] But as blood group compatibility became increasingly visible, it also became apparent that families could not always provide suitable donors. Hospitals soon developed more expansive social and administrative practices for procuring blood, and "sharing" came to include a wide range of exchanges. In the United States much blood was provided by (male) professional donors, who received payment for bleeding.[30] Some hospitals kept local lists of donors; others depended on commercial agencies that maintained professional donor registries and commanded high procurement fees. Still others relied on the American Red Cross, which operated a system of free donation. The New York Blood Transfusion Betterment Association of the 1920s attempted to reform and standardize transfusion in that city, offering only moderate payments and subjecting donors to health screens. In Chicago in 1936, the first blood "bank" did not depend on financial remuneration but required a patient or family member to "pay back" a prior withdrawal in blood.[31]

In Paris, the French government stepped in to limit what donors could be paid. Their program of "minimal compensation" was intended to offset the time and trouble of volunteers.[32] There, L'Oeuvre de la transfusion sanguine d'urgence (Emergency Blood Transfusion Service) kept an orderly system of donor cards and enforced stringent blood grouping and regular health checks. In Russia, individual hospitals—even individual surgeons—had their own ways of procuring blood. Some used free volunteers, others paid for it, and others used a combination of paid and unpaid.[33] Russian transfusion specialists held numerous meetings to try to institute a standardized system of donorship, but they were unable to persuade higher authorities to sanction the plan. This mixture of methods may have been a driving force that led to exceptional innovations in the Soviet Union: some researchers experimented with transfusions using blood from cadavers, and others focused on technologies of storage.[34]

In Britain, the London Red Cross Blood Transfusion Service emphasized the moral significance of freely given blood, a notion that it exported widely to other countries.[35] London transfusion organizer Percy

Oliver rigorously vetted his donors for good character as well as health. He insisted that donation was in the service of humanity and deliberately minimized contact between donor and recipient in case familiarity obscured the broader moral objective.[36] Oliver supplemented his lists by appealing to established organizations that already fostered a sense of civic duty. Blood donation flourished in an era of volunteerism, drawing on existing volunteer organizations such as the Rover Scouts.[37] As blood traveled further, and donor lists lengthened, transfusion produced a bureaucracy that grew from and expanded along the lines that connected families, communities, hospitals, and existing volunteer organizations.

This variation in the policies and politics of procurement was maintained against a backdrop of continued national and international debate over the curative powers of citrated versus fresh blood, the safety of preservatives, and the effectiveness of different types of equipment. But although transfusion organization differed between countries, overall trends in administrative practice are clear. First, hospitals became increasingly vigilant about checking the groups of donor and recipient. Second, hospitals expanded their pools of available donors by sharing lists. Private agencies, independent organizations, and philanthropic societies began organizing "panels"—that is, lists of willing donors—who served multiple hospitals. These institutions increasingly took on the work of mediating interactions between donors and hospitals. During the 1920s, the number of hospitals served by the London Red Cross increased exponentially. In Paris, L'Oeuvre de la transfusion sanguine d'urgence carried out only a few hundred transfusions a year in the late 1920s but many thousands in the 1930s. As these administrative systems expanded, the Soviet Union saw the earliest efforts toward a countrywide service. During the early 1930s, the highest government officers began responding to pressure for a centralized transfusion service, and a system, centered on Leningrad, came into operation.[38]

Third, blood could be preserved for longer and could travel further. Whereas donor lists had been limited by the distance donors could travel, during the mid-1930s, storage innovations allowed blood to move where donors could not. The widespread use of refrigeration was pioneered by the Soviet Union's new national infrastructure for transfusion. On the bloody battlefields of the Spanish Civil War, transfusion pioneer Frederic Duran Jordà used fridges and mobile refrigerated vans.[39] The Soviet Union and Spain took seriously the virtues of stored blood—in part because it was well suited to a socialist, centralized administration, and in part because of desperate need.[40] By the late 1930s, transfusion specialists across Europe were learning that fridges and anticlotting chemicals could maintain the therapeutic value of blood for up to two weeks.[41] The bureaucracy of transfusion in all places expanded as the storage capacity and mobility of blood gradually increased. Technological developments in Spain, in particular, would make a strong impression on those building a wartime transfusion infrastructure in Britain.

As the bureaucracy of transfusion expanded, and blood storage became more reliable, blood group tests became an essential and routine component of transfusion. Blood grouping required careful training—textbooks were full of detailed instructions—but expertise was spreading fast. As blood groups became more visible, so blood grouping antiserum became a crucial substance circulating between hospitals and transfusion centers. In 1930, the importance of blood groups was publicly acknowledged when Landsteiner was awarded the Nobel Prize for Physiology or Medicine and cast as the true hero of the therapy. Meanwhile blood group nomenclatures were becoming standardized internationally, a move driven by the first International Congress of Blood Transfusion in 1935, and the second in 1937.[42] At the latter, blood groups comprised a quarter of the whole program, given equal billing with discussions of blood storage and donor management.[43]

In summary, as new practices of blood transfusion proved their therapeutic power after the First World War, biochemical differences in human blood became clinically meaningful. Wartime experiments on anticlotting techniques allowed the movement of blood between bodies via syringe, making transfusion safer and easier. What had begun as a fragmented infrastructure for managing donors and blood had coalesced into a large-scale bureaucratic enterprise. As a consequence, the classificatory power of blood groups took on a new and vitally important meaning in surgery. The routine flow of blood between bodies made identifying blood groups crucial to the safety of transfusion. Blood groups were now also abundantly available as biochemical markers of human difference. The routine testing of blood groups, and the lists of blood groups in files and records, came to support new research on biochemical difference and politically urgent forms of social identity.

Genetics and Forensics

During the first three decades of the twentieth century, radically different cultures, nations, governments, and social movements sought to apply theories of heredity to reproductive policy and practice.[44] There was particularly keen interest in precisely how human mental and physical characteristics were inherited, but the specific pattern of inheritance of particular human traits—mental defect, intelligence, criminality—proved to be vexingly complex.[45] This was especially frustrating to scientists who were following the flourishing research programs of chromosome mapping on fruit flies and crop plants.[46] So it was tantalizing when in the 1910s serologists and microbiologists Ludwik Hirszfeld and Emil von Dungern first convincingly raised the possibility that blood groups were inherited via simple Mendelian laws.[47] Although the underlying genetics remained unclear for almost two decades, Hirszfeld and von Dungern established the ABO blood groups as some of the first human Mendelian traits.

Then, in the early 1920s, the abundance of blood group data generated by the transfusion services caught the eye of Felix Bernstein—mathematician, actuary, and head of the Institute of Mathematical Statistics in Göttingen, who was keenly interested in material to which he could apply new mathematical techniques for human genetics. Engaging with the *Vererbungsmathematik* (mathematics of inheritance) of German doctor and statistician Wilhelm Weinberg, Bernstein was fascinated that a simple equation might be used to estimate the population allele (gene) frequencies of a human trait from the frequencies of its observable characteristics (phenotype). Weinberg's methods offered a way of testing whether (and how) a trait was inherited via Mendelian laws (although Bernstein offered important revisions to these).[48] Historian Pauline Mazumdar explains that the theory was ready and waiting; now blood groups provided the data to which such theory could be applied and tested.[49] Bernstein applied Weinberg's methods to the blood group results of 20,000 people—from populations as diverse as Korea, India, Norway, and Madagascar. In 1924 he published his results, arguing that the ABO blood groups must be inherited via a single locus with the alternative alleles A, B, and O.[50] As Mazumdar puts it, Bernstein's paper resulted in "a global dust storm of Mendelian algebra" as serologists and geneticists tested Bernstein's methods using the abundance of blood group results that had accumulated in medical settings.[51]

Legal applications of Bernstein's conclusion followed swiftly.[52] Forensic science in Europe was becoming increasingly institutionalized, and the new genetic understanding of blood group inheritance suggested that blood groups might be used to resolve legal disputes over paternity.[53] During the same year that Bernstein persuasively demonstrated his model of ABO inheritance, German serologist Fritz Schiff presented the lecture "Blood Group Diagnosis as a Forensic Method" to the Medico-Legal

Society of Berlin.[54] Ten years earlier, Italian serologist Leone Lattes had shown that blood group tests could be carried out on weeks-old blood-stains, raising the possibility of blood group serology in forensic analysis. Now, Schiff showed that blood grouping could be used for another purpose: to resolve contested paternity. He explained to his audience that if the blood groups of a mother and child were known, then this narrowed the possible blood groups of the child's biological father. Thus, the courtroom soon became one of the settings in which blood groups achieved wide attention. In 1926 a blood group test changed the outcome of a courtroom trial for the first time. Within a year, serological testing had been requested for several hundred paternity cases in Germany.[55]

Schiff's publications received a wide readership, including in English-language journals.[56] His work generated discussion in medical societies across Europe about the use of blood group tests in establishing nonpaternity. By the end of the 1920s, blood groups were admitted in courtrooms in Sweden and Norway. In Italy they became allowed in 1931, in Ireland in 1932, in the United States in 1935, and in Britain in 1939.[57] By then medical experts were convinced of the use of blood grouping in paternity determination, although their wide acceptance by magistrates and juries took far longer. Nevertheless, the courtroom became an arena in which blood groups came to be understood as offering irrefutable evidence of fatherhood.[58] These were the first settings in which family relationships were legally defined by blood group.

Interest in the inheritance of blood groups reinforced the notion that these biochemical entities might underpin heritable differences between people. The spread of transfusion coincided with the intensification of racial nationalism across Europe. While governments considered new laws for the control of migration and reproduction, doctors and serologists carried out studies to correlate blood group categories with race, class, and nation.[59]

Race and Identity

Since Landsteiner's initial findings at the beginning of the century, serologists had been investigating associations between blood groups and characters such as skull and face shape, fertility, physical fitness, constitutional type, and skin, eye, and hair color. The First World War had produced the conditions not only for refining transfusion but also for the first systematic attempt to bring together blood group results collected from people of different national, religious, and racial identities. Ludwik Hirszfeld and his physician wife, Hanna Hirszfeld, worked as army doctors in the Aegean port of Salonika (now Thessaloniki), where various Allied troops had retreated from German forces. Bringing together their diagnostic serological expertise with their first attempts at blood transfusion, they took the opportunity to test the blood groups of troops and local populations that represented what they saw as "distinct" national or ethnic categories.[60] Their compelling results were published in the *Lancet* and *L'Anthropologie*; the *British Medical Journal* reportedly considered them too striking to be believed.[61] The Hirszfelds seemed to have shown that different national, racial, and religious populations had varying frequencies of the A and B blood groups. To capture this, they assigned a single "biochemical race index" to each "nationality" that indicated the ratio of A to B carriers in each population.[62] Amid fierce debates about which biological characteristics best distinguished races and nations, the Hirszfelds' index suggested that blood groups could sweep away contested racial classifications based on skulls, body shape, and skin color.[63]

Racial nationalism intensified and tied together a growing emphasis on the eugenic circumscription of national ideals with turbulent disputes about territory. A passion for serological race science studies rippled across Europe, the United States, the Middle East, and Japan. Serologists and doctors carried out their own tests but also mined donor lists in attempts to correlate blood group frequencies with the categories of race and nation. For the Hirszfelds and many of their readers, the unit of analysis was now no longer the blood groups themselves, but rather the frequency or ratios of blood groups within populations. Calculating a population frequency or ratio required that researchers circumscribe a population by nationality, race, ethnic group, religion, age, mental characteristic, or disease category. Having defined a population group in this way it was straightforward to calculate the ratio of A to B groups (or some other ratio), or the frequency of A (for example) in the population. Differences between populations were often slight and depended on the ways that the populations had been circumscribed, although sometimes the calculation of ratios and indices amplified those numbers. Methods were contested; there were plenty of disputes over what kinds of groupings might be understood as biological (or not)—ratios included the "biochemical race index," "racial index," blood group frequency, blood group ratios, and a wide array of other graphical methods based on these.[64] But the bigger message was that blood groups offered a newly scientific way of articulating human difference.

These population studies meant different things in different places. In German-speaking lands, much (but not all) research on blood groups and race was fashioned as an anthropological field that coalesced around the Deutsche Gesellschaft für Blutgruppenforschung (German Society for Blood Group Research).[65] Established in Vienna in 1926 by anthropologist Otto Reche and surgeon Paul Steffan, the society was firmly embedded in a *völkisch* ideology: as sometime editor of the journal *Volk und Rasse* (*People and Race*), Reche strongly advocated the notion that the human race of the future should be bred from the peasantry with long historical roots in German soil. Establishing the *Zeitschrift für Rassenphysiologie* (*Journal for Racial Physiology*), Reche and Steffan made serology serve a racial ideology that promoted the superiority of the Nordic stock. In numerous surveys and maps they forced a connection between blood and soil: the A blood group supposedly dominating Western Europe, and B prevailing in the Slavic East.[66] Reche worked alongside several senior Nazi officials through his work for *Volk und Rasse*, while the Deutsche Gesellschaft counted among its members luminaries from across the German academic world.[67]

Efforts to classify people serologically were by no means only a German phenomenon. Blood groups were made to function in discourses about race across Europe and beyond. Over Germany's eastern border, where central European nations were being reconfigured after the breakup of the Austro-Hungarian Empire, anthropologists and doctors attached different nationalist narratives to blood groups. As in Germany, anthropologists in Romania and Hungary envisaged the concept of *Volk* in physical and biological (as well as cultural and linguistic) terms, and they mobilized craniometry and serology to articulate the racial histories of their respective nations. Unlike members of the Deutsche Gesellschaft für Blutgruppenforschung, however, Hungarian and Romanian anthropologists defined their national types through a precise *mixture* of races, although they were no less anxious about the preservation and propagation of their respective national identities.[68]

The picture was also complex in the Middle East. Following the collapse of the Ottoman Empire, different ethnic and religious communities and Middle Eastern nation-states struggled for political sovereignty and international recognition. In an era of fervent nationalism, anthropologists invented or erased whole population groups by marshaling the anthropometric measurements of bodies, heads, skeletons, and skulls. Some also embraced the blood group techniques of the Hirszfelds, first pioneered in the former Ottoman region of Salonika. Blood group and anthropometric measurements were mobilized to consolidate or refute population groupings in Egypt, Turkey, Syria, and Palestine, identifying some groups as admixed and others as racially pure.[69]

In other countries, anthropologists hitched blood group research to imperial agendas. In Paris, the Centre d'études des groupes sanguins (Center for Blood Group Research) at the Pasteur Institute became a central institution for anthropological blood group research, headed by Nicholas Kossovitch, who had worked with Ludwik Hirszfeld during the First World War.[70] At the Pasteur Institute researchers focused especially on the populations of France and its colonies in West Africa and Madagascar.[71] British medical journals published data from grouping tests collected in hospitals and clinics in former or current parts of the European empires.[72] Transfusion arrived in Japan in 1919 as interest in "pure blood" was intensifying along with ambitions for imperial expansion.[73] US curiosity in blood group distributions was often directed toward indigenous communities, minority communities of color, and immigrant populations.[74] Meanwhile, the Soviet Union institutionalized the field in the country's territories by establishing a commission to study blood groups.[75]

Blood groups were taking the world of race science by storm—but they were contested, with many anthropologists deeply skeptical that that they were superior to anthropometric measurements.[76] The more population data that accumulated, the more complex the picture became. Some researchers were puzzled to find that apparently unrelated populations had near identical blood group frequencies. Others were dismayed that differences between populations were sometimes very slight and often bore little relationship to political borders. In Germany, notwithstanding the supremacist tenor of Reche's and Steffan's program, the reception of blood group work among Nazi audiences was mixed.[77] Many blood group frequency maps did not offer a clearly defined picture of racial types and so were not easily reconciled with a vision of an Aryan race. Nor did the Nazis, once they were in power, make much practical use of blood groups, generally preferring classification based on traditional physical characteristics.[78]

Several prominent geneticists and serologists reflected on the apparent failures of blood groups to clarify race. US geneticist Lawrence Snyder admitted that their applications to anthropological problems had been rather "vague," though he insisted that blood groups should still be studied alongside "pigmentation, hair form, cephalic index, and the rest."[79] In 1935, US immunochemist William Boyd and anthropologist Leyland Wyman reflected on why blood groups had not lived up to expectations. Marshaling contemporary theories of population genetics, the authors explained that while blood groups did not clearly correlate with the established "races," they promised to be valuable for subtler probing of the history of human migrations.[80]

Despite these worries, the quantity of blood group frequency data for racial, geographical, religious, and other human groupings continued to rise. In the late 1920s, some authors began synthetic compilations in books, which included *Blood Grouping in Relation to Clinical and Legal Medicine* (1929) by Snyder, *Individuality of the Blood* (1932) by Lattes, and *Blood Groups* (1939) by Boyd. These typically preserved the social groupings reported in individual studies, many of which had been carried out in the course of medical practice. From a French doctor using transfusion to treat malarial anemia in a hospital in Syria, to US immunologists carrying out serological tests in indigenous public schools in Oklahoma, on-the-ground judgments about the social identities of patients and donors in local surgeries and hospitals were written into published papers, then reproduced as such studies were incorporated into books and worldwide maps. Such publications argued that differences between human groups could be articulated biochemically. Because of where and how these blood groups had been collected, these human groupings had a granularity that was organized by the social and political boundaries that structured the interwar world.

Data was so abundant that by the early 1930s, more than fifty researchers across the world were actively publishing on national and racial distributions of blood groups. Historian William Schneider estimates that by 1939, this had resulted in more than 1,200 papers describing original research on the geographical, racial, or national distributions of blood group frequencies. Combining data from transfusion donors, hospital patients, prisoners, and military recruits, this amounted to tests on an estimated 1.3 million people.[81]

"Medicine, Biology, Anthropology"

The Second International Congress of Blood Transfusion, in Paris in 1937, marked two decades of remarkable changes in the understanding and meanings of blood. Opening the congress, the French Minister of Health, Marc Rucart, explained to delegates that the miraculous therapy of transfusion had only been made possible by the discovery of the blood groups. Only months before the congress, Nazi planes had demonstrated the deadly consequences of aerial bombardment over the Basque village of Guernica, and the Spanish Republicans were already making remarkable advances in transfusion technology.[82] It had taken two decades for transfusion to become a widely practiced therapy for shock related to blood loss, and it was now dramatically proving its worth in the face of the destructive technologies of modern warfare.

Yet Rucart declared that blood groups had ramifications "far beyond the scope of medicine, for the field of biology and anthropology."[83] By drawing special attention to their "anthropological" and "biological" significance, he was alluding to the profusion of studies that had now been published on the blood group frequencies of different racial and national populations, and to the idea that blood grouping would reform the science of human heredity. As the minister put it: blood groups had not just practical significance but also deep scientific value. His words, and their backdrop of growing fear of a new war in Europe, captured the hope invested in blood transfusion and the promises that blood grouping held for understanding human biological difference and identity.

Although by now the knowledge of blood groups could be heralded as having made transfusion possible, and Landsteiner was widely celebrated as their discoverer, their path to that status had not been so simple. Blood groups were not readily available objects that could be seen and handled. They had been brought into being by the improved preservation and expanded mobility of blood, and by technologies that included syringes, bottles, and donor lists. They depended on the cooperation (or coercion) of hundreds of thousands of people, who were recruited in contexts of political tension and war. Blood group population frequencies were shaped by the institutions in which blood was drawn, and by the everyday assessments and prejudices through which doctors and anthropologists shuffled and sorted people into social groupings. In an era of nationalist-inflected eugenics, blood group data could be deployed for strikingly different political agendas, and by amplifying or flattening population categories, they could be marshaled to support diverse national histories.

The frenzied interest in blood groups in continental Europe and the United States had almost passed Britain by during the 1920s. But in the early 1930s, they came to the attention of a small community of influential British geneticists who put them to work in another project: one that was no less nationalist, and was profoundly eugenic, but was also (for some) distinctively antifascist. This politically diverse group of scientists made blood groups into central objects for the reform of human heredity. They used blood groups to frame a new vision for "modern genetics."

TWO

Reforming Human Heredity in the 1930s

In December 1934 the president of the Royal Anthropological Institute wrote to Julian Huxley asking him to support a proposed "racial survey of Britain." The president, Edwin Smith, hoped that anthropological measurements of people across the country would "throw light not only on our history but also on sociological and medical questions."[1] Huxley agreed, but he told Smith that it was "extremely important" to have "the assistance of men versed in modern genetics." Huxley recommended long-standing friends and colleagues J. B. S. Haldane, Lancelot Hogben, and R. A. Fisher for this task, adding that he himself was writing "a little book on the topic of racial problems" because he was "rather appalled by the lack of appreciation among anthropologists of modern genetic work."[2]

When Huxley wrote of bringing "modern genetics" to surveys of human populations, he had blood groups in mind. Haldane, Hogben, and Fisher were all members of the Human Genetics Committee, established in 1932 by Britain's Medical Research Council (MRC). Interest in blood group research had been almost entirely absent from Britain during the 1920s.[3] But at its very first meeting, the MRC committee seized on recent European research and outlined how blood groups might reform the study of human heredity.[4] The committee members believed that with their clear-cut genetic inheritance, blood groups would provide a crucial reference point for pinning down the inheritance of more complex traits, such as "mental defect" or "intelligence," and might even lead to the first maps of human chromosomes.[5] As Fisher put it, blood groups could give human genetics a "a solidly objective foundation, under strict statistical control."[6]

As well as the study of human inheritance, members of the Human Genetics Committee sought to apply blood groups to the genetics of human populations. Fisher and Haldane were developing mathematical techniques for modeling the dynamics of small-effect Mendelian genes in populations—a field that would later become known as "population genetics." Haldane had recently become fascinated by the notion that the geographic diversity of blood groups might offer a way of studying human migrationary history. Meanwhile, these model genetic traits would prove useful in 1930s discourses about the British nation. Hogben and Huxley, in particular, believed that blood groups could be useful tools in their antifascist promotion of social equality and a democratic world order. Huxley, a prominent advocate for science and its application to social reform, outlined some of the moral lessons of blood group genetics in his "little book on racial problems," *We Europeans* (1935).

This chapter outlines why these scientists singled out blood groups in talking of modernity, reform, and race. It explores the disciplinary contexts in which blood group research was performed, the technologies through which it circulated, and the political uses to which it was put in 1930s Britain.

Heredity in Britain

In the 1920s, a community of intellectuals engaged in the problems of social reform and biology felt growing frustration. Commitments to eugenics underpinned interest in human heredity, and almost all of the scientists who claimed to study genetics believed that eugenic measures were essential for the long-term future of society.[7] But British eugenic research was apparently not keeping up with advances elsewhere. The Eugenics Education Society (later "Eugenics Society") counted plenty of prominent intellectuals among its supporters but never had as large a membership as similar organizations in other countries. It used public lectures, films, and the journal *Eugenics Review* to spread the message that the control of human heredity was of central civic concern, but it did not sponsor a great deal of research.[8] Eugenics was only weakly institutionalized among British universities: there were no university departments or institutes devoted to the subject. Moreover, the research that the Eugenic Society did sponsor tended to rely on the collection of family data on disease or social traits, and it generally took the observation that a defect ran in families to be evidence of genetic heritability.[9] To some, these methods looked very out of date. By that time many biologists were beginning to understand the problem of estimating genetic inheritance to be far more complex; German mathematicians were developing sophisticated corrections to account for small sample sizes and observation bias.[10] Several British biologists were concerned that the methods advocated by the Eugenics Society lagged behind the sophisticated mathematical techniques being developed elsewhere.[11]

The only British academic institution devoted to the sustained study of heredity and its eugenic implications was Karl Pearson's Department of Applied Statistics at University College London, which was home to the Galton Laboratory. Pearson dominated statistical theory in Britain, having invented the standard formula for the correlation coefficient, as well as the chi-squared test for estimating the goodness of fit between observation and theoretical prediction. With a bequest from the Victorian-era anthropologist and statistician Francis Galton for a laboratory dedicated to eugenics, Pearson drew to University College researchers from Scotland, continental Europe, the United States, India, and Japan. Thoroughly unconvinced that Mendelian genetics could offer meaningful insights into variation and evolution, Pearson and his colleagues used and developed "biometric" methods: that is, they collected data on continuous traits and applied statistical techniques to probe their inheritance. Pearson was a committed eugenicist but was disdainful of what he called Eugenics Society "propaganda." Styling his work as "scientific" and "mathematical," he cofounded and edited the journal *Biometrika*, which promoted the study of biological statistics.[12] Later he also founded the *Annals of Eugenics*, which he initially devoted "wholly to the scientific treatment of racial problems in man."[13] Most of the work of Pearson's department was published in these journals, and his group dominated the science of human heredity in Britain.[14]

But by the end of the 1920s, the questions and methods of human heredity research were shifting. Not only were some professional geneticists impatient of the Eugenics Society's methods, but, to some, Pearson's fierce resistance to Mendelian genetics was beginning to look outdated.[15] During that decade, both Haldane and Fisher had published significant work on the mathematics of selection in populations, showing how evolution might be modeled for traits inherited via Mendelian laws. Nevertheless, it was not yet at all clear how these techniques might be applied to the genetics of human populations.

Responding to this state of affairs, the MRC established a new commit-tee that would advise and direct funding for research on human genetics.[16] Pressure had been building for some time for a new British institution devoted to the science of human heredity. For example, Cambridge ge-neticist Charles Hurst had been lobbying the MRC for a new "Bureau of Human Genetics."[17] Although circumspect about Hurst's proposals, MRC secretary Walter Morley Fletcher and London School of Economics (LSE) director William Beveridge—social reformer and enthusiastic member of the Eugenics Society—agreed to convene a meeting of physicians, sociolo-gists, geneticists, and anthropologists to discuss the proposal.[18] Following the meeting, Hogben and Haldane persuaded Fletcher and Beveridge that rather than creating a new institute, the MRC would do better to recruit a group of experts who could advise and direct funding for human genetic research across the country.[19] Beveridge agreed, and the MRC formed its new Human Genetics Committee. It invited as members Fisher; Haldane; Hogben; Julia Bell, editor of *The Treasury of Human Inheritance*; Lionel Pen-rose, researcher at a mental hospital in Colchester; and Edward Cockayne, pediatrician at Great Ormond Street Hospital in London.

Their first meeting was held at the MRC offices in February 1932, where discussion was dominated by blood groups and the problem of genetic linkage.[20] Following Bernstein's sensational paper describing the genetics of the ABO groups, the German mathematician had recently published a new article describing a method for determining linkage be-tween the ABO blood group locus and another human trait, using infor-mation from just two generations.

Genetic linkage is a measure of the likelihood of the cosegregation (coinheritance) of two traits. That likelihood of cosegregation offered a way of estimating the approximate distance between specific genes on a chromosome: a basic genetic "map." Chromosome mapping had been a major objective of genetics since Thomas Hunt Morgan's fruit fly lab of the 1910s—but whereas *Drosophila* had only four pairs of chromosomes, humans had twenty-four (or so it was thought at the time; we now know the number is twenty-three), and so far only three blood group loci were known (ABO and the less important MN and P), severely limiting their mapping potential.[21] The Human Genetics Committee researchers were thrilled by Bernstein's advances, but they knew that many more blood group loci needed to be found before they would have a reasonable chance of establishing linkage to disease traits.[22] Nevertheless, things looked promising: Haldane had recently introduced Fisher to the work of geneticist Charles Todd, who was studying serology in poultry at the MRC research institute in Hampstead. Todd had identified a large number of new blood groups in chickens, and the geneticists took this as an indication that many more might be discovered in humans, considerably expanding the potential for linkage mapping.[23]

Fired up by this possibility, all five researchers agreed that blood group serology offered the most promising avenue for reforming human heredity and applying it to eugenic problems. They declared that they would "welcome the fullest possible extension of Todd's work" and that every opportunity should be taken to promote "the study of serological differences among men and animals."[24] Writing to Todd after the meeting, Fisher conveyed the committee's view that serological work would "lead to a greater advance, both theoretical and practical, in the problems of human genetics than that expected from *any further work on biometrical or genealogical lines*" (my italics).[25] As Hogben put it in a report after the meeting, the blood groups offered the distinct possibility of a "chromosome map of the human species."[26]

One strand of this proposed research was an urgent need to collect blood group data alongside data relating to disease, with the aim of tracking their degree of coinheritance. Julia Bell had worked for many years collecting pedigree data at University College. Both a mathematician and medical doctor, she was overseeing Karl Pearson's long-running project on human disease and mental disorders for the *Treasury of Human Inheritance*, a massive, multivolume compendium of pedigrees.[27] Open to new methods in genetics, Bell began integrating blood group data into her work, collecting material herself from medical records and textbooks, families and physicians.[28]

Bell's interests in pedigrees overlapped with those of Penrose, who was enthusiastic about the potential for blood groups to disentangle the roles of heredity and environment in mental disorders. Penrose had been licensed in medicine in 1919, and in 1930 he had obtained a post at the Royal Eastern Counties Institution, a hospital for people suffering from mental conditions. He believed strongly in the social and political utility of human genetics, but he was concerned about the biases and assumptions of the research methods promoted by Britain's Eugenics Society. Penrose was deeply invested in the clinical prognostic value of genetics and, in developing Bernstein's methods, he offered a technique for estimating linkage from family genotypes of a single generation. Supported by the Pinsent-Darwin Trust and the MRC, he embarked on a large-scale study of the clinical and genetic aspects of mental defect. The Colchester Survey, which ran from 1931 to 1938, involved 1,280 patients, one of the largest studies of human inheritance to date.[29] Penrose outlined a detailed methodological vision of new standards for research on mental health disorders in his book *Mental Defect* (1933).[30] Although he did not incorporate blood groups into the Colchester Survey, Penrose tested the blood groups of 1,000 patients and remained closely engaged with serological research for several decades.[31]

Hogben incorporated blood group genetics into his program to give the social sciences a mathematical basis. Newly appointed to the chair of social biology at the LSE, he believed that the science of human heredity was in desperate need of reform. Hogben had been a committed socialist since school and envisaged a society transformed by widespread understanding and appreciation of science.[32] Appalled by the racial prejudice he had witnessed among scientists while working in Cape Town, he arrived at his position back in London as part of a program set up by Beveridge to import quantitative methods developed from biology into the social sciences at LSE.[33] Hogben was by no means against eugenics per se—he could see many valid reasons to be concerned about the future evolution of humankind—but he led an attack on what he saw as the uncritical and biased methods of the Eugenics Society.[34] Hogben was already putting the final touches to his *Genetic Principles in Medicine and Social Science* (1931) when he discovered Bernstein's work on blood groups. Hurriedly revising his proofs, Hogben added an extra chapter that explained that blood groups were both "an encouragement for the belief that human genetics may be made an exact science, and . . . an object lesson to those who are disposed to construct pretentious hypotheses on the basis of isolated pedigrees."[35] In Hogben's view, building a new genetic approach to human heredity on a foundation of blood groups would prevent its use to support racial and class prejudice.

Meanwhile, Haldane and Fisher were helping to establish a strand of mathematical genetics that dealt with the dynamics of Mendelian genes in large populations. In 1930, Haldane was holding down three professional positions, as a lecturer in biochemistry at the University of Cambridge, as "officer in charge of genetical investigations" at the John Innes Horticultural Research Station at Merton in South London, and as Fullerian Professor of Physiology at the Royal Institution, London.[36] A socialist since his student days, Haldane wrote extensively about science and its application to politics and everyday life, and by the 1930s, his articles, lectures, and broadcasts had made him a well-known public figure. Haldane's significant mathematical contributions included work on detecting genetic linkage, and an extensive series of papers on artificial and natural selection published between 1924 and 1934.

Fisher worked at Rothamsted Experimental Station in Harpenden, north of London, where he was developing new methods for the design of experiments on fertilizers and crop yields and for the statistical analysis of large quantities of data. Fisher also had considerable freedom to pursue a wide range of genetic research projects, directing breeding experiments on mice, snails, and poultry and writing a series of articles on the mathematics of evolution, culminating in *The Genetical Theory of Natural Selection* (1930).[37] A committed eugenicist, he wrote extensively on the implications of genetics for the future of society and of humankind, and he devoted almost half of the *Genetical Theory* to human heredity and evolution.[38] Haldane's and Fisher's "population genetics" would later underpin the notion that genetics was fundamental to a synthetic understanding of biology and its evolution, a view powerfully promoted by Huxley in *Evolution: A Modern Synthesis* (1942).[39] If blood groups were amenable to the tools of population genetics, then humans might be brought into this scheme.

The members of the MRC Human Genetics Committee had diverse political commitments. Fisher was a political conservative, advocating a kind of eugenics that would promote the interests of elites and certain sections of Britain's middle class.[40] Haldane, Hogben, and Huxley were all left leaning and believed that eugenic principles should be utilized to bring about greater social equality.[41] But all the committee members were united in their desire to reform human heredity. Large quantities of data, rigorous methods for selecting subjects, and the use of mathematics for asking questions about linkage represented a style of human genetics quite different from that currently advocated by the Eugenics Society or promoted by Pearson. With knowledge of blood groups in hand, Hogben, Haldane, Fisher, Penrose, and Bell believed they could put human heredity on a Mendelian footing.

British Race Science

Newly confident that they could finally turn the study of human genetics into an "exact" science, several members of the Human Genetics Committee also made interventions in the meanings and scope of race. Haldane had recently become fascinated by the notion that the geographic distribution of blood groups could offer insights into human migrationary history. In 1931 Haldane delivered a lecture to London's Royal Institution called "Prehistory in Light of Genetics," in which he introduced his British audience to the now abundant overseas work on the geographical and racial distributions of blood groups. But rather than focusing on their potential for race classification, Haldane explained that a gradient in the frequency of any one blood group allele represented a trace of past human migratory history. Presenting an isoline map showing the worldwide distribution of blood group B, he argued that it suggested strongly that it corresponded to a "migration outwards from Central Asia in prehistoric times" (figure 2.1).[42] Where blood group population data had previously served arguments about race classification, now it bore traces of humankind's historical past.[43]

In claiming that the geographic distribution of genetic diversity could reveal human prehistory, Haldane cited Russian geneticist Nikolai Vavilov, who had recently visited London and presented Soviet research on genetic geography.[44] Speaking at the Second International Congress for the History of Science a few months earlier, Vavilov had argued that hot spots of plant genetic diversity indicated the centers of origin of plants, and might correspond to prehistorical human settlements.[45] Haldane presented to his Royal Institution audience one of Vavilov's maps alongside his blood group map, arguing that they both demonstrated the concurrent historical migration of people and agriculture. More fundamentally, they demonstrated that genetics might become an indispensable tool for understanding human prehistory: "While the work presented here is far from complete, I think that it has now progressed so far that no anthropologist who wishes to take a large view of human origins can possibly neglect it."[46]

Haldane would later elaborate this to claim that different kinds of genetic character could resolve history at different temporal scales. Using geology as a comparative metaphor, he explained that the frequency distributions of some characters—"such as pigmentation"—yielded information about the recent past, "just as the recent and Pleistocene deposits tell of recent glaciation, volcanism, and so on." Others, like blood groups, had not apparently been shaped by evolutionary selection (although on this point Fisher disagreed) and so had the potential to "give information of a more fundamental character on racial structure, just as do the Paleozoic rocks on geological structure."[47] The historical scale that could be resolved from such data depended on whether a character had been shaped through evolutionary selection or not. Genetics had the potential to yield multiple layers of human history.

We have already seen that Karl Landsteiner and colleagues were attempting to use blood group data to recover the evolutionary relationships between humans and primates (chapter 1). This was different: Haldane's excitement over blood groups was about their potential to illuminate human population genetics and recover histories of settlement

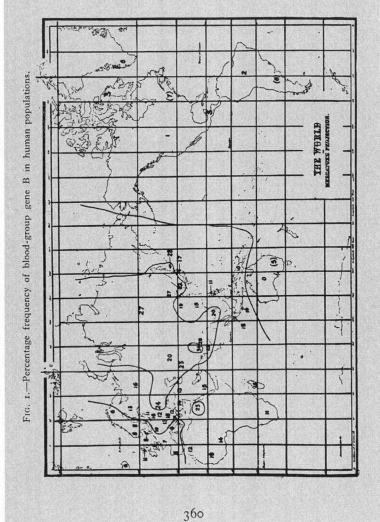

Fig. 1.—Percentage frequency of blood-group gene B in human populations.

THE WORLD
MERCATOR PROJECTION.

2.1 A contour map presented to the Royal Institution by J. B. S. Haldane and published in his 1931 paper "Prehistory in the Light of Genetics." Haldane superimposed on a Mercator projection map of the world numbers indicating the percentage frequency in local populations of "blood-group gene B." He used these numbers to estimate the positions of isolines, which connected geographic regions with equal allele frequencies. Haldane intended the isolines to suggest a diffusion of allelic variation across geographic space, which, in turn, would indicate historical human migration. From Haldane, "Prehistory in the Light of Genetics" (1931), 360. The Royal Institution, London/Bridgeman Images.

and migration. This was evolutionary biology at a local, incremental level. Population genetics was about resolving the mechanisms of evolutionary change, rather than resolving deeper species relationships.

Haldane enthusiastically promoted genetics in academic debates about the meaning and scope of race science. In 1934 he joined the Race and Culture Committee organized by the Royal Anthropological Institute and London's Institute of Sociology to "clarify the term 'race'" and to "consider the significance of the racial factor in cultural development."[48] The committee sought to determine "how far particular races and populations are actually linked with particular cultures"—in essence, did "race" determine "culture"?[49] The committee consisted of anthropologists, anatomists, archaeologists, and sociologists, with widely divergent interests—ranging from racial determinists George Pitt-Rivers and Reginald Ruggles Gates to left-wing pacifist Herbert Fleure.[50] The resulting twenty-four-page pamphlet, *Race and Culture* (1936), gives insights into the disputes over race in Britain and indicates Haldane's success in persuading his colleagues that genetics had something to offer.[51] One of the definitions of "race" given in the pamphlet was distinctly genetic:

A Race is composed of one or more interbreeding groups of individuals. . . . It is a biological group or stock possessing in common an undetermined number of associated genetical characteristics by which it can be distinguished from other groups, and by which its descendants will be distinguished under conditions of continuous isolation.[52]

But when Haldane claimed that race was "genetical" and that blood groups offered a reasonable way of moving forward in the study of human diversity, what was it that he was arguing *against*?

In Britain during the 1930s, "race science" was not a single coherent field. Race was studied by comparative anatomists, physicians, geographers, anthropologists, and physicians. In practice, this meant evaluating and measuring a wide range of human characteristics, of both living people (body measurements; head shape; skin, hair, and eye color; customs; personality; reaction times) and cadavers and human remains (skull measurements, excavated artifacts). These studies depended on surveys by teachers and doctors, and on large numbers of artifacts and bones collected by colonial officials and missionaries overseas, as well as from graves unearthed around the United Kingdom. In Britain such studies were deployed in debates about human origins and evolution, human migratory history and racial identity, and territory and empire. By the early twentieth century, prodigious quantities of these kinds of research were published in the journals of the Royal Anthropological Institute, *Biometrika* and *Annals of Eugenics*.[53]

These last two were overseen by Pearson, whose department at University College produced a large quantity of research on race, alongside his work on heredity. Pearson had inherited his pursuit of "types" from his patron Francis Galton; he developed techniques for reducing numerous measurements on his extensive collections of skulls to mathematical indices that he believed could encapsulate the abiding characteristics of a race, and offer insights into racial history and early migration.[54] Pearson was a positivist and a socialist, and he was radically nationalist, believing that the state's interests could be advanced through economic and military competition with other nations. For Pearson a "civilization" demanded "the struggle of race with race, and the survival of the physically and mentally fitter race," as he declared in a lecture and pamphlet published as *National Life from the Standpoint of Science* (1901). For Pearson, and many who shared his social-Darwinist beliefs, there was much to be gained from a detailed understanding of the racial "types" under the imperial power of the British government.[55]

The study of racial types was an important strand of research in interwar Britain.[56] But beyond typology, there were other ways of studying race. Specifically, many geographers studied human physical characters in relation to landscape and applied their studies to debates about town and country planning, human population migration, and negotiations over political boundaries in Britain and Europe.[57] Their concern with population dynamics, mixing, and landscape cohered neatly with Haldane's claims for blood group maps.

The Race and Culture Committee included prolific geographer and anthropologist Herbert Fleure, who advocated a style of nationalism that was closely focused on the lives of the British people. Twenty years younger than Pearson and committed to a kind of eugenics more closely aligned with the political left, Fleure did work that was emblematic of British interwar geography's interest in race. While Pearson and his colleagues tended to study race using skulls, Fleure was an assiduous surveyor of living populations.[58] Originally trained as a zoologist, he was professor of geography and anthropology at the University of Wales at Aberystwyth, and a prolific author of books and articles on race and "human geography" for both academic and nonspecialist audiences.[59] Between 1909 and 1940, much of Fleure's research consisted of detailed surveys of Wales and the Isle of Man, in which he combined the collection of archaeological data with extensive detailed anthropometric measurements on living people.[60]

Fleure was committed to "human geography," a strand of research that investigated the relationships between the physical characteristics of humans, their culture and traditions, and the landscape.[61] Deeply influenced by fellow geographer and sociologist Patrick Geddes, Fleure was devoted to the Regional Survey movement, which had aimed to cultivate in British citizens an interest in, and affinity with, the local landscape.[62] The First World War had given new impetus to that project, having wrought, in another geographer's words, "catastrophic" disturbances to the "equilibrium of civilization and the physical environment."[63] Fleure argued that varied environments around the British Isles had shaped the physical, social, and spiritual characteristics of different social groups; it was an approach that in many ways resembled some of the *völkisch* principles advocated in Germany during the same period.[64] But he was deeply critical of racial theories used to veil insidious political propaganda, even insisting that the term "race types . . . should not be used without great reserve in scientific discussion."[65] Nevertheless, he believed that the careful study of physical characteristics could yield important information about the races and movements of the past. He focused his own work on "fringe" or "remote" populations, which he considered epistemologically valuable because they were assumed to have stayed in one place and therefore to have preserved traditions and types far older and more informative than those in other regions of the country.[66]

Thus, Fleure and other human geographers emphasized the study of living people, the geographical distributions of their characters, and the continuous dynamics of populations. To Haldane, blood group frequency data likewise proved that human populations were dynamic and subject to selection, drift, and other genetic forces. That geography was one of the principal academic discourses on race goes some way toward explaining why Haldane could persuade colleagues on the Race and Culture Committee that genetics was central to race.

"We Europeans"

While Haldane pushed for a genetic race science at the Royal Institution and Royal Anthropological Institute, Huxley was particularly vocal in bringing this message to audiences beyond the academy. Like Fleure, Huxley believed that there were great political dangers to existing "popular thinking" about racial types, and he argued that a scientific approach was necessary for reform. Genetics was revealing that animal, plant, and human populations harbored an astonishing and previously unseen genetic diversity. Moreover, mathematical population genetics had apparently shown that natural selection did not eradicate that genetic variation. Rather, at equilibrium, gene ratios remained stable, generation after generation—racial purity was not natural, but diversity was. Huxley and colleagues saw an opportunity to use genetics to speak out against fascism—to use science to promote social justice.[67] Huxley made these arguments about genetics and race in the popular British book *We Europeans: A Survey of Racial Problems* (1935). Coauthored with anthropologist Alfred Haddon and sociologist and demographer Alexander Carr-Saunders (and republished as a Penguin paperback in 1939), *We Europeans* sought to counter a "vast pseudoscience of 'racial biology'" that served "to justify political ambitions, economic ends, social grudges and class prejudices."[68] The book explains that the "geographical distributions of the blood groups" offer a "new approach" to race, one that disavows the fixity of races and instead shows just how mixed "we Europeans" really are.[69] The phrase "we Europeans" was intentionally vague, and it tacitly underlined a distinction between the European nations and their imperial territories.[70] But it also sought to convey a message of unity in the context of intensifying territorial disputes unsettling the continent. The uses to which *We Europeans* put blood groups were no less political than those described in chapter 1: in its antifascism the book was deeply nationalistic, citing a range of social statistics that compared Britain favorably to Germany. Strikingly, though, the book's moral argument was that genetics not only offered a politically neutral race science but also had lessons about a future democratic world order, an assertion that (as we shall see later) Huxley and others rearticulated with more force on an international platform after the Second World War.[71]

In 1936, the year after *We Europeans* appeared, the British Association for the Advancement of Science convened a "lively" meeting on the topic of "genetics and race," recounted in the *Manchester Guardian* in an article entitled "Nazi Conception under Fire." The session was a fiery public conversation between Huxley, Fleure, Pearson's loyal colleague Geoffrey Morant, and right-wing eugenicist Reginald Ruggles Gates. Fleure, now fully subscribed to Huxley's view, enthusiastically declared that "race types" could now be explained using "the modern science of genetics." Fleure claimed that Mendelian theory showed how the "characters of ancestors" could reappear generations later in their descendants. The term "race," he believed, had too fixed a meaning to capture the phenomenon explained by Mendelian genetics. Huxley backed him up, declaring, "Race could only be defined genetically, and if we tried to define it in terms of any culture or nation we could not do so in any scientific terms." Fleure went on to disavow the term "race" entirely. Genetics shows us that the world is too mixed to depend on such categories, he said; the term was "obstructing anthropological progress. It prevents us from asking the right questions."[72]

So where some interwar serologists and anthropologists in several countries had mobilized blood groups to serve narratives of nationhood and belonging, from the 1930s Huxley and colleagues used them to disavow fascist commitments to race purity. One conclusion to be drawn from this is that blood groups were strikingly flexible. In an era of nationalist-inflected eugenics, they could be deployed for remarkably divergent political agendas, whether in the pages of *Zeitschrift für Rassenphysiologie* or *We Europeans*. In Britain, it was the emphasis on deep human history that resonated with anthropologist–geographers like Fleure, who later became an influential postwar advocate for the use of blood group genetics to study human diversity. Fleure's enthusiasm for blood group genetics particularly highlights how blood groups powerfully appealed to those who wanted a dynamic, politically engaged, geographical race science that could claim to be unshackled from the prejudicial excesses of the past.

The Galton Serological Laboratory

The Human Genetics Committee's ambitions for blood group genetics were most clearly articulated by a new laboratory that Fisher established in the mid-1930s.[73] In 1934, Daniel O'Brien of the Rockefeller Foundation traveled to London to discuss with the MRC how the foundation might best support research into human genetics, and especially mental disease.[74] He met with several members of the MRC Human Genetics Committee, including Fisher, who laid out to O'Brien a novel program of laboratory-based research that would use blood groups to study human heredity. One year earlier, Pearson had retired from University College at the end of a long and immensely productive career. In a decisive shift of emphasis, the college appointed Haldane as professor of genetics on a part-time basis, and Fisher as professor of eugenics—making him also director of the Galton Laboratory and editor of *Annals of Eugenics*.

From the outset, Fisher found himself up against the college's limited financial resources, and he was grateful for the chance to discuss funding with O'Brien. He enumerated the ways in which blood group research could benefit the study of human defects. Identifying blood group markers that cosegregated with disease loci could potentially help identify people carrying disease-causing variants. Establishing genetic linkage between blood groups and human disorders offered the potential to map genes responsible. Fisher expressed the hope that some gene variants responsible for "human anomalies" might themselves be detected via serological tests. Moreover, serological methods might help not only to elucidate medical disorders but also to detect factors that "exert a positively beneficial influence on health, intelligence, artistic appreciation, sensory discrimination, longevity etc."[75]

The notion that blood groups would help to identify people carrying disease genes particularly appealed to the MRC and the Rockefeller Foundation. It cohered perfectly with interests within in the MRC, which had recently established a Mental Disorders Committee. It also impressed Rockefeller officials. O'Brien underlined how valuable the work was, assuring his directors that the study of "mental defectives" was the "most obvious follow-up of these serological genetic studies."[76] The foundation was so impressed by these potential applications of Fisher's proposed work that they decided to grant him the funding from the medical sciences rather than from the natural sciences program, with the expectation that Fisher would in some way tether "fundamental genetics" to medicine.[77] The Rockefeller's enthusiasm signaled a firm belief that the study of mental traits represented a major practical application for blood group genetics.

In April 1935, the Rockefeller Foundation approved support for the new Galton Serological Laboratory—funding that would be administered by the MRC.[78] To run the lab, Fisher appointed George Taylor, a medically trained serologist who had worked in the pathology department at the University of Cambridge for six years.[79] A few months later, Fisher and Taylor hired research assistants Aileen Prior and Elizabeth Ikin, and the medically qualified Robert Race, who had been working as an assistant pathologist at the Hospital for Consumption and Diseases of the Chest in Brompton.[80] Because this was the first dedicated blood grouping lab in Britain, Taylor sought advice and resources from well-established serological laboratories in Europe and the United States. In 1936, he spent several months in the Retsmedicinsk Institut (Forensic Institute) of Copenhagen University, learning new techniques in their lavish new laboratories and ordering up-to-date equipment for Fisher. University College granted Fisher part of the refurbished animal house for keeping immunized rabbits, which were suitable for making some grouping sera. He was also given half of the former department museum for his blood grouping equipment, which included a refrigerator, centrifuge and hot-air oven. The serological work took place alongside Fisher's numerous other genetic projects, including a substantial program of research on the mathematics of population genetics, as well as selection experiments on animals.[81]

Over the next five years the Galton Serological Laboratory, collaborating with Hogben, Haldane, Bell, and Penrose, developed methods for investigating blood groups and human disease, and collaborated with hospitals and general practitioners.[82] Alongside the ABO system, new groups had recently been discovered. Karl Landsteiner and Philip Levine in New York had defined the P blood group in 1927, and the MN groups in 1928. The British researchers carried out their own hunt for new serological reactions. Fisher, Penrose, and their colleague John Fraser Roberts carried out blood group surveys on patients in two major mental hospitals, apparently finding "a very remarkable series of reactions of blood" among these individuals. This seemed to point to a "genetical factor" found only among the "mentally deficient," which was exactly the kind of thing that the Rockefeller had been hoping for.[83] It looked as though they had identified a serological marker relevant to mental disease, and Fisher's results were sufficiently promising for the MRC to provide money for two new assistants to work on the case.[84]

Meanwhile, down the road at LSE, Hogben was applying Bernstein's mathematical techniques to a pedigree of Friedreich's ataxia, a disease causing progressive damage to the nervous system. He followed the families of patients from hospitals in London, visiting them at home and testing their blood groups.[85] At the Galton, Ikin and Prior were collaborating with physicians around the country to find evidence of linkage between blood groups and a range of diseases. Taylor worked with doctors on hereditary eye conditions, with Penrose on phenylketonuria, and with hematologist Janet Vaughan of the Hammersmith Hospital on acholuric jaundice.[86]

Alongside this work on linkage, another strand of the Galton Serological Laboratory's research was to adapt mathematical methods of population genetics to the study of human populations. Collecting data from colleagues, students, and friends, Taylor and Prior surveyed the ABO and MN blood groups of just over 400 unrelated people. Although there had been a few studies of blood groups in Britain to that point, including one on forty families in Glasgow, this London study was the most extensive blood group survey ever carried out in Britain. As Taylor and Prior described in a three-part paper published in *Annals of Eugenics*, their aims were severalfold. One was to explain the techniques deployed for grouping, especially the titrations used for determining the newly discovered M and N groups. Another was to generate Mendelian human data for testing new mathematical techniques for probing inheritance. Taylor and Prior also used both family and population data to test and elaborate the techniques described by Bernstein and modified by Fisher, and they concluded that the blood groups of "England" agreed with "the accepted genetic theory."[87]

Thus, during its first few years, the Galton Laboratory developed and institutionalized some of the new standards for human genetics envisaged during the early meetings of the Human Genetics Committee. Departing radically from Pearson's biometric program, its approach was intended to make human heredity Mendelian. Publicly underlining this shift, Fisher changed the slogan of the *Annals of Eugenics* from "A Journal for the Scientific Study of Racial Problems" to "An International Journal of Human Genetics." With its emphasis on data abundance and mathematical analysis, the Human Genetic Committee would seek to elevate eugenics from an uncritical reliance on pedigrees. Above all, the Galton Serological Laboratory would put blood group research in Britain on an institutional footing that would be sustained for the next twenty years.

From Skulls to Blood

In May 1935, Fisher wrote a frustrated letter to Pearson about the skulls, skeletons, and skins still taking up space in the museum rooms at the laboratory, two years after Pearson's retirement. Fisher pleaded that Pearson clear his private collection, to make room for the fridge, centrifuge, and oven that Fisher had recently purchased for a blood grouping laboratory. He implored Pearson to "make arrangements during this spring or summer for housing this material elsewhere as it is certain that the museum will be increasingly needed for other purposes."[88] Blood groups versus skulls, skins, and skeletons: the dramatic contrast drawn by this correspondence between long-term combatants Fisher and Pearson highlights not just to the novelty of blood group genetics but also its local reformist meanings.

Fisher's phrasing speaks volumes about his vision for "modern" human genetics. In direct opposition to Pearson's research program, but in keeping with the vision of the MRC's Human Genetics Committee, the Galton Serological Laboratory promoted standards for human heredity research that could now claim to be Mendelian. The local displacement of Pearson's specimens by Fisher's refrigerators represented a shift from a science of race based on skulls and skeletons to the study of living people, their geographic distribution, and the population genetics of their blood groups.[89] While Pearson specifically wrote that he did "not believe that measurements on living people are of much value," Fisher bemoaned that the durability of skulls had meant that they were collected and stored in absurdly large numbers.[90] He argued that, by contrast, living material had huge advantages: "the sex is known" and "blood relationships" are known, "as are nationality, language, religion and social status." Most important of all: "The student of living measurements can choose his material and be sure of getting enough of it."[91] Over the next five years, Fisher would find he could get more than enough. In 1939, the Galton Serological Laboratory received an appeal for practical work that Fisher had not anticipated: to become part of the Emergency Blood Transfusion Service in anticipation of a new war.

THREE

Blood Groups at War

In October 1939, R. A. Fisher and George Taylor published an appeal in the *British Medical Journal* (*BMJ*) to workers in the newly founded Emergency Blood Transfusion Service (EBTS). For the past five years the two men had been running the Rockefeller-funded Galton Serological Laboratory at University College London, from which they collaborated with other geneticists to investigate the genetic relationships between diseases and blood groups. With the outbreak of war, the lab was coopted by the EBTS to provide antisera (that is, testing reagents) to the new wartime depot laboratories. As the Rockefeller Foundation suspended its funding, the MRC reorganized the lab's responsibilities, moving it to Cambridge and renaming it the Galton Serum Unit.[1] A little dismayed by this turn of events, Fisher nevertheless saw a remarkable opportunity to couple his genetic research to the needs of the EBTS. The *BMJ* appeal authored by Fisher and Taylor was entitled "Blood Groups of Great Britain" and entreated blood depot medical officers to send them the blood group results of volunteer donors. These records would apparently constitute valuable "genetical and ethnological data," which "not only [will] . . . throw light on points that require very large numbers for their elucidation, but will open up the field, at present wholly unexplored, of the homogeneity or heterogeneity in respect to blood groups of the population of these islands."[2] In other words, EBTS records might yield important information about the genetics of the British people.

Fisher and Taylor published their appeal only one month into the war. The lab was by then at the center of a network of wartime depots. The EBTS already had an extensive infrastructure composed of apparatus (bottles, tubes, needles, sterilizers, refrigerators), transportation (crates and vans), institutions (the Post Office, blood depots, volunteer donor organizations, hospitals, and laboratories for blood processing and testing), as well as lots and lots of paper. The last took the form of index cards, lists, letters, enrollment forms, donor cards, and labels, and it had devoted to it a veritable army of clerical staff responsible for managing and directing people and blood. In response to their appeal, Fisher and Taylor were rapidly inundated with lists of blood group results and donor enrollment forms, and with those they launched two research programs that they pursued throughout the war. Blood transfusion and human genetics had become institutionally linked for the first time, a coupling that would endure for two decades.

The last chapter described how Fisher and his colleagues on the Human Genetics Committee developed theoretical and ideological commitments to blood group genetics. This chapter is about how these were put into practice at a time of war. This is, in part, a story about how research is shaped by infrastructures: that is, the apparatus, institutions, materials and social practices that provide the resources for doing science. The infrastructures that condition science can be hard to see.[3] British wartime blood group genetics is unusual because the infrastructural change was so decisive. In just a couple of years, transfusion shifted from a set of small-scale donor systems with few formal institutional connections between them, to a nationwide wartime service underpinned by the routines and procedures for the mass storage and management of blood.[4] In part a response to the alienating nature of these new technologies, donor recruitment propaganda figured blood donation as a humanist contribution to the war effort.[5] The 1939 national remodeling of transfusion put in place not only new technologies for moving blood and paper but also novel centers of expertise. As the Galton Serum Unit became one such center, its workers turned EBTS institutions, materials, and cultures of exchange into unparalleled resources for studying blood groups. Recruitment programs, cooperative depot directors, clerks, a reliable paper-based administration, and a postal service were transformed into resources for human genetics.

This is also a story about what can be done with paper. The EBTS infrastructure was held together by a system of donor registry that chimed with the era of paper-based citizenship ushered in by the 1939 National Registration Act. Like ration books and identity cards, color-coded donor cards became another way of managing and monitoring individuals.[6] Clerks used letters to summon donors to specific places at particular times to give blood. Index cards noting blood group and general state of health defined the value of each donor (figure 3.1). Labels on bottles

3.1 A photograph of "the office at work": the North West London Blood Supply Depot, housed in the Slough Social Centre, taken between 1940 and 1943. Janet Vaughan (in glasses) presides over female clerks sorting registrations and calling up donors. To the left is a large map on which the hospitals supplied by the depot are marked. The Slough depot was responsible for the blood supply of the northwest quarter of London, which included Basingstoke, Buckingham, and Aylesbury. Vaughan recalled that an essential part of her job was to visit all of those hospitals, check that they had the right supplies of blood and apparatus, and bring them news of the latest developments in transfusion techniques. Made as part of a series of publicity photos for the Emergency Blood Transfusion Service, the photo underlines the broad geographical reach of this community service and the atmosphere of busy concentration created by its workers.
Reproduced with the kind permission of the Bodleian Libraries, University of Oxford.

directed blood to the correct bodies. Letters between Galton Serum Unit researchers and depot directors determined the kinds and quantities of data that would reach Cambridge. Antisera defined the wartime relationship between unit researchers and transfusion officers, and paper mediated those relations.

Wartime Transfusion Infrastructure

Technologies of blood preservation remodeled transfusion in Britain from a series of local donor panels and collection sites into a large-scale, modern, nationwide service. As historian Nicholas Whitfield has ana-lyzed in rich detail, this was also a shift from one-to-one giving to the large-scale collection and cold storage banking of blood.[7] Plans for the wartime service began shortly after the crises of August and Septem-ber 1938, when German troops invaded Czechoslovakia. In November that year, prominent London hematologist and socialist activist Janet Vaughan convened an informal "subcommittee" of medical practition-ers in her Bloomsbury flat to discuss advances in blood transfusion, es-pecially those being developed in Spain and the Soviet Union.

Vaughan was an established expert on the diseases of the blood, which were only gradually becoming consolidated into the medical discipline of hematology. She was author of the standard text *The Anaemias* (1934) and was in charge of blood medicine and transfusion at the Hammer-smith Hospital.[8] Immersed in a progressive left-wing community, with links to the Bloomsbury Group, Vaughan was director of the Holborn and West Central London branch of the Spanish Medical Aid Commit-tee, which organized the provision of medical supplies to Republicans fighting in the Spanish Civil War. In her memoir, she recalled how the "committee met night after night in a small attic room up many dark stairs . . . trying in vain to keep track of all the many conflicting left wing organizations in Spain." For Vaughan, supporting the Spanish Civil War became "the great opportunity to stand against fascism. . . . I walked in poster processions through London streets; I spoke on soap boxes at street corners and in huge public meetings in town halls."[9] It also brought to her fascinated attention the critical importance of technolo-gies for blood storage.[10]

The Spanish Republicans were now able to store blood for more than two weeks using glass bottles with added anticoagulant, keeping it mo-bile and a few degrees above freezing using refrigerated vans.[11] In 1938, de-termined to test some of these techniques, Vaughan and her colleagues had made large numbers of transfusion sets and begun extracting blood from donors. At the time of the Munich Crisis in September 1938—when many in Britain first became convinced that the country would go to war—she and her colleagues at the Postgraduate Medical School had been told to prepare for up to 57,000 casualties in London. On that oc-casion, though, she used all the collected blood for preservation experi-ments.[12] In Vaughan's own recollections, a friend quipped, "The only blood lost at Munich was what Janet collected at Hammersmith."[13]

During the spring of 1939, Vaughan continued to convene meetings in Central London at her own house in the "heart of pacifist Bloomsbury."[14] Although the "subcommittee" still had no official status, participants now included doctors from the major London hospitals and MRC representatives. They debated how many blood depots London would need, where they should be located, to what regions and hospitals they would provide blood. Structurally, they envisaged the service to be integrated with a planned set of nationwide wartime medical services. The Ministry of Health and the Department of Health for Scotland were developing plans for an Emergency Medical Service, which would organize and administer the activities of hospitals.[15] Alongside this, the MRC was to run the Emergency Public Health Laboratory Service—that is, a network of laboratories equipped to identify and monitor epidemics.[16] Between them, the Emergency Medical Service and Emergency Public Health Laboratory Service bound into a national system hospitals and laboratories that had previously been under local management.

The MRC was also to administer the proposed "Emergency" Blood Transfusion Service, as it was now being called to cohere with these other nationwide organizations.[17] Like them, the EBTS was built in part on existing transfusion institutions—donor panels and small-scale transfusion centers. Vaughan's committee decided that the wartime service should comprise a network of "empaneling centers" for registering and testing donors; a transfusion "panel" was simply a list of volunteer donors recorded on index cards. Mobile and static bleeding stations would extract blood from donors of the right group. Central depots would refrigerate the blood in glass bottles and distribute it to hospitals in refrigerated vans.

The question of where best to position the principal stores of blood was not an easy one to answer. The organizers anticipated that the transportation systems in London might be thoroughly disrupted during emergency situations, with the potential to seriously hamper the movement of blood, donors, and patients. Judgments about where depots should be located sought to balance distance with road communication; for the London area this took into consideration how the movement of blood might be impeded by lack of transport across the river Thames.[18] Eventually the subcommittee decided that London would be served by four depots outside Central London—in Luton, Slough, Sutton, and Maidstone. The Army Blood Transfusion Service, which was to be responsible for transfusion overseas, would be based in a former hospital maternity ward in Bristol, within reach of the southwest coast. Each depot was to be equipped with autoclaves, stills, hot-air ovens, a generating plant (in case of electricity failure), a cold room, mobile refrigerators, couches, dressing trolleys, sterilizers for instruments and dressings, and needle sharpeners. In a striking confluence of technologies for dairy and blood

3.2 Photograph of Mr. Hennington sharpening needles at the North West London Blood Supply Depot in Slough, taken between 1940 and 1943. The young man, with a neat haircut, casual jacket, and tie, sits next to a large window and uses a sharpening stone on a lathe. Puncturing a donor's skin to extract blood required sharp needles, which the depot then sterilized and reused. The photograph was one of a series of publicity photos for the Emergency Blood Transfusion Service—a series that also included images of bottle washing and sterilizing equipment—and helped to convey the message that the material infrastructure of transfusion relied on a wide variety of paid and volunteer labor. 21 × 16 cm.
Reproduced with the kind permission of the Bodleian Libraries, University of Oxford.

products, the subcommittee decided that blood would be kept in the "ordinary pint milk bottle" (by United Dairies) and transported in a fleet of Walls refrigerated ice cream vans.[19]

Each depot was to be directed by a medically trained officer working with a handful of other medics, nurses, and volunteers to bleed donors, give transfusions, clean and sterilize equipment, sharpen needles, assemble transfusion kits, and work in depot laboratories (figure 3.2). Female assistants were to be hired as clerks or trained in blood grouping; unskilled volunteers were to help with administrative work and drive vans and ambulances. Vaughan recalled, presumably to illustrate the commitment of her volunteers, that one of her drivers was a titled seventy-year-old woman who "always wore several strings of pearls and a toque [small hat] rather like Queen Mary."[20] The committee also re-solved that during quiet periods, the depots would carry out research; thus, many depots also employed two or three research assistants.[21] This was coordinated centrally by an MRC committee, soon known as the Blood Transfusion Research Committee, which kept the depots and the armed forces abreast of new research. This included studies on the storage of blood, on grouping techniques, and on the transfusion of plasma— that is, the liquid part of blood, which could be freeze-dried and shipped overseas to aid the fighting forces.[22]

Vaughan's subcommittee agreed that the EBTS would only collect therapeutic blood from donors classified as group O, making it (at least theoretically) unnecessary to group recipients under emergency condi-tions. It estimated that each depot would need to test 20,000 people to obtain 8,000 or 9,000 of these "universal" group O donors. This meant recruiting and training a lot of people in the techniques of blood group-ing. This was where the Galton Serological Laboratory, still at its original location in London, first became involved in the plans.

"Very Large Numbers"

During the planning stages of the EBTS, Vaughan asked George Taylor, head of the Galton Serological Laboratory, whether he would take charge of making antisera and training blood groupers for the service. Vaughan's link to Taylor and Fisher was of many years standing. The British he-matologist community was small, and hematology a young field. A few years earlier, Vaughan had collaborated with Taylor in a research project on linkage between blood groups and jaundice. Vaughan believed Tay-lor to be one of Britain's most experienced serologists, and he became Vaughan's main authority on the technical aspects of blood grouping. For his part, Taylor was deeply committed to participation in wartime blood transfusion; he had already volunteered his expertise to those performing transfusions at the nearby University College Hospital, and he attended every EBTS planning meeting.[23] Vaughan assigned Taylor particular responsibility for training "girls" in the practices of group-ing blood. Serological work was routine and repetitive, but it was also skilled and required dedicated focus, all qualities considered well suited to women. Taylor was a serious-minded character, giving Vaughan all the more delight in her reminiscences that she could "always remem-ber [him] . . . the English authority on blood groups, saying with great solemnity: 'you must also enroll girls to determine these blood groups and this should be done at once; it is not easy to procure young girls!'"[24]

The prospect of devastating aerial bombardment in London meant that rapid donor recruitment was crucial. This took the shape of a vigorous publicity campaign to frame the act of donation as a contribution to the war effort. Recruitment looked significantly different than it had just a few years earlier.[25] In the 1920s and 30s, the London Red Cross Blood Transfusion Service had restricted its selection of donors to people it deemed to have a high moral character. Blood was then transfused directly from donor to patient, so donation and transfusion was an intimate transaction. Now, demand for blood was far higher, but disembodied blood could be kept for longer and travel longer distances. Blood depots, which stored hundreds of bottles of donated blood, relaxed demands on donors to be in particular places at particular times. Whereas the London Red Cross Blood Transfusion Service had restricted its selection of donors to people with specific morals and temperaments, the EBTS opened the door to a far larger constituency, selecting only on age, general health, and blood group.[26] Vaughan's subcommittee initially gave depots responsibility to recruit and manage donors, but the MRC soon began coordinating appeals for blood, in newspapers, on the radio, in cinemas, in pamphlets, and at local institutes, as well as in medical journals (figure 3.3). On July 3, 1939, London's "empaneling stations" opened. Volunteers were asked to register and have a small quantity of blood taken for blood group testing—"a mere prick of the finger is all that is necessary"—and were assured that they would only be called upon to give blood in the event of an emergency.[27] The campaign was phenomenally successful; barely a month after recruitment began, the *Times* announced that the service had registered its first 100,000 donors.[28]

Blood was needed not just for transfusion but also to prepare reagents for testing. Stable sources of such substances depended on reliable human donors. Serum, the fluid component of blood, contains antibodies and other proteins; antiserum was blood serum that contained known specific antibodies that could be used for grouping. So while the EBTS used group O blood for transfusion, it needed A and B blood for making antisera—preferably high-titer group A and B blood, that is, blood with high serum concentrations of the respective antibodies. When recruitment had begun in mid-1939, the MRC had sought to address the problem of recovering sufficient quantities of high-titer A and B sera. Taylor himself had suggested that the Galton Serological Laboratory

3.3 "Your blood can go on active service": A double-page spread of a four-page leaflet entreating potential donors to enroll in the service. The front page (not shown) reads "YOUR BLOOD can save the life of someone, somewhere," and depicts two soldiers attending to a wounded colleague, helping him to a cigarette. The inside pages, designed with black, blue, and red type, narrate the active life of donated blood in the war effort: "Blood has been on Active Service at Dunkirk, in the Battle of Britain, at home in the Blitz. Dried blood has been dropped by parachute to save lives in Malta. It has saved lives in ships at sea, lives of men of the Royal Navy and the Merchant Navy in Russian, Atlantic and African convoys. It has been flown over the Himalayas to China. Wherever there has been fighting—in North Africa, Egypt, in Libya, in Tunisia—blood has saved lives." The third page of the leaflet comprises an enrollment form that volunteers could fill out, fold, and mail. Emergency Blood Transfusion Service, ca. 1943. 26 × 21 cm.
Wellcome Collection, London, GC/107/1. Reproduced under Crown Copyright/Open Government License.

could produce and distribute the reagents.[29] The MRC also considered an offer by the London Red Cross Blood Transfusion Service to collect blood that would be turned into antisera by the pharmaceutical behemoth Burroughs Wellcome & Company, but the MRC decided to keep its serum requirements independent of commercial interests.[30] Instead it supported Taylor in building up the necessary supplies of boxes, ampuls, and equipment for filling them with sterile antiserum.[31]

Taylor was by now carrying out antiserum provision and blood grouping training for the EBTS and so was carving out a central role for the Galton Serological Laboratory in wartime transfusion. The MRC proposed that if hostilities began, the lab would move to Cambridge and would be staffed by Taylor, Robert Race, and fellow research assistants Aileen Prior and Elizabeth Ikin, along with two laboratory "attendants," Douglas Keetch and George Tipper. They were to be housed in the Cambridge Department of Pathology, which would also become a central node for the regional Emergency Public Health Laboratory Service.[32] MRC physiologist Alan Drury had already worked for some years as a full-time researcher in the Department of Pathology. Through Drury, the MRC established the department as central to its wartime work, giving space not only to the Serum Unit but also to part of the Emergency Public Health Laboratory Service (including a *Streptococcus* laboratory) and to the MRC's Serum Drying Unit for processing freeze-dried plasma.

In its new incarnation, the central responsibility of the Galton Serum Unit was to make large quantities of antiserum, so the lab needed a reliable source of group A and B blood. Arthur Landsborough Thomson of the MRC had initially wondered whether mental health patients would be a good source, thinking perhaps of the extensive blood group tests that Penrose had already carried out at the Royal Eastern Counties Institution. Writing to Laurence Brock of the Board of Control for Lunacy and Mental Deficiency (part of the Ministry of Health), Thomson explained how difficult it was "to find people who will be stationary in a time of emergency."[33] Brock vetoed that plan because they decided that patients could not give proper consent ("It would clearly not be an easy matter to explain to most mental patients what is wanted of them and why").[34] Other proposed sources of human serum included MRC employees themselves, and Taylor received a considerable quantity of blood from people working at the National Institute of Medical Research in North London.[35] In the end, the choice of Cambridge as the lab's new home was in part owing to its proximity to students (as Taylor described them, "healthy young adults willing to be bled").[36] As demand later outstripped available student donors, the Galton Serum Unit sought additional stable sources of blood. Taylor persuaded the Royal Air Force to allow him to determine the blood groups of men undergoing flying training at the southwest seaside town of Torquay in Devon. After a difficult time negotiating with senior officials of the Air Force, Taylor found this cohort of young men to be a "splendid source of grouping serum."[37] Finally making official this commitment to donation, the Air Force agreed to stamp the men's blood groups on the reverse sides of their identity disks.[38]

Grouping sera were stored frozen, and Taylor advised the laboratory recipients of the antisera that the ice drawers of domestic refrigerators perfectly served the purpose. Taylor's own stores of what he described as "high-grade grouping serum" were backed up by a reserve supply kept at −10°C at the Worcester and Midlands Ice Company in Gloucester, to mitigate against catastrophic loss.[39] Over the next few years, individual depots gradually became self-sufficient in antisera (using A and B donors from their own panel lists), but not before the manufacture of grouping reagents had put Taylor's laboratory firmly at the center of the new nationwide health service.

As the war went on, the lab's responsibilities would extend beyond the supply of antisera to giving opinions on puzzling samples of blood and distributing expertise on "blood-grouping problems," as well as authorship of technical publications on blood grouping.[40] Taylor later explained that the lab's practical responsibilities made it possible to establish channels through which other kinds of expertise could be circulated around the country: "As a result of supplying serum, examining and giving opinions on troublesome samples of blood, and being consulted about blood grouping problems in general, we are in touch with very large numbers of civilian and service workers interested in blood transfusion."[41] These "very large numbers" would become precious to Fisher. The lab's wartime setting, and its routine responsibilities for making, testing, and distributing antisera, resulted in a new resource for the study of human genetics.

"Actual Laboratory Work"

Three days before war was declared, Janet Vaughan was at the North West London Blood Supply Depot in Slough, on the premises of a local social club, when she received "a laconic telegraph from the Medical Research Council to 'Start Bleeding.'"[42] The managers of local factories were asked if they could release their employees for bloodletting, while "loudspeaker vans were sent out into the streets asking donors to repost at the depot."[43] Vaughan recalled that factory directors "immediately arranged to drive their employees to and from the blood depot while whole families came together from country villages on their bicycles."[44] As for the transfusion workers themselves, Vaughan remembered the scene with characteristic attention to the theater of the situation:

The medical personnel drove the Walls Ice Cream vans with their refrigerators down from the Mount Royal Depot and our improvised organization moved into action. That

Sunday morning we stood in the Social Centre bar in our white coats with the locals, to hear Chamberlain state we were at war, and then we went back to our bleeding.[45]

At almost the same moment, on August 29, 1939, Fisher's staff at the Galton Serological Laboratory in London were ordered to pack up and drive to their new institutional home at the Department of Pathology in Cambridge. As the renamed Galton Serum Unit busily took on its new responsibilities, Fisher himself remained in his job at University College, although the college soon evacuated its premises in Central London. Furious at his employers for forbidding researchers access to the laboratories (as he complained in a letter to the *Times*), Fisher negotiated the temporary use of rooms at the Rothamsted Experimental Station, north of London and about forty-five miles from Cambridge, where he had worked before his professorial appointment (figure 3.4).[46] Fisher moved to Rothamsted, taking with him many of his experiments, two members of the staff, and his Millionaire calculator. This was the first commercially successful machine that could perform direct multiplication; it was later described by Fisher's daughter Joan Fisher Box as making a "noise like an old-fashioned threshing machine."[47] Other University College colleagues to move there included geneticist J. B. S. Haldane (now Weldon Professor of Biometry) and zoologist Helen Spurway, who were both carrying out statistical work for the Royal Air Force, the Army, and the Ministry of Aircraft Production.[48]

Conditions in the plant pathology building at Rothamsted were crowded, and Fisher shared his office with his secretary, Barbara Simpson, and other Galton staff. Simpson and genetical assistant Sarah North became responsible for much of the work of sorting blood group records.[49] No longer tied to his serological laboratory, Fisher remained in close contact with Taylor and others at the Galton Serum Unit by mail. In regular reports to the Rockefeller Foundation, Fisher insisted that his "old group" was "working together as an unbroken unit."[50]

Just days before war broke out, Fisher had presented data from 58,000 donors at the International Congress of Human Genetics in Edinburgh. But he wanted more. The blood group records of the wartime transfusion services offered a stunning opportunity for collecting abundant numbers of blood groups. As well as his appeal in the *BMJ*, Fisher wrote directly to the depots to ask for their grouping totals. Although he was no longer officially part of the Galton Serum Unit, Fisher framed his requests for data as though they represented an exchange for the antisera that the unit was supplying. He carefully reminded depot workers of the "large quantities of testing fluids" (i.e., antisera) being supplied by

71

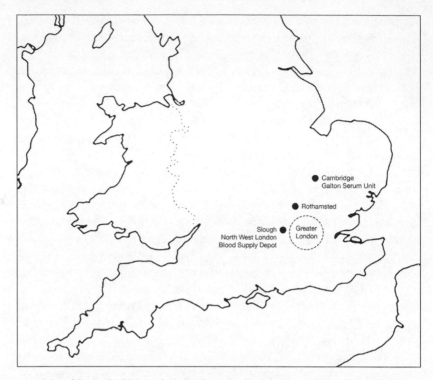

3.4   Map of the south of Britain, depicting the approximate locations of the major wartime institutions in this chapter: Galton Serum Unit in Cambridge (60 miles north of central London), the Rothamsted Experimental Station (30 miles northwest of London), and the North West London Blood Supply Depot in Slough (20 miles west of London). The proximity of these institutions facilitated collaboration between Janet Vaughan, R. A. Fisher and Barbara Simpson, and George Taylor and his colleagues; while their distance from one another made necessary the exchange of paper correspondence on which this story is based.
Map by the author.

Taylor, before asking whether they might be willing to send "grouping totals" to Rothamsted.[51] Fisher was negotiating a system of exchange: by mobilizing his connection to the unit he turned the essential and practical provision of antisera into a resource for garnering materials for research.

Just like Vaughan's clerks, Fisher's workers were dependent on vast quantities of paper. At the depots, secretaries and administrative assistants managed hundreds of thousands of donor records. Each record held an array of information about a volunteer: name, address, telephone number, general state of health, past serious illnesses, and whether under national service obligations. Once a donor's blood had been selected for transfusion, clerks marked the donor's enrollment card with the date on which the blood had been taken, the result of the Kahn test for syphilis, and, in relevant cases, the concentrations of A and B antibodies.[52] The donors themselves received a corresponding card containing some of that information.[53] Testifying to the importance of blood groups to the identity of the blood, each bottle carried a label that was color-coded according to the donor's group. This label was stamped with an index number that matched the filed index card of the donor; the bottle also had a tie-on label carrying the same number. After a bottle of blood had been transfused into a patient, the tie-on label was removed from the bottle and returned to the depot along with information on how the patient had responded. Reflecting on the growing transfusion service in a report to the Ministry of Information in 1940, Vaughan noted how these paper technologies allowed blood to be tracked between donor and recipient:

Each bottle carries a coloured label appropriate to the blood group for the contained blood[,] stamped with an index number similar to [the] filed index card of the donor, and also a tie-on label carrying the same number. The latter label after a transfusion is given, is removed [from] the bottle and returned to the depot stating whether the blood was in a satisfactory state when received and whether the result of the transfusion was good. In this way, each depot is able to keep a complete record of the number and success of the transfusions given. In the event of any difficulty occurring it is possible by means of the number on both bottle label and donor's card to check up any possible fault in technique which might account for the difficulty.[54]

Paper technologies for marking and tracing blood meant that each depot was able to keep a complete record of the transfusions. Labels maintained a chain of reference between donor, blood, bottle, and patients. They allowed depot workers to investigate donors whose blood had adverse effects on patients. An article in the *BMJ* reported the day-to-day activities of Vaughan's Slough depot: within three months of beginning recruitment, they had 15,000 potential donors on their books. The article exclaimed admiringly: "The mere card-indexing of such a number would be regarded as a very serious business indeed for the average city office, but in this depot they seem to take it in their stride."[55] Administrative paperwork was essential to the smooth functioning of the depots.

It was also highly desirable to the scientists at Rothamsted and Cambridge. Even by the time Fisher's and Taylor's *BMJ* letter had appeared, 32,000 forms had arrived for processing.[56] The letter had asked transfusion workers to send results in the form of totals: frequencies of A, B,

O, and AB, summed separately for the two sexes. Many correspondents obliged, although some just sent the original forms and left it for the Rothamsted workers to do the counting.[57] Simpson, Fisher's secretary, did much of the practical paperwork of duplicating the forms, arranging their "transmission by rail, car, etc.," sorting index cards, calculating grouping totals, and ensuring "avoidance of confusion between different batches—some perhaps from different centres."[58]

In a report to the Rockefeller Foundation, Fisher pointed to the serendipitous institutional alignment of genetic research and the Galton Serum Unit's war work. He explained that the EBTS had brought into his orbit centers "where quite unprecedented numbers of the British population are being grouped for transfusion purposes." He emphasized how hard he and his colleagues had worked to gain access to this network, being "at pains to maintain these contacts." Fisher marveled that the transfusion enrollment forms yielded not just blood group data but also characteristics such as "sex, age and surname." He remarked that these paper resources also engaged him and Simpson in administrative labor, including the maintenance of correspondence relationships, the reception and dispatch of parcels, and careful attention to detail to prevent errors.[59]

As well as capturing blood group results and ancillary information, Fisher and Simpson did further work to make the blood group records yield genetic data, transposing the totals into allele frequencies. Owing to the dominance of certain alleles, it was impossible to tell without family data whether an individual of phenotype A, for example, had the genotype AA (meaning both the relevant chromosomes carried the gene for group A) or AO (meaning the individual had one gene for group A and one for group O). But at the population level a new kind of data could emerge. A principle of population genetics—the Hardy-Weinberg equilibrium—stated that in a large population, the allele frequencies of a Mendelian gene (for example, the alleles A, B, and O) could be predicted from phenotype frequencies.[60] Using population totals, Fisher and Simpson could use the Hardy-Weinberg principle to estimate the underlying allele frequencies. In other words, genetic data was made by sorting, counting, and organizing cards and mathematically transposing population results. Writing to the MRC, Fisher referred to Simpson's "sorting, counting and compiling" as "actual laboratory work."[61] Pointing to a more general feature of human genetics during that period, Haldane echoed Fisher's allusion to paperwork as laboratory work in 1941 when he remarked that "statistical methods" had "replace[d] the various technical devices, such as milk bottles and etherizers, which are familiar to the Drosophila worker."[62] Human genetics was about practices using paper.

This was not the first time that Fisher and his colleagues had carried out blood group surveys. In the mid-1930s the Galton Serological Laboratory had grouped over 3,500 families and unrelated individuals in London with the express intention of testing the application of new mathematical population genetic techniques to human subjects.[63] But the sample size of this earlier survey was rapidly dwarfed by those achieved when the laboratory started grouping blood transfusion volunteers. After the first EBTS donor drive, Fisher and the unit had produced data from 58,000 people, and by February 1940, from more than 100,000.[64] Having officially recruited Air Force men to provide blood for grouping, George Taylor and Robert Race continued to make the complex journey from Cambridge to Torquay every two weeks—a "long drag across country in trains sometimes unheated, delayed or crawling through air raids"—which by December 1940 yielded 10,000 Air Force forms typed by the Galton Serum Unit itself.[65]

It was not just raw numbers that were important to Fisher; he was interested in *where* in Britain the records were from. Although regional EBTS depots were not officially established until July 1940, provincial transfusion services were functioning well before this, and during these early days Fisher received results from Scotland, Wales, and the north of England. With blood group records flooding in from all around the country, Fisher began to see the regional results as "ethnographical" data.[66] He signaled his new interest in allele frequency distributions when in November 1939 he asked the University College library to subscribe to the *Zeitschrift für Rassenphysiologie* (*Journal for Race Physiology*), which was published by the Deutsche Gesellschaft für Blutgruppenforschung (German Society for Blood Group Research).[67] Characterized by a *völkisch* ideology, the *Zeitschrift* was one of the principal international journals for comparative work on blood group frequencies and race, with over half its articles by international contributors.[68]

Thus the mobility of the donor records and the institutional structure of the wartime transfusion services shaped the parameters and methods of Fisher's research. A mere three months into the war, he authored with Vaughan his first "ethnographical" paper, "Surnames and Blood Groups." Using the records from Slough—an industrial town to the west of London—Vaughan and Fisher reported that people with distinctively Welsh surnames ("Davis, Edwards, Harris, Jones, Lewis, Morgan, Phillips and Roberts") had significantly different blood group frequencies from the rest of the area's cohort. The authors suggested that this difference could be explained by an influx of workers associated with recent industrial development.[69] Vaughan herself spoke of Slough's quality as a "'frontier town' in the American sense of the word." Reflecting on her work with Fisher, she noted that the city "had grown up after the first world war, round a vast new trading estate, full of migrant workers."[70] This recent migration could apparently be traced in the blood of donors.

In the early 1930s, Haldane had been enthusiastic about the use of blood group data to recover historical migration. Fisher and Taylor now made a first systematic attempt to correlate the geographical distribution of blood group frequencies across Britain with narratives about British history. They presented a much-expanded study to the Pathology Society in Cambridge in February 1940. Shortly afterward, they published the research in *Nature*, under the title "Scandinavian Influence in Scottish Ethnology." The researchers reported a continuous gradient in the A:O blood group ratio from northern to southern Britain. Much as in the large body of racial blood grouping work that had accumulated in the 1920s and '30s, the authors reached for narratives of racial history and stories of invasion and mixing, which were even more relevant now that Britain was at war. Taylor and Fisher compared their observed frequencies to those found in other European populations and evoked an account of British history that put special emphasis on the Viking conquest and settlement in Shetland, Orkney, and the Scottish mainland. Against the expectation that people from the north of England and Scotland would have "a greater infiltration of Scandinavian blood," the authors claimed to be taken aback when they found that frequencies in Northern Europe more closely resembled the ratios in the south of England, and that "no Continental population . . . comes near to the Scottish ratio."[71] Iceland was the only European country with results like those from Scotland. The scientists' explanation was that the blood group constitution of the Continental Northern Europeans had itself changed since Viking times, "presumably," as they suggested, "by infiltration from Central and Eastern Europe." According to this new historical account, Scotland could remain "Scandinavian." Alluding to contemporary German narratives of an "Eastern" threat to "Nordic" blood, Fisher commented to Vaughan, "Almost topical, isn't it!"[72]

The war made possible the recruitment of blood donors via appeals to a national war effort; it also allowed Fisher to spin a nationalist genetic narrative about Britain. Just as people were persuaded (by the EBTS) to have blood taken from their bodies, so depot directors were persuaded (by the geneticists) to lend their valuable donor records. Despite the apparently serendipitous confluence of the interests of the Galton Serum

Unit with those of the EBTS, Fisher and his colleagues invested considerable time and work in transforming routine wartime practices into resources for making genetic knowledge. They established several overlapping informal systems of exchange: antisera for donor records and specimens for serological expertise. Another, less successful, component of this system was the exchange of statistical expertise for checks on blood grouping technique. This last element would reveal professional tensions between depot workers, geneticists, and lab serologists about the right way to test blood.

Statistics as Serology

Early in the war, Fisher made a bold intervention in the serological work of the depots: he suggested using a basic tenet of population genetics to monitor the accuracy of their blood grouping tests. "Population genetics" was just beginning to be understood as a subfield in its own right, one devoted to the dynamics of evolutionary change at a genetic level. The Hardy-Weinberg principle (which stated that allele and genotype frequencies in a large, randomly mating population stand in a fixed relation to one another) was regarded by Fisher and like-minded colleagues as an essential instrument in the population geneticist's toolkit. Deviations from Hardy-Weinberg (ascertained using the chi-squared test) could be used to evaluate assumptions about the genetic basis of a trait, such as the numbers of loci involved and the dominance relationships between alleles.[73] It might also indicate that a population was under selection, or that the population had recently been through a bottleneck, or that the sample was inadequate. Extending this last category, Fisher proposed to use Hardy-Weinberg to monitor the reliability of a technician's blood grouping technique.[74] He reasoned that inexperienced technicians or out-of-date reagents could result in consistent biases in the reading of blood group results. Early on, Fisher and his colleagues had noticed "anomalies" among the 32,000 results coming in from London.

Fisher made his Hardy-Weinberg protocol function as part of the exchange between practical war work and acquiring resources for genetic research. He told Taylor that "the detection of anomalies in grouping frequencies is one of the most useful by-products of the collections we are making."[75] Forced to justify the salary of his secretary, Simpson, to the University College authorities, who were intent on sacking anyone not perceived to be contributing to the war effort, Fisher implored the

MRC to plead his case, explaining that through Simpson's work and their statistical checks on the transfusion data, "numerous anomalies and discrepancies in the frequencies of blood groups obtained have been brought to light."[76] Fisher was eager to help his country: a member of the Home Guard, he was deeply disappointed not to serve in an official capacity, and he saw his analysis of the blood donor records to be an important contribution.[77] Thus, in letters to colleagues, Fisher began commenting on the credibility of the blood grouping results arriving from depots, and comparing these to the results published internationally on the worldwide distributions of blood groups. Encountering a particularly obstinate set of records from Scotland, he told Taylor, "The Glasgow series practically knocks me flat. . . . There are nearly 15% B's among the males, and 26% in the female list."[78] This, he observed wryly, could "give rise to alarming ethnological speculation, as it probably could not be paralleled nearer than Northern India."[79] Really, he was suggesting there was something seriously wrong with the grouping techniques of the Glasgow depot. In carrying out these checks, Fisher emphasized to the MRC that he and Simpson were making a "positive contribution to the efficacy of the Blood Transfusion Service."[80]

We know from his involvement with the MRC Human Genetics Committee that Fisher had larger-scale ambitions to establish the authority of a new kind of human genetics, and we might see his application of statistics to transfusion data as an attempt to extend the jurisdiction of genetic principles to new domains. Haldane had implied that mathematical techniques were essential laboratory tools for genetic research. Now Fisher went further, turning techniques based on the Hardy-Weinberg equilibrium into part of a routine protocol for the clinical laboratory. In Fisher's eyes, the mathematical tools of population genetics had universal applicability.

If this was self-evident to Fisher, it was less obvious to the depot medical officers. Statistics featured regularly in more specialist biology and mathematics journals, but trained physicians were unlikely to be familiar with the chi-squared test, still less the Hardy-Weinberg principle. Given the long and fraught relationship between medical statistics and clinical judgment, Fisher found that he had to tread carefully.[81] He commented to Taylor that it was "not for me to meddle in such matters." Wondering to Taylor how he should communicate his concerns about the strange results from Scotland, he suggested he would diplomatically disclaim "all technical knowledge of serological work."[82] He added, alluding to the tensions in Northern Ireland, which were particularly keenly felt in Glasgow, "I will have a shot at trying to do something tactful about the

Glasgow man, but I feel almost as guilty as if I were planting an I.R.A. bomb in his Laboratory."[83]

Conversely, some medical officers already had misgivings about techniques advocated by Fisher's colleagues at the Galton Serum Unit. To group blood, most transfusion centers used the simple "tile" or "slide" method, in which a lab worker mixed blood samples and antisera on sections of glass or porcelain and checked for agglutination by eye after a few minutes. The unit, by contrast, preferred the more reliable "tube technique," where the reagents were mixed in test tubes, left to stand for two hours, then examined under a microscope. But not only did the latter use more equipment and take much longer, it required an expert eye to transpose degrees of agglutination into binary data, which suited Taylor and his colleagues but was unwieldy for some blood depots.[84]

Unit workers were so alarmed by the results yielded by the depots that they began to advocate the use of better techniques. Vaughan was an easy convert; early in the war she had checked her depot's protocols by retesting 5,000 donors using the tube technique. Others were harder to convince. Taylor used the networks through which his unit circulated antisera to try to persuade depot workers to use the tube technique. Publishing his favored protocol in the *Journal of Pathology and Bacteriology*, Taylor suggested to the MRC that it might be a "good plan to send [copies of the paper] to all the people who apply to us for grouping serum."[85] He had some success in persuading depots, and a debate started up in the pages of medical journals. A regional blood transfusion officer in Cardiff advocated Taylor's position when he wrote to the *BMJ* declaring that although it meant "extra work in the laboratory," he had "completely discarded" grouping on slides "in favour of tubes in view of Dr. G. L. Taylor's article." Drummond cautioned readers, "Some years ago I observed two fatalities due to incompatible blood transfusion. . . . Both were due to . . . faulty grouping by the slide technique."[86]

Others in the transfusion service did not welcome being lectured about laboratory technique by an institution clearly far removed from the front line of clinical work. In 1941, the Blood Transfusion Research Committee, chaired by Alan Drury, drew up a report on the techniques of blood grouping, which he then circulated to various depot workers before publication.[87] The ensuing correspondence between Drury, Vaughan, and Taylor hints at a general impatience with the techniques advocated by the Galton Serum Unit. Committee head Drury had a keen sense of what was appropriate for the transfusion setting.[88] He suggested that the details of the tube technique, which he tactfully confirmed was "very reliable," should nevertheless perhaps be left out of the memorandum and published separately.[89] Consequently, in the first draft of the memorandum, Taylor's technique got only a short mention, although his advocates had won a fuller account of it by the time the galley proofs were ready.[90]

After sending proofs to specialists around the country, Vaughan received some scathing criticism back from the depots. Taylor had drafted the section on the problematic topic of "cold agglutination"—whereby the external conditions of the reaction caused red-cell clumping even when the two blood samples were the same—but his account was evidently complicated, and several people thought it should be considerably simplified for the clinician.[91] Brigadier Whitby, a long-serving army man, also a "brilliant clinical pathologist" and now head of the Army Blood Transfusion Service, told Vaughan that Taylor's section was "simply appalling" because it "savour[ed] of the laboratory and [did] not solve the bedside difficulties."[92] As far as Whitby was concerned, "the bedside" (though perhaps more accurately the transfusion clinic) demanded a practical sensibility entirely lacking in those doing laboratory research. The controversy over the "Blood Group Memorandum" was evidently severe, and Vaughan did her best to smooth things over, telling Drury that she had "written to Whitby (very sweetly)" and that she was "supplying sedatives by telephone and letter to an almost hysterical George Taylor."[93]

Wartime blood group records were (on the whole) successfully shared between people with different interests. But efforts to make those exchanges work, and the ensuing controversies over technique, reveal some of the diverse commitments of different communities. That Whitby could decry the Galton Serum Unit's techniques as inappropriate for "the bedside," while Fisher could regard the unit staff as "the only professionals in the country," speaks of conscious regard on both sides for the domains of the clinic and research laboratory. The unit itself occupied hybrid ground: it was a place of expertise and specialist serological reagents and a nationwide hub of genetic research.

Sharing Blood and Data

The wartime story of blood transfusion and genetics is about the alignment of two modernizing projects dedicated to mass data. One was a long-standing program that sought to turn the study of human heredity into "modern genetics." The other was the modern, bureaucratic EBTS:

a planned service dedicated to the organization of large numbers of people and quantities of blood on an industrial scale. The conditions for new infrastructures and technologies for managing blood were made by an array of materials and techniques: anticoagulants, cold storage, milk bottles and ice cream vans, record cards, index cards, labels. The array of paper technologies involved was coopted to serve a research program that dealt with large quantities of atomistic population data. This techno-scientific system included professional, social, and political labor. Clerks and secretaries dealt with transfusion paperwork; Ministry of Health propaganda experts procured volunteers; donors yielded blood; nurses, bottle washers, needle sharpeners, van drivers, and transfusion doctors moved blood between bodies. Paper records could be used and reused for different purposes. Organizational (paper) tools for the transfusion services (donor records) were transformed in Rothamsted into the raw data of human inheritance (blood group totals). Blood groups functioned as classificatory devices in transfusion centers and as diagnostic tools in hospitals, while at the Galton Serum Unit they were phenotypes that could be made to yield patterns of genetic diversity.

The material properties of paper mattered. It could be handled by laypeople as well as clerks and scientific professionals, so transfusion enrollment forms could be cut out from newspapers, and donors could carry around personal reference cards. Paper inscriptions could be modified and added to over time. The supreme mobility of paper was enhanced by a dedicated infrastructure for its circulation: the Post Office. The availability and material dimensions of paper meant it could accumulate in large quantities and be managed using sorting technologies such as card-index systems. Labels tied onto bottles could be taken off after transfusion, sent back to the depot, and used to produce a record that linked donor and transfusion outcome. And the material properties of paper—flat, cheap, adaptable—meant it could mediate between different purposes.[94] Paper records were sortable, mobile, and repurposable, making them valuable to scientists in pursuit of data. These administrative tools made the wartime blood transfusion service unexpectedly productive for genetics.

The partial centralization of the transfusion services and the formal alliance between those services and the Galton Serum Unit turned the latter into a passage point for sera, blood samples, and grouping results, which they found exceedingly rewarding.[95] Depots did not always align with the unit's interests—regional centers were not obliged to use the same grouping techniques, and they gradually became self-sufficient in antiserum—but Taylor's lab did manage to carve out its authority as the professional center of expertise through which data and specimens flowed throughout the war. Fisher's fortuitous position within this system for the procurement and movement of blood meant he could align his research program with a civic commitment to national service. That authority did not go unchallenged. Fisher's attempts to monitor blood grouping technique using statistics was only partially successful. The next chapter analyzes how blood groups were negotiated and shared between scientists, doctors, transfusion specialists, and others.

FOUR

The Rhesus Controversy

During the war, R. A. Fisher repurposed depot donor records to study human genetic diversity. Using the same links with depots, he and the Galton Serum Unit were also deeply engaged with a second wartime research program: namely, on the genetics of the Rh (Rhesus) blood groups, a serological system discovered in the United States in the early 1940s.[1] Soon after this, the Rh groups had rapidly become the topic of intense clinical interest when they were shown to be responsible for a condition called "hemolytic disease of the newborn," or "erythroblastosis fetalis," a severe and often fatal form of anemia suffered by some newborn babies. Here was a blood group system with dramatic ramifications beyond transfusion. Fisher and his colleagues realized that the genetics of Rh could be crucially important to human health: the testing and management of the Rh groups could save the lives of newborn babies. The Galton Serum Unit's research on Rh became deeply consequential for the careers of unit workers; in particular it would establish Robert Race as a leading international specialist in blood group genetics. The unit's Rh research was also the focus of a highly visible controversy about blood group nomenclatures that would consume researchers, doctors, and transfusion specialists for over a decade.

The protagonists of that controversy were Alexander Wiener, who had codiscovered the Rh groups with Karl Landsteiner in the early 1940s, and Fisher and Race, who later proposed a hypothesis about the serology and genetics of the system and a nomenclature to go with it. Wiener felt that in making these proposals, Fisher and Race had overextended the available knowledge about Rh, and he took every opportunity to resist their proposed nomenclature. But it caught on among many transfusion workers in Britain and generated heated disputes about the virtues and meanings of different blood group symbols. As researchers on both sides of the Atlantic probed the Rh groups in detail, the serological system became far more complex than the ABO or MN groups had been. Over the next decade, the system expanded and both the Wiener and Fisher-Race nomenclatures multiplied.

For Wiener, Race, and Fisher, this was a dispute about the way that the immune system (and its genetics) functioned. But for many doctors, clinical pathologists, immunologists, geneticists, and anthropologists, nomenclatures mattered for other reasons. In countless letters to journals, these constituencies weighed in with their views on the proposed nomenclatures and suggested alternatives. Meanwhile, official bodies in the United States and Europe held international meetings to standardize the Rh nomenclatures one way or the other, although in some respects the dispute was never resolved, and modified versions of both nomenclatures continue to be used to this day. Those historically revealing materials make visible some of the varied practical functions of nomenclatures. They also show how blood group research was shared and contested by people with different interests. They offer a snapshot of serological "knowledge in transit," as new understanding was made through communication and practice with paper and verbal tools, as well as though their miscommunication and resistance.[2] Nomenclatures also provide a unique vantage point from which to view the strengthening relationship between blood transfusion and genetics as the Galton Serum Unit delved into research that would have powerful consequences for prenatal and postnatal care.

Early History of Rh

The Rh blood groups were announced in 1940 by Landsteiner and his younger colleague Wiener. Two decades earlier, Landsteiner had fled economic chaos in Vienna to take up a position at the Rockefeller Institute in New York. There, he continued work on problems of immunity and blood groups, including the characteristics of human blood in relation to that of other primates.[3] Eliciting help from the superintendents of several New York zoos, his research program extended studies of "serological taxonomy" begun by Cambridge parasitologist George Nuttall.[4] Landsteiner attempted to use blood groups to construct detailed genealogies of the "lower apes," the "anthropoid apes," and humans. Alongside his research, Landsteiner was medical consultant to the New York Blood Transfusion Betterment Association, and he began collaborating with Alexander Wiener, geneticist and blood group specialist at the Serological Laboratory of the Office of the Chief Medical Examiner of the City of New York.[5] With the intention of searching for new blood groups, the two men started a project to raise antibodies against various kinds of primate blood and test those sera against human samples. Thus, injecting blood from rhesus monkeys into rabbits and guinea pigs (to "raise" anti-rhesus antibodies in these animals), Landsteiner and Wiener isolated an "anti-rhesus" serum that could detect a novel antigen, Rh, in some human blood samples. If a sample agglutinated with this serum, it was called Rh-positive (Rh+); if not, it was Rh-negative (Rh–).[6] They soon found evidence that this anti-rhesus serum (anti-Rh) could also be found in samples derived from patients who had suffered deadly hemolytic reactions following transfusion.[7] The inference was that these patients were Rh– individuals who had been inoculated with Rh+ blood.[8]

Soon after their initial discovery, Philip Levine, physician collaborator of Landsteiner and now serologist at a hospital in Newark, New Jersey, found that the Rh blood groups were responsible for erythroblastosis fetalis in some newborn babies.[9] Levine showed that this occurred when an Rh– mother gave birth to an Rh+ baby and that any subsequent pregnancies by the same mother would be affected by the condition.[10] Levine conjectured that an Rh+ fetus could "immunize" its Rh– mother during pregnancy, causing the mother to make antibodies against the fetus's own blood, and endangering future babies. Here was a blood group system that had very immediate ramifications for mothers, babies, and their families.

Across the Atlantic, these striking clinical effects particularly caught Fisher's attention. If mothers and babies could carry different Rh alleles, with fatal consequences, then how had the Rh polymorphism survived in evolutionary terms? Researchers in the new field of population genetics would have expected this variation to have been selected out of human populations. Fisher was deeply interested in how selection operated on genetic loci, and he had long insisted that the ABO blood groups were under selection (most other researchers believed that they were selectively neutral). Eagerly seizing on this new human Mendelian trait as a fascinating topic of research, he asked Taylor to purchase two rhesus monkeys from the superintendent of the London Zoo, and he was delighted when the animals were delivered to the Galton Serum Unit in Cambridge in

January 1942.[11] The lab was now in a position to isolate the same anti-Rh serum that Landsteiner and Wiener had made in New York. They injected rhesus blood into guinea pigs and rabbits to raise anti-Rh antibodies, and they used that anti-Rh serum to test human blood for the presence of the Rh group. Fisher's first human subjects were lab members, who, as expected, showed distinct agglutination reactions to the sera. Soon, Fisher and Race were able to obtain clinically relevant samples by recruiting mothers and babies from Addenbrookes Hospital, Cambridge, which at that time was just across the road from the Galton Serum Unit.[12]

When Landsteiner and Wiener had first defined the Rh groups in 1941, they had seen two clear categories, "Rh+" and "Rh−." Reporting the incidence of the blood groups in families, they reached the conclusion that the Rh groups were inherited via a single gene with two variants (or "alleles"), one dominant to the other. Following the conventions for genetic nomenclature at the time, Landsteiner and Wiener denoted alleles using italicized two-letter symbols (*Rh*) and represented the dominance relationships between alleles using upper- and lowercase first letters (*Rh* and *rh*). Thus, they inferred that individuals who were Rh+ had either the genotype *RhRh* or the genotype *Rhrh*, while those who were Rh− had the genotype *rhrh*.[13]

So far so familiar: The Rh groups and their allele variants were denoted with alphabetical symbols, with italics and upper- and lowercase letters indicating characters and properties of the genes. However, as Fisher and his colleagues began working on Rh genetics in 1942 they found that the agglutination reactions that they observed did not always result in the same simple grouping patterns. They noticed some striking patterns of agglutination, indicating new kinds of Rh antibodies and new blood groups. As these observations became more and more numerous, the researchers struggled to absorb new patterns of agglutination and inheritance into the existing blood group nomenclature.[14]

Wiener, too, was finding the system more and more complex. For the first couple of years, the Galton Serum Unit corresponded regularly with the Wiener lab, exchanging results in advance of publication. In 1943, Fisher was appointed professor of genetics in Cambridge, positioning him to work even more closely with the unit. Researchers on both sides of the Atlantic grappled with the increasingly elaborate patterns of agglutination and their inheritance. They were making their own antisera locally and carrying out tests on different blood specimens—so sometimes consensus was hard—but by 1944, the two labs had defined the set of alleles Rh_1, Rh_y, Rh', Rh_2, Rh_0, Rh'', and *rh*. However, in June 1944, the harmony between the groups was disrupted when Race published a paper in *Nature* arguing for a radically new mechanism of Rh inheritance. Though the paper was authored by Robert Race, the new genetic proposal was Fisher's, who had reportedly worked it out on a napkin in a Cambridge pub.[15] Fisher proposed that instead of a single gene with many possible allele variants (which was the model of the ABO and MN systems), the Rh locus consisted of three genes, tightly linked on the chromosome, and each with two possible alleles. In this scenario, the *combination* of alleles would usually function as an allelic unit. On the basis of this proposed genetic architecture, Fisher and Race suggested a complete revision of the nomenclatures used to denote the Rh antisera, genotypes, and blood groups. They denoted the three pairs of alleles *C/c*, *D/d*, and *E/e*, "chosen to avoid confusion with any so far used." They called this choice "arbitrary," but the letters C, D, and E were presumably chosen to follow from the A and B of the ABO nomenclature.[16] Whereas in the earlier phase of research, scientists had understood there to be single antigens, Fisher and Race now saw combinations of three antigens, corresponding to three pairs of alleles.

Historian Pauline Mazumdar has explained how the rival nomenclatures of Wiener and Fisher reflected fundamentally different interpretations of how antibodies and antigens interact, and how this related to a much longer-standing dispute over biological specificity.[17] In outline: Wiener believed that individual antigens differed by gradual degrees, and that an antigen (and therefore a blood group) was defined through its slightly different reactions with a range of antisera; as he put it, "a single antigen molecule can react with several antibodies of different specificities."[18] Fisher, by contrast, assumed that the interactions between antibody and antigen did not differ by degree but were "all or nothing." He believed that the specificity of antibody-antigen binding was generated by a triplet combination of antigens.[19] Wiener and Fisher had distinct views of how the immune system created specificity and different pictures of how Rh genetics was organized. In Wiener's system, each gradually varying antigen corresponded to one of a range of alleles (Rh_1, Rh_y, Rh', Rh_2, Rh_0, Rh'', and *rh*) that were all variants of a single gene, while Fisher and his colleagues saw three separate (though tightly linked) loci each associated with a pair of possible alleles (*C* and *c*, *D* and *d*, *E* and *e*). For Fisher, the relationship between allele, antigen, and blood group was direct and simple: a single allele variant corresponded to a single antigen, which reacted with a single antibody.[20] The Fisher-Race CDE terminology emphasized the unit-like character of the three antigens, the three separate alleles that corresponded to them, and their combinatorial specificities (*CDE*, *CDe*, *CdE*, *Cde*, *cDE*, *cDe*, *cdE*, and *cde*). Therefore, the symbols used to denote the Rh groups referred to

TABLE 1
The Rh Series of Allelic Genes*

Designation of genes†				Reactions with Rh antisera			Reactions with Hr antisera		
1 Preferred	2	3	4	Rh'	Rh''	Rh_0 (standard Rh)	Hr' (standard Hr)	Hr''	(Hr_0)
rh	rh	rh	rh	−	−	−	+	+	(+)
Rh_0	Rh^0	Rh_0	Rh^0	−	−	+	+	+	(−)
Rh'	Rh'	Rh'	Rh'	+	−	−	−	+	(+)
Rh_1	Rh^1	Rh_0'	$Rh_0{'}$	+	−	+	−	+	(−)
Rh''	Rh''	Rh''	Rh''	−	+	−	+	−	(+)
Rh_2	Rh^2	Rh_0''	$Rh_0{''}$	−	+	+	+	−	(−)
(Rh_y)	(Rh^y)	$(Rh'{''})$	$(Rh'{''})$	(+)	(+)	(−)	(−)	(−)	(+)
Rh_z	Rh^z	Rh_{12} or $Rh_0{'''}$	$Rh_0{'''}$	+	+	+	−	−	(−)

* Does not include the intermediate genes. Reactions, genes and antisera enclosed in parentheses have been predicted but not yet encountered.

† These do not represent different nomenclatures but merely variations of a single method of designating the genes.

Name of serum:	Anti-Rh_1	St	Anti-Rh Standard	Anti-Rh_2	Not yet found	
Antibody present:	Γ	γ	Δ	H	δ	η
Genes						
Rh_z CDE	(+)	(−)	(+)	(+)		
Rh_1 CDe	+	−	+	−		
Rh_y CdE	(+)	−	(−)	+		
Rh' Cde	+	−	−	−		
Rh_2 cDE	−	+	+	+		
Rh_0 cDe	−	+	+	−		
Rh'' cdE	−	+	−	+		
rh cde	−	+	−	−		

Those reactions not yet determined serologically are given in brackets.

4.1   As researchers defined new Rh groups, they altered names and symbols to help organize observed patterns of agglutination. The two tables convey the complexity of the Rh groups by the mid-1940s, and also the sense of uncertainty and conflict over the names used to denote antisera and antigens. (a) Table from a paper by Alexander Wiener in 1945, which includes the names of the new antisera Hr' and Hr". Comparing the + and − reactions along the table shows that the symbols Hr' and Hr" were meant to indicate that these antisera produced the opposite reactions to antisera Rh' and Rh". From Wiener, "Theory and Nomenclature of the Hr Blood Factors" (1945). (b) Table from the paper by Robert Race that first described Fisher's new hypothesis. Here, the authors propose reciprocal allele names (C and c, D and d, E and e) to reflect contrasting serological reactions. One can look vertically down the table to compare the + and − reactions for antisera Γ and γ corresponding to the putative alleles C and c. From Race, "An 'Incomplete' Antibody in Human Serum" (1944).

(a) Licensed by the American Association for the Advancement of Science.

(b) Reproduced under license from Springer/Nature.

different objects—for example, where Wiener still saw one allele (e.g., Rh_1), Fisher and Race saw three (e.g., CDe). In this sense the two nomenclatures defined different Rh systems (see figure 4.1).

Excitingly for the Cambridge geneticists, the combinatorial features of Fisher's new theory and nomenclature also generated graphic hypotheses.[21] The three tightly linked genes would (occasionally) be able to recombine, so the researchers were able to predict the existence of two rare alleles that had not yet been defined. Fisher argued that rare recombination events between loci would be expected to have quite specific effects on the Rh frequencies among the general population. He predicted that his theory might be confirmed by surveying thousands of blood samples for evidence of the very rare combinations, an experiment perfectly suited to a laboratory with strong links to medical services. Sure enough, to Fisher's immense satisfaction, one of those predicted alleles was soon discovered by an EBTS depot director, who collaborated with Race and Taylor to investigate Rh inheritance.[22] Over the next months and years, the Galton Serum Unit sifted through specimens sent to them from doctors and hospitals around the country, confirming and elaborating these remarkable new features of the Rh system.

"Wanted: Anti-Rh Sera"

The Galton Serum Unit had access to a great many blood samples. George Taylor and transfusion expert Patrick Mollison, who worked at the Southwest London wartime depot, appealed to doctors for specimens. In a letter to the *BMJ*—"Wanted: Anti-Rh Sera"—they asked doctors for samples of blood from any mother who had recently given birth to a baby suffering from erythroblastosis fetalis. They added that they would also "be grateful for brief clinical notes," as such information was crucial for interpreting samples.[23] If the Rh system was responsible for the condition, then affected mothers would have antibodies against the blood of their Rh+ babies. Samples from these mothers would serve as both reagents and case studies; indeed, some became sources of new kinds of Rh antibodies.[24] In these ways, certain mothers and their families became prized not just as research subjects but also as resources for providing reagents to investigate the properties of the Rh groups.[25]

Private correspondence shows that Fisher initially had to work hard to persuade even his closest allies to take him seriously. After all, his triplet-gene theory was extremely bold: a tightly linked cluster of genes had never been described for humans, and had been described only once in any organism.[26] In a letter to his friend Daniel Cappell, head of the Dundee depot in Scotland, Fisher confessed that while the theory had "grown on Race . . . Taylor still [had] his doubts."[27] Nevertheless, British physicians and transfusion specialists soon began using the new "CDE" nomenclature. This testifies to the strengthening authority of the Galton Serum Unit as a passage point for blood samples and expertise. The Rh groups were quickly becoming significant to the practice of medicine in British transfusion depots and hospitals, and in the medical offices of general practitioners (GPs). To procure blood samples and case details the researchers elicited the collaboration of GPs and pathologists. In turn, doctors pursued their patients and family members for blood samples, although this was proving particularly challenging during wartime. As Race put it, "men and women are away from home in the Forces; travelling is difficult; and, with doctors so fully occupied, personal visiting is practically impossible."[28] Nevertheless, many GPs and pathologists throughout Britain managed to obtain samples, which they sent to the unit or to Mollison's depot in Surrey. Within the year, unit researchers were working with doctors in Scotland to publish a report on fifty individual cases of the condition, citing the help of twenty-nine GPs around the country.[29]

The Rh system was consequential enough for clinicians that research was not confined to those in the orbit of the elite Galton Serum Unit. The Wellcome Bureau of Scientific Research began to keep in stock rhesus monkey blood with which researchers could inoculate guinea pigs to make the relevant antisera.[30] Doctors and pathologists in different parts of the United Kingdom tested the Rh blood groups of blood donors to obtain population estimates of frequencies of Rh+ and Rh− individuals.[31] The medical journals *Lancet* and *BMJ* published their first accounts of Rh in January 1942, and they were soon recommending that any transfusion recipient who had been pregnant or who had received a previous transfusion should be given only Rh− blood.[32] So urgent was the response to the Rh discovery that from late 1942, many depots in the United Kingdom began specifically testing for and supplying Rh− blood for Rh− women, although testing antisera were initially hard to come by, and doctors relied on cross-matching (the direct mixing of donor and recipient blood).[33] From 1943, many EBTS depots tested donors for Rh− and Rh+ and reserved stocks of Rh− blood for pregnant women and patients needing repeat transfusions. Soon expectant mothers (and some husbands) began to be routinely tested for the Rh blood group as part of antenatal care. Rh− individuals were issued with "special group cards," signaling to any medical professional that they were never to be given

Rh+ blood. Such cards, Vaughan later reported, also served to educate "both the public and the general practitioner in the importance of Rh tests in maternity cases."[34] The Rh groups were becoming embedded in the administrative structure of transfusion and other medical services.

But although the Fisher-Race nomenclature caught on quickly in Britain, the Wiener nomenclature was also popular. Alexander Wiener had considerable authority in the United States, and—working at the New York Medical School and in the Blood Transfusion Division at the Jewish Hospital in Brooklyn—he had access to a great many blood samples. Thus, as the Rh blood groups became highly visible over the next decade, the Rh antisera, groups, genotypes, and antigens were denoted using two principal, competing terminologies: the Wiener nomenclature and the Fisher-Race nomenclature. By the late 1940s, the discrepancy was sufficiently serious that authorities in the United States and Europe organized major international meetings to resolve the issue, while scores of scientists and clinicians continued to write to journals to debate the virtues of the two nomenclatures and suggest new ones.[35]

Previous accounts of the Rh controversy are right to claim that at its heart, this was essentially a dispute over how immunological reactions worked, and therefore how best to explain Rh inheritance.[36] But the published literature generated by this dispute also includes numerous reflections on the utility of nomenclatures and how these might serve different purposes. The Rh controversy gives us a different way of looking at the places in which blood groups were used, and at some of the negotiations that went on between these settings. These reveal a world of practice in which nomenclatures served different purposes in the research laboratory and clinic.

Sharing Nomenclatures

Rh blood samples, antisera, and data records were circulated between laboratories and hospitals and among geneticists, immunologists, clinical pathologists, and blood bankers. From the early 1940s, the number of institutions invested in Rh blood testing increased steeply. As well as the UK blood transfusion depots, various US laboratories were established, in large part to deal with the emerging complexities of Rh. The growing list of Rh antigens did not all produce the same kinds of agglutination reactions, so various blood grouping techniques competed as the most appropriate for diagnosing the Rh groups. Reagents for Rh testing (sera and red cells) were derived from human blood and were far harder to come by than those for the other known blood groups. In the United States in the 1940s this led to the establishment of several sites that became devoted to the production and circulation of antisera, such as the Rh Typing Laboratory in Baltimore, Louis Diamond's Blood Grouping Laboratory in Boston, and Philip Levine's research laboratory at the Ortho Research Foundation in New Jersey.[37]

Researchers in these varied settings tried (and struggled) to make nomenclatures reflect their priorities and commitments. One advocate of the Fisher-Race system, Dundee depot director Cappell, argued that names of the antisera should refer to the (prior) names of the antigens and their alleles.[38] The earliest version of the Fisher-Race nomenclature had denoted antisera using the Greek letters Δ, H, Γ, and γ; Cappell suggested that the antisera be renamed using Latin letters to make clear that they related to "the elementary antigens of the Rh complex with which they react." In Cappell's scheme the antisera would be called "anti-C," "anti-D," "anti-d," and so on. He explained that this new system had "the further merit that it is easily adaptable as knowledge advances," with "knowledge" here clearly taken to mean genetic knowledge.[39] Fisher, Race, and their geneticist colleagues rapidly accepted and developed the terminology, privileging reference to the allele names.[40]

Others argued that genotype terminology should reflect the actions of the antisera. In 1944, John Murray, an assistant pathologist at the Middlesex Hospital and a collaborator of Fisher and Race, suggested in *Nature* that the Rh antisera should be assigned the numbers 1, 2, 3, and 4, and that genotypes might then be given names that reflected the patterns of agglutination with those antisera. This, he explained, would mean that "at a glance it may be seen exactly with what sera the cells have been tested."[41] In Murray's scheme, various antigens would therefore be denoted: Rh^{136}, Rh^{126}, Rh^{123}, and so on, with supernumerals indicating the sera with which the antigens reacted. So although Murray accepted the Fisher-Race assumption of a 1:1:1 mapping between antibody, antigen, and allele, his nomenclature privileged the clinical significance of antisera. For a pathologist trained in serological techniques, the agglutination reactions were diagnostic tools. In other words, workers in different settings—geneticists, clinical pathologists, and physicians—favored nomenclatures that elevated their own practices. The different emphases given to nomenclatures revealed the divergent concerns of different practitioners.

Thus, some nomenclatures were contested because of what they represented, others because of how they were used. In 1948, Arthur Mourant, Race, and their hematologist colleague Mollison authored the MRC's memorandum "The Rh Blood Groups and Their Clinical Effects" for an audience of physicians; there, they introduced a range of "short symbols" for the Rh genotypes, most of which were "based on those by Dr A. S. Wiener." For example, "cde/cde" was given the shorthand "rr," and "cDE/cdE" was reformulated as "R_2R."[42] Physicians apparently found these shorthand symbols so useful that Race and Ruth Sanger included them in the first edition of their textbook *Blood Groups in Man* (1950), describing them as "necessary," "convenient in conversation," and "much used."[43] These uses included private laboratory scribblings. In the lower section of the notes shown in figure 4.2, Race worked out the genetic map distance between the genes C, D, and E using genotype frequencies within families.[44] In separate calculations in the top part of the page, he denoted genotypes using the shorthand symbols. In another example, a photograph of a donor card published in the *Daily Express* (during a search for a rare kind of blood) explained that the desired blood needed to be "O Rh″" blood of the genotype "cdE/cdE."[45] In the world of serological genetics, sometimes a single nomenclature was not sufficient, even on a single page.

One reason why both nomenclatures were deemed "necessary" was that in different clinical or scientific settings, the competing nomenclatures presented various typographical challenges, depending on whether they were made using typewriters, with printing presses, or by hand. When a nomenclature committee convened by the US National Institutes of Health (NIH) attempted to standardize the names of Rh antisera on bottle labels, it noted that the Fisher-Race nomenclature was easier to handle typographically than Wiener's. The latter, they decided, involved "complications, both typographical and genetic, of subscripts, superscripts, numbers, primes, and other symbols." In the committee's view, the Fisher-Race nomenclature was, by contrast, "simple and direct, both typographically and genetically."[46] The implication was that subscripts, superscripts, and primes presented problems for the typewriter and printing press.

Similarly, when a correspondent from Sweden wrote to Arthur Mourant to ask him to check a series of Rh blood grouping results, he sent two documents with identical data: one using a typewriter, the other written by hand.[47] The typewritten sheet deployed the Fisher-Race nomenclature (upper- and lowercase letters), while the handwritten sheet used the Wiener-modified shorthand (with sub- and superscripts), Apparently, for some people, subscripts and superscripts could be articulated in handwriting with greater ease, and upper- and lowercase letters more easily on the typewriter.[48] Indeed, in some circumstances the Fisher-Race nomenclature was potentially hazardous: the instability of handwriting meant that the

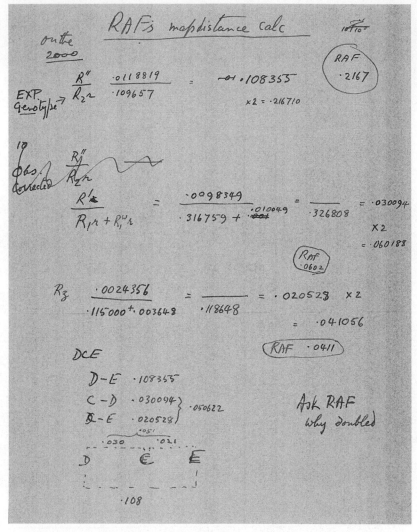

4.2 Robert Race's handwritten calculations in correspondence with R. A. Fisher as the former worked on new estimates for the structure of the Rh locus in February 1947. He deployed both the CDE nomenclature (bottom left) and the Wiener-modified shorthand nomenclature (top left). Race appears to have been comparing two methods for estimating the frequencies of Rh allele combinations (*cde*, *Cde*, *cdE*, etc.) in a population. These notes were likely preparation for Mollison, Mourant, and Race, "The Rh Blood Groups and Their Clinical Effects" (1948). 26 × 23 cm.
Wellcome Collection, London, SA/BGU/F.1/1/1. Copyright Medical Research Council. Reproduced with the kind permission of the Medical Research Council, as part of UK Research and Innovation.

uppercase "C" and lowercase "c" might generate dangerous ambiguities, especially if lower- and uppercase were not displayed side by side.[49] Unambiguous reading was an essential practical consideration. The Michigan geneticists Edward Ducey and Robert Modica, who supported Fisher's triplet-gene theory, proposed a modified version of the Fisher-Race nomenclature whereby the names of the antisera included only lowercase letters. By removing capital letters from the antisera, Ducey and Modica reserved them for antigens and also replaced the uppercase/lowercase denotation of alleles with a system of primes. This would preserve the reciprocal relations between antigen and antiserum but preclude confusion in the "reading of these terms."[50]

Meanwhile, the Rh nomenclatures also had to be articulated orally in the genetics laboratory, and especially in the clinical laboratory and hospital. Ducey and Modica addressed this issue when they argued that their version of the Fisher-Race nomenclature would prevent confusion in speech. Elsewhere, researchers complained that the dependency of the Fisher-Race system on upper- and lowercase alleles made the spoken articulation of genotypes extremely cumbersome. The antisera names were not too difficult; in a British film about grouping technique released in 1955, the narrator simply refers (phonetically) to "big-d" (i.e., D) antiserum.[51] But genotypes were problematic, since the single genotype *cde/CdE* would have to be said as "little-c, little-d, little-e over big-c, little-d, big-e." Dallas hematologists Joseph Hill and Sol Haberman, working in a clinical pathology lab and a hospital blood bank, respectively, declared, "One of the most frequent criticisms that have been made of the Fisher-Race notation is that it is difficult to use verbally." They suggested a whole new system for speaking these nomenclatures, which, they argued, would be "useful to the laboratory worker as well as to the clinician." This involved articulating only the uppercase allele and preceding it with the terms "homozygous" and "heterozygous." This meant pronouncing the genotype *"Cde/CDe"* as "homozygous C, heterozygous D," leaving the homozygous "e" silent.[52]

Soon, textbooks began flagging the huge importance of spoken articulation. Recall that following conventions in genetics, both nomenclatures used italicized letters to distinguish alleles from their corresponding antigens; the American textbook *Blood Transfusion* (1949) gave precise instructions for making that distinction in speech without the italicization available to printed inscriptions. Describing Wiener's nomenclatures, for example, the textbook declared that "in vocalizing, the [genotypes] are distinguished from the [blood group] names by the lack of *h*'s in the gene characters," and advised readers that on paper it was best to drop the "h" from the *Rh* symbols.[53] The pressures that speech exerted on nomenclatures were powerful enough for textbook authors to attempt to standardize their articulation.

In short, although the Rh controversy was ostensibly a dispute about genetic and immunological theory between elite researchers, the choice of nomenclatures had practical ramifications that reached deep into other parts of the transfusion community. In the hands of serologist-geneticists, Rh groups were made more elaborate with increasingly subtle genetic architectures, while for doctors and transfusion workers they led simpler existences. Both interests were powerful enough to prevent complete standardization of terminology. Different nomenclatures had different virtues in different settings, and no single nomenclature was sufficiently flexible to function smoothly in all of them.

The Controversy Fades

One puzzling feature of the Rh controversy is why numerous official meetings failed to standardize even the nomenclatures for labels affixed to bottles of antisera. For example, the conveners of the 1948 NIH meeting reluctantly admitted in the journal *Science* that they had been "forced to the conclusion that for the present a compromise must be made," recommending that "the Wiener terminology appear first . . . followed by the Fisher-Race terminology in parentheses."[54] This ruling—upheld in 1953 by the Food and Drug Administration—has often been seen as a missed opportunity to resolve an urgent problem.[55] Some histories attribute the failure to resolve the dispute in favor of the Fisher-Race nomenclature to Wiener's forceful personality and tenacious hold over the US authorities. But we might ask, instead, why was there not more pressure from clinicians to standardize labeling?

Part of the answer may have to do with the fact that usually only two of the Rh groups, Rh+ and Rh− (D and d antigens), caused incompatibility with any serious medical effects, so that to many clinicians only these binary designations mattered.[56] After 1942, books and medical journals had kept clinicians abreast of the exciting developments in genetic research on Rh, but by 1950, interest was waning. An amniocentesis diagnostic tool had become available for detecting fetal hemolytic disease, and Rh testing was fully routine in medicine.[57] By then, textbooks often emphasized that for clinicians the complex genetics of the system were largely irrelevant. The *BMJ* reminded its readers that despite the exciting nomenclatural controversies, "only one, D (big D), is of great clinical importance."[58] Mollison's *Blood Transfusion in Clinical Medicine* (1951) carefully explained the Fisher-Race genetics but then stated, "Fortunately . . . a simple subdivision of human beings into Rh positive and Rh negative . . . is sufficient for routine clinical purposes."[59] Mollison was a close and sympathetic colleague of Race, but he declared in a *BMJ* review of *Blood Groups in Man* (1950) that it was "useless to pretend any longer that what is interesting and even fundamental in the field of blood groups is necessarily of the slightest interest in clinical medicine."[60] By then, the Rh system was straightforward for clinicians—the blood groups were prognostic tools rather than complex genetic traits. There was little pressure to resolve their nomenclatures one way or the other.

Meanwhile, the fevered excitement among geneticists had also calmed. By the mid-1950s, Fisher's triplet-gene theory had become generally established as the more accurate description of the genetics of the Rh system, even though the predicted anti-d serum never materialized. At the end of that decade Fisher's work would spark a productive new research program at that University of Liverpool that would focus on Rh genetics and erythroblastosis fetalis and eventually lead to a way of preventing the condition.[61] The broad acceptance of the Fisher-Race nomenclature in Britain was partly due to the publishing success of its key advocates. Race, Mourant, and Mollison authored several MRC memoranda on Rh grouping for physicians, while Race and Sanger's *Blood Groups in Man* (1950) eventually ran to six editions, becoming the standard textbook on blood grouping for clinicians, geneticists, and anthropologists.[62] By the mid-1950s in Britain, the Fisher-Race nomenclature had largely (but by no means wholly) won out, and editors had ceased to publish Wiener's diatribes against it.[63]

Between Transfusion and Genetics

The Rh debate was initially the province of a relatively small community arguing about immunology and its genetics. But it drew in clinicians, pathologists, and serologists, many of whom were far removed from the central protagonists. This was in part owing to the authoritative practical position of the Galton Serum Unit within the EBTS network. The unit was the primary wartime institution for the production and circulation of standardized antisera, and therefore a passage point for samples and blood grouping results, as well as a center of expertise for difficult grouping problems. Wiener benefited from a similar position: in New York, he circulated antisera to and from colleagues, and at the end of the 1940s he was granted an NIH license to make and distribute reagents. Thus, both labs at the center of the dispute offered practical services further afield, and they could export their systems of labeling to other settings. The scope of the controversy over immunological and genetic nomenclatures testifies to the far-reaching authority of those labs into the clinical realm.

In the context of the Rh dispute, nomenclatures were epistemic things: they functioned as provisional scientific objects and might consolidate into tools for making new knowledge.[64] But this is also a story about ontologies—that is, how blood groups were made, known and used.[65] As they moved—literally, between laboratories and hospitals, and disciplinarily, between fields of research—the Rh groups shifted from being diagnostic or prognostic tools to being complex genetic traits. They had different theoretical meanings to the serologists and geneticists who studied them. They also had varied practical functions: nomenclatures were used to organize serological results and genetic experiments on paper; they were also made to function as labels.

Overlaid onto all of this, the Rh groups had to be articulated through different media: notebook jottings, typewritten letters, published papers, and speech all constituted different kinds of inscriptions, and no single nomenclatural system was flexible enough to work in all settings. A symbol that was simple to write on a blackboard was not always so easy to *say*. The circulation of even the most mundane and abundant nomenclatures could be problematic, and by the late 1940s, textbooks had even begun to stipulate which nomenclatures were appropriate for which contexts. Users had loyalties to different practices and objects, so that nomenclatures could have distinct meanings in different settings. The failure by official bodies to standardize nomenclatures was a consequence of the fact that no single system performed in all situations.

The story of the Rh nomenclatures unfolded just as genetics was becoming visible to physicians. Before the 1950s, few doctors considered genetics relevant or useful for their work. But research on the Rh groups helped that picture to change. Knowledge about human inheritance began to enter the medical literature, albeit in a carefully mediated form. In 1950, the British Medical Association added a "medical genetics" section to their annual meeting, and that same year, the *Lancet* explained, "The most obvious practical application of human genetics to medical practice is, of course, the analysis of the blood-groups in relation to hemolytic disease of the newborn."[66] The Rh groups became a conduit of genetic terminology into the clinical realm.

Outside the world of human and clinical genetics, researchers studying inheritance in other organisms were alerted to the fascinating complexities of Rh, and some worried that blood group genetic terminology might be a barrier to engagement. Oxford geneticist E. B. Ford, a close ally of Fisher's, argued that "confused terminology" was a "potent factor in preventing many geneticists from including the blood groups within their sphere of interest."[67] In a private letter to Race about a specific publication he exclaimed, "Why ever are the authors using such an unfortunate and wholly ungenetic nomenclature?" Ford implored Race: "Why, my dear Rob, does this still go on? General geneticists outside blood grouping cannot understand it, and it prevents them from taking an interest in serology."[68] US geneticist Herluf Strandskov worried in the *Journal of Heredity* that this lack of conformity might put geneticists off from taking human research seriously.[69] Ford and Strandskov both worried that this lack of standardization would hinder research on human genetics.

Race was more sanguine. When revising *Blood Groups in Man*, he resisted Ford's plea to standardize blood group nomenclature in line with genetics, remarking to a colleague, "I think our knowledge of blood groups and of the nature of the gene is growing, or perhaps shrinking, so rapidly that any sort of laying down the law about notation would be a nuisance—whichever way a decision went."[70] Race implied that nomenclatures fix understanding and that standardization would too quickly close down subtleties that were still emerging.

The Rh episode offers another vantage point from which to view the strengthening relationship between blood transfusion and human genetics in the 1940s. In *Lords of the Fly* (1994), historian Robert Kohler gives a rich account of an infrastructure and social practices assembled for making and mapping genetic difference—in that case using the fruit fly *Drosophila melanogaster*.[71] Like the *Drosophila* geneticists, Fisher, Race, and other midcentury serological geneticists established a robust social and material infrastructure to produce and map human difference, but in their case they used and intervened in a public health service, which meant aligning themselves with medical workers, patients, and donation customs and bureaucracies. There was push and pull between these fields and communities. Following nomenclatures offers a route to understanding these professional dynamics.

The MRC appreciated this remarkably productive use of the transfusion infrastructure. By the end of the war, the Galton Serum Unit, working with the EBTS and hospital workers, had made astonishing progress in understanding new immunological complexities, with far-reaching implications for neonatal care as well as transfusion. The MRC was so impressed by these productive collaborations that they chose to preserve these links in peacetime. The war had created a set of relationships between the state and public health that would endure. As blood transfusion was brought into the administration of the new National Health Service, it would expand and consolidate a serological infrastructure for a brave new postwar era.

FIVE

Postwar Blood Grouping 1: The Blood Group Research Unit

"Blood Group researches are intimately bound up with Transfusion Research," wrote Medical Research Council official Arthur Landsborough Thomson to a colleague as the war drew to an end. In ongoing discussions about where the Galton Serum Unit should be placed in peacetime, he noted that "opportunities for clinical contacts, co-operation with Blood Supply Depots and research workers interested in the field appear . . . favorable in London."[1] Following the wartime successes of the unit in disentangling the complex Rh (Rhesus) system, MRC officials wanted its scientific activities to remain coupled as closely as possible to the transfusion services. The MRC decided that the genetic research should be brought into proximity with a broader range of transfusion activities, with maximum access to "clinical material and records."[2] London's concentration of hospitals, GPs' offices, and depots promised rich resources for serological genetics.

As a result of those discussions, the MRC decided to reconstitute the Galton Serum Unit in the form of two new laboratories at the Lister Institute of Preventive Medicine, an institution that was devoted to research on public health and was located in a pink-and-red-brick Victorian building in the West Central London borough of Chelsea, on the northern banks of the river Thames. The first of these new MRC laboratories was the Blood Group Research Unit (henceforth, "Research Unit"), which was to be directed by serologist and geneticist Robert Race, who had formerly been in charge of the Galton Serum Unit. The Research Unit was to be a small lab dedicated to the serological properties and genetic inheritance of the blood groups. The second of these was the Blood Group Reference Laboratory (henceforth, "Reference Laboratory"), which would make and distribute antisera.[3] Chapters 7 and 8 discuss the Reference Laboratory in more detail; here I focus on the Research Unit, and the way that it used the donors, samples, and distributed infrastructure of the National Blood Transfusion Service (NBTS) to study the serological properties and genetic inheritance of blood groups.

By the end of the war, a sizable bureaucracy organized around regional depots kept track of nearly a million donors. Refrigerated vans, glass bottles, paper labels, sterilizing autoclaves, and a substantial telephone network kept disembodied blood moving between collection site, depot, and hospital (figure 5.1).[4] The therapeutic properties of blood had changed since 1939. It was now much more than just an emergency substance for treating shock. Transfusion was becoming a routine part of elective surgery and an important component of neonatal care. The days of indiscriminate transfusion of group O blood were long past. The remarkable unfolding of the intricate Rh system had drawn attention to the complexities of blood groups, while also highlighting the dangers that transfusion still posed. Now, surgeons took greater care over group compatibility and routinely tested recipients as well as donors. Crucially, they also performed "cross-matching" tests before surgery, which meant mixing the samples of blood from donor and recipient to check for hitherto unknown incompatibilities.[5] As the peacetime NBTS got underway, the country also embarked on an ambitious new project of nationalized health care. Within a couple of years, the transfusion service was brought under the administration of the National Health Service (NHS), which tightened the links between hospitals and transfusion depots. This national structure, and the new emphasis on the specificities of blood groups, intensified the highly distributed scrutiny of blood as it passed from donor, to depot, to blood bank, to patient. Using this infrastructure during the postwar decade, the Research Unit defined new blood group variants and systems, with important clinical consequences for transfusion patients and exciting possibilities for genetics.

Blood groups were still the best-understood human genetic traits, and they were certainly the only human characters for which genetic data was so abundant. Back in 1930 they had served as a model for what human genetics could be—mathematically informed and amenable to be-

5.1   Photograph of the cold room of a National Blood Transfusion Service depot in North
London, 1951. Single donations of blood, in conspicuously labeled glass bottles, have
temporarily separated into two visible layers. The bottles are stored in stacked metal racks
at 4°C, waiting to be moved to the blood banks of local hospitals. One of several
publicity photographs; photographer and commission unknown.
Photograph copyright TopFoto.

ing recorded in very large numbers. During the war this vision had
been fulfilled as Fisher's colleagues repurposed the EBTS donor records
for research. Now, studies of the many new blood groups contributed
to a growing interest among London researchers on human genetics.
Much of this work was centered at University College, where J. B. S.
Haldane still worked and where Lionel Penrose now directed a world-
famous program of research on the inheritance of complex human
traits at the Galton Laboratory.[6] The Research Unit's efforts to find new
blood group loci were vital to these linkage studies. The more blood

groups that were known, and the better their inheritance was defined, the clearer the path to understanding the genetics of complex traits.

This is a story of postwar biomedicine built on relationships between laboratories, depots, and hospitals, as well as public health authorities.[7] Blood and its labels moved between donor, doctor, serologist, researcher, and patient, and, in doing so, threaded together the organization of transfusion, blood grouping techniques, and processes of serological discovery. In this way, a clinical infrastructure and protocols of routine surveillance were used to build lifesaving knowledge about blood.

Organization

Officials involved in the wartime EBTS had begun planning the postwar NBTS in 1943, alongside discussions about a new system of social security and the national organization of health infrastructure.[8] Postwar transfusion was to be organized regionally. During the war, the country had been divided into fourteen areas, each served by regional hospital boards.[9] Initially the NBTS comprised ten regional transfusion centers and two London blood supply depots.[10] Each was run by a medically qualified regional blood transfusion officer, later "regional transfusion director," or RTD, the term I use here. Scotland and Northern Ireland had separate services, but their directors kept in contact with centers in England and Wales.[11] After the war, the MRC handed blood transfusion over to the Ministry of Health (MoH), but in 1948, in accordance with the NHS Act, management of regional transfusion centers was passed down to NHS regional hospital boards. The peacetime transfusion service, then, was made to function within the NHS.

RTDs were responsible for all operations at the regional centers, which included testing new donors, organizing mobile bleeding units, preparing plasma, and distributing sterile transfusion equipment.[12] RTDs were also responsible for supervising the transportation of blood to hospital blood banks, where carefully labeled bottles were refrigerated, ready for transfusion. Hospital pathologists oversaw these banks and supervised the systems and checks for testing blood and delivering it to doctors and surgeons. Regional centers typically issued local hospitals with quantities of blood to be stored until it was needed, and hospitals returned unused blood once it had expired. RTDs liaised with hospital pathologists to investigate puzzling blood grouping tests or adverse reactions to transfusion. The system, then, emphasized routine communication between hospital and depot. Alongside these responsibilities, RTDs were expected to carry out research, which many did with enthusiasm—some taking an interest in improving transfusion technique, others investigating serological problems.

Officials worried that the regional structure of NBTS management would result in a loss of uniformity of general policies relating to donors, techniques, equipment, and uniforms. RTDs agreed that they should try to maintain national standards, and they helped to author pamphlets that were updated regularly.[13] RTDs also routinely met at MoH headquarters under the chairmanship of William Maycock, the consultant advisor on blood transfusion. Nevertheless, NTBS officials did not attempt to enforce total uniformity, as volunteer recruitment was judged to be sensitive to local culture and institutions of donation. Indeed, in some parts of the country the organization of donor panels had remained independent of the NBTS. In those places, the Red Cross, local churches, or enthusiastic individuals continued to run their own recruitment services, as they had before and during the war.[14] Since RTDs were expected to coordinate carefully with those local panels, the MoH appointed a number of "donor panel liaison officers" to oversee their smooth administration across the country.[15] MoH officials consulted liaison officers on issues such as the design and organization of filing systems and index cards, the publicity strategies of recruitment campaigns, and the design and administration of donor certificates and badges.[16] This attention to regional variation in the needs and proclivities of donors was reflected in NBTS recruitment propaganda. Regional liaison officers became closely involved in publicity, including expensive promotions such as films and TV spots.[17] The MoH recognized the importance of regional variation and local familiarity in matters relating to donor recruitment and wanted to strike a balance with efforts toward national standardization.

Notwithstanding this emphasis on regionality, specialist laboratory expertise on blood groups, blood products, and transfusion technologies remained concentrated in London. Just as during the war, the MRC continued to oversee research on plasma fractionation and drying, transfusion technique, blood coagulation, and serology and genetics—much of it at the Lister Institute. That organization already consisted of eight departments doing work related to public health, covering topics such as bacteriology, experimental pathology, and nutrition.[18] The Lister had been a major site for antiserum and vaccine production during the war, and from 1943 had been directed by Alan Drury, former chair of the wartime Blood Transfusion Research Committee. Several departments worked on some aspect of blood science, from the serological analysis of bacterial antigens, to the properties of clotting factors, to the effects on blood of malnutrition and jaundice. The MRC now added to this the two new blood grouping labs: Race's Research Unit and Mourant's Reference Laboratory, which carried out the "research" and "practical" aspects of blood grouping, respectively. Both labs were expected to work closely with another MRC institution, the Blood Transfusion Research Unit, directed by Patrick Mollison at Hammersmith Hospital. During Mollison's tenure as director of the South London Blood Supply Depot, he had published important work on the storage of red cells, and his postwar unit (in a small room next to the hospital's obstetric ward) was devoted to research on blood preservation and on improving the care of newborn babies affected by Rh incompatibility.[19] The Lister Institute also housed the MRC Blood Products Research Unit—an incarnation of the wartime Serum Drying Unit that had been based in Cambridge—which was steadily contributing to improvements in the storage capacity of whole blood and the expanding uses of blood fractions. The Blood Products Research Unit tested and prepared dried plasma and large quantities of the blood products fibrinogen, fibrin, and thrombin for use in hospitals. In the building too was a lab devoted to the biochemistry of blood groups: that is, the chemical structure of red cells and antigens, and the mechanisms of antibody-antigen binding.[20] This last was run by chemist and serologist Walter Morgan, who had been brought into the orbit of the Galton Serum Unit during the war and continued to work closely with Race and Mourant.[21] Together, the Lister Institute labs managed the provision of antisera and carried out research into blood groups and their genetics, the practicalities of transfusion, and the technologies of storage and movement.

London, then, was the site of multiple kinds of scientific expertise on blood groups, specifically, their genetics and their use in transfusion. This arrangement fulfilled the NBTS planners' vision: although transfusion was to be regionally organized, it was still an experimental and potentially dangerous technology and offered potentially hazardous products, which had been responsible for a number of deaths with causes that were not fully understood.[22] This was a cutting-edge and still relatively little-understood practice, the highly distributed organization of which was overseen by metropolitan experts.

Blood Grouping Technique

The smooth and safe operation of the NBTS depended on the practices of blood grouping. Throughout this distributed system, routine tests were carried out in a variety of institutions. Beyond the depot and specialist

laboratories, tests were now routine in hospitals. And where blood grouping had seemed straightforward in the 1920s, it now involved a far
greater range of techniques. The two main categories of blood group test
were the "tube" technique, which mixed antisera and samples in test
tubes arrayed on a wooden block, and "slide" technique, which employed a white tile or slide on which to mix blood. Blood groupers might
also use capillary tubes, centrifuges, combinations of antisera, or spe-
cially treated cells. The startling array of methods was in part because
different blood groups required subtly varying methods of detection—
different antigens and antibodies had to be treated in different ways
to make agglutination visible. As new blood groups were discovered,
methods were adjusted and refined. Moreover, the choice of technique
depended on institution and circumstance, on how quickly a test had
to be carried out, and on what kind of training the serologist had, as well
as how many and what kinds of blood groups were being tested. Moreover, there were no strict rules on who could carry out blood grouping.
The first formal qualifications for laboratory technicians in pathology
were issued in the late 1940s, but few blood groupers held these. And
although general practitioners and junior doctors likely had cursory
knowledge of blood grouping, not all had much practical ability.

Disputes over techniques bring into sharp focus the significance and
meaning of blood grouping tests and offer insights into the dynamics
of authority among the different parts of Britain's transfusion service.
We have seen that during the war, the Galton Serum Unit experts had
clashed with hospital workers over methods—at that time, over the relative virtues of the tube versus the slide technique. Postwar, grouping
tests were more numerous and practiced in many new places, and disputes were more public. Preeminent experts on blood grouping worked
at the Lister Institute's Reference Laboratory and Research Unit, and at
regional centers, where serologists had extremely stringent standards for
technique, organization, and demeanor.

The Lister Institute serologists had their own typewritten guide to
blood grouping with the enigmatic title "Their Life a General Mist of
Error, or, Hints to Blood Groupers."[23] This document offered instructions
on how to organize and read arrays of blood grouping reactions and to
prepare controls.[24] Its tone was stern and familiar and its instructions exceptionally detailed. Use of the pronouns "she" and "her" indicates that
blood grouping technicians were expected to be women. Both Race and
Mourant often employed their young female technicians straight out of
school, reasoning that the Lister Institute was one of the only places in
the country that could give appropriate specialist training.[25]

Lister staff were fastidious about organization and behavior. "Hints to Blood Groupers" included a lengthy section on "general serological behaviour," which stipulated, "The pencil must be reasonably sharp and the rubber [eraser] found. (If later the rubber has to be looked for or the pencil sharpened a dangerous diversion will be caused.)" It explained sternly how to organize and label samples, and it offered advice on how to minimize confusion at the laboratory bench by arranging tubes in wooden support blocks.[26] The document also provided meticulous instructions on how to write labels, how to handle abbreviations, and how to support and orient tubes (which should "be raised . . . so that the mark on the pipette is not lost in deep darkness but is visible against the white background of the back of the label"). It stipulated how reactions carried out at the laboratory bench should be written into the notebook, and it went into great detail about how to transpose agglutination results onto paper. Expanding on that theme, the document repeatedly implored lab workers to pay careful attention to their handwriting, at various points stipulating that the technician should "label the tubes of cells slowly, so that they can be read," or "write down slowly all the details on the serum tube," and "write legibly on all tubes."[27] The message was that accurate results vitally depended on careful inscription.

The document also paid careful attention to cultivating the correct atmosphere for accurate blood grouping work. It was "essential that there should be no talking and that the door should be shut," and if you were scoring the tests, you should "never turn around to welcome anyone until you have studied, and recorded, in your own time, the slide you have made." Under the heading "General Serological Behaviour," the document explained further how workers should interact with one another in the room in which tests were being done. It directed, "If you want to speak to a person working try to see when it will cause least disturbance," and "If you cannot see ask her to tell you when she can speak and pretend to be interested in something else in the room; do not stand within the worker's vision."[28] Above all, the atmosphere had to be calm and measured. This last requirement was underlined by other experienced blood groupers. For example, the Sheffield RTD Ivor Dunsford authored a highly regarded handbook of blood grouping technique that warned that tests should be carried out "in an unhurried atmosphere as free as possible from distractions."[29] The Lister document "Hints to Blood Groupers" also underscored the preeminent authority and responsibility of the Research Unit and Reference Laboratory; blood grouping was a serious business, and the instructions ended with a severe caution:

"Our groupings are accepted without question because of our prestige, remember the responsibility of this for an error could easily be lethal."[30]

During wartime, transfusion had been carried out using only "universal" (group O) blood, which meant that donors, rather than recipients, were subject to the overwhelming majority of grouping tests. But as the uses of blood expanded, doctors made greater efforts to match donor and recipient. While hospitals tended only to test the ABO and Rh groups, they also carried out cross-matching tests as a final check on compatibility.[31] Thus, as surgeons and obstetricians relied increasingly on transfusion, a wider range of people were expected to know the principles and practices of blood grouping, and these came to include general practitioners, nurses, hospital doctors, and midwives. There were still debates about the relative efficacy of the slide and tube techniques—the latter tended to be used in depots, and both were used in hospitals—but this had mostly settled down, with a consensus that different techniques were useful in different circumstances. The 1955 color film *Blood Grouping*, made at the "Group Laboratory" of London's Mile End Hospital, demonstrated to junior doctors and house officers both the slide and tube techniques used in routine testing.

Serological experts did, though, entreat doctors to leave blood grouping to NBTS professionals where possible.[32] In the *Lancet*, *BMJ*, and medical textbooks, Mourant, Race and the RTDs particularly cautioned doctors of the dangers of grouping in a hurry or under stress. They explained that GPs should arrange for pregnant women to be tested at the local transfusion service, and that geographically isolated doctors should obtain proper training from their RTD. If a patient had to be tested quickly, the greatest danger was apparently an inexperienced serologist attempting blood grouping in a rush.[33] "Under emergency conditions, when the temptation to do the test will be the greatest, the danger of error will also be the greatest," wrote Mourant. In that situation a doctor would do better to wake the "irate pathologist" from his bed and compel him "to travel many miles to the hospital."[34] Indeed, in an apparent attempt to discourage general practitioners from doing the tests, Mourant's instructions explained the "principles and not the practice," for which he referred the reader to specialist MRC booklets.[35] He instead focused on methods for collecting, packing, and transporting specimens. Mourant exerted control by withholding instructions.[36]

The divergent interests of the Lister scientists and RTDs, on the one hand, and hospital pathologists, on the other, clashed most visibly in the mid-1950s in a dispute over the introduction of a new technology for blood grouping: the "Eldon card."[37] In 1955, British surgeon James Rice-Edwards reported in the *BMJ* a technique developed in Denmark, in which pieces of card were impregnated with patches of dried A, B, and Rh antisera.[38] To use the cards, Rice-Edwards explained, a lab technician simply added small drops from a blood sample, and agglutination would become apparent in less than a minute. One of the advantages of the technique, Rice-Edwards claimed, was that the cards themselves could be filed as a permanent record, removing the possibility of clerical errors. He judged the procedure quick and foolproof, and it would be easy, he thought, for hospitals to keep supplies of cards at hand in the emergency department or on the wards.[39]

Hospital doctors seem to have found this idea appealing, but it broke all the RTDs' rules of blood grouping, and they generated a storm of protest. Over the next year, the *BMJ* and *Lancet* published a flurry of letters debating the virtues and vices of the card technique. Transfusion officers called it "unsafe" and worried about the "disastrous" absence of adequate controls.[40] Hospital physicians countered that the "simple" paper test would prevent dangerous delays in testing and greatly facilitate the work of qualified serologists.[41] They complained that, by opposing the Eldon card, transfusion-service experts were obstructing the development of techniques better suited to a hospital setting. For NBTS officers, hospital physicians were "court[ing] disaster" by contemplating shortcuts.[42] When the expert RTDs discussed this issue at their next London meeting, many considered that the "apparent simplicity of the method [was] its most serious disadvantage, since this would lead inexperienced or untrained persons to use it with a false sense of security."[43] *"Its most serious disadvantage"*—seemingly worse that any inefficacy—was that the new method might tempt the "inexperienced" to try their hand at blood grouping.

The dispute illustrates the disruption wrought when blood grouping expanded into new domains.[44] The Eldon cards provoked a fight over the control of blood grouping in light of divergent professional interests. On one side, RTDs and their colleagues at the Lister asserted the absolute necessity of superior serological expertise.[45] On the other, hospital workers accused NBTS officers of willful ignorance of the pressures of the hospital setting.[46] Even a regular ally of the NBTS, the distinguished Oxford clinical pathologist Margaret Pickles, who worked closely with Mollison, chastised the organization for its stance: "It reveals a curiously uncritical attitude when two directors of the National Blood Transfusion Service are prepared to make sweeping statements on a blood grouping technique—the Eldon card—without having tested it carefully in their own laboratories."[47] For general practitioners, blood grouping was a diagnostic tool among several valuable techniques in clinical pathology. In hospitals, it was one among many often-precarious surgical protocols. For those within the NBTS and Lister Institute, the safety of the whole service depended on test accuracy.

However, the serological testing at the Lister Institute was not just about managing the therapeutic consequences of transfusion. Serological samples also became resources for building new knowledge about blood. At the Research Unit and the Reference Laboratory, puzzling donor and patient blood from all over the country took on new lives and new meanings.

Identifying New Blood Groups

Notwithstanding the rigorous discipline of serological testing, the Research Unit was by all accounts a lively, scientifically engaged, relaxed place to work. In the 1930s, Race had trained as a medical student at St. Bartholomew's Hospital in London, and after working as an assistant pathologist for a couple of years, he had responded to an advertisement in the *BMJ* for an "assistant serologist" at Fisher's Galton Serological Laboratory. Between 1937 and 1939, Elizabeth Ikin, Aileen Prior, and George Taylor had trained him in serological techniques, and he had worked with them on blood group linkage and gene frequency distribution. During the war, Race continued this work in the Galton Serum Unit, becoming a highly visible expert on Rh and other blood groups.[48]

In establishing the Research Unit, Race hired two research assistants. One was Sylvia Lawler, a medically qualified serologist who had also worked with the Galton Serum Unit in Cambridge.[49] The other was Ruth Sanger, a zoology graduate from the University of Sydney.[50] Sanger had been on the scientific staff of the New South Wales Red Cross Blood Transfusion Service when she first became interested in Rh genetics. With a desire to deepen her knowledge, her director had arranged for her to apply to the Research Unit. Traveling on one of the first postwar passenger ships to England, Sanger joined the unit in 1946, arranging for the work she carried out in the lab to be submitted as a PhD thesis to the University of London.[51]

Race saw the remit of the unit to include not just the search for new blood groups—although this was crucial for the management of therapeutic blood—but also the genetic analysis of those blood groups. He was keen not to lose sight of the grand genetic ambitions that he had for

blood groups: "In the short view these antigens are of practical importance as the cause of haemolytic disease of the newborn and of transfusion reactions; in the long view their main importance lies in their ability to label human chromosomes."[52] Genetic linkage was still a central
contribution that blood groups could make to human genetics. The unit
was small and tended to have only three to five members at any one
time, but over the next ten years it built up a formidable reputation.

The unit's twofold aims depended ultimately on the work of the RTDs
around the country as they encountered and investigated puzzling transfusion outcomes or unexpected cross-matching tests. There were always
two sides to any serological equation. Red-cell agglutination was the
consequence of a reaction between an antigen (attached to the surface
of red cells) and an antibody (a soluble protein in the fluid component
of blood). In the early days of blood grouping, this kind of analysis had
been relatively simple: only one blood group system was known (the
ABO system), and only four blood groups were associated with that system: A, B, AB, and O. But by the late 1940s, many new blood groups had
been discovered, long-established systems had become far more complex, and analyses demanded more numerous and subtle reagents and
protocols.[53] When RTDs sent intransigent samples to the Lister for further
testing, the Research Unit and Reference Laboratory worked together to
investigate.[54]

The Lister Institute laboratories were in a good position to carry out
such tests because they had extensive "panels" (systematic arrays) of refrigerated red cells, each containing mixtures of surface antigens, which
they could use to probe blood for antibodies. If an antibody agglutinated
a new combination of those red cells, those antibodies would be frozen
and kept for future testing. Unlike antisera, red cells (antigens) could not
be frozen and needed to be replenished regularly. The red-cell panels at
the Research Unit and Reference Laboratory tended to comprise samples
donated by staff of the Lister Institute or by loyal local donors, although
as time went on they built up far broader resources.

During the course of such testing routines, members of the Research
Unit and Reference Laboratory kept a vigilant eye for evidence of new
blood groups. If it looked as though they had defined a new antigen, the
Research Unit staff would follow up with donors with that antigen and
attempt to test as many of their family members as possible. This work
was productively recursive. As they carried out these very investigations,
the Lister labs accumulated diverse examples of antisera and characterized more and more antigens. The broader the array of sera sent to the
Lister laboratories for investigation, the wider the range of antibodies

they accrued in their freezers, and the better the resources they had for probing unknown samples of blood. This, in turn, consolidated the expertise of the Research Unit and Reference Laboratory, within Britain and overseas.

To keep interesting samples flowing through the institute, the Research Unit scientists built generative social relationships with transfusion doctors, patients, and donors, who yielded blood samples for investigation. The archived correspondence from Race and Sanger and their interlocutors is striking in its warmth, humor, and liveliness. For example, in Britain one of Race's regular correspondents was Ivor Dunsford, the RTD for the Yorkshire city of Sheffield, who coauthored the highly regarded *Techniques in Blood Grouping* (1955). Race frequently investigated Dunsford's samples and kept him abreast of new discoveries. Dunsford told Race how gratified he was to be involved in the fast-paced developments in serological genetics: "I often feel like a lonely lighthouse keeper in a sea of problems, as you can well imagine with between 2000 and 3000 samples per week passing through the laboratory."[55] Dunsford would process thousands of routine blood samples and would select and pass on to Race and Sanger the most unusual and interesting specimens (figure 5.2). Although it felt lonely to Dunsford, this monitoring work yielded some stunning finds for the Lister researchers, who happily credited this collaboration in coauthored papers.[56] When individuals like Dunsford sent samples to London, Race and Sanger investigated, often encouraging correspondents to publish or coauthor papers with them.[57]

As well as friendships and collaborations within the NBTS, Race and Sanger also cultivated relationships with transfusion workers overseas. Although the NBTS was productive for Race, he did admit frustration that depot directors did not share more materials with him. Many RTDs were deeply interested in serological research and, having "done all the donkey work" (that is, the routine serological testing), they "naturally want[ed] to tackle the more interesting and complex bloods" themselves—but they were often so overworked that they could not find the time. Race complained to a US colleague that despite the rich material available, "very little indeed penetrates the filter of the Depots."[58] From the early 1950s, US blood banks helped to fill that gap. In the United States, numerous independent blood providers and private blood banks worked alongside the American Red Cross.[59] Among the archives of the unit, some of the richest series of letters are to and from Amos Cahan, of the private Knickerbocker Blood Bank in New York City, and Louis Diamond, who hosted Race for several months at the Blood Grouping Laboratory in Boston and was also medical director of the (now extensive) Red Cross National Blood Program.

5.2 Robert Race and Ruth Sanger at the Blood Group Research Unit, ca. 1950. In this staged photograph, Race and Sanger are at a laboratory bench by a window, in front of a light box and microscope. Both are wearing smart casual clothes and not laboratory coats, perhaps signaling their seniority in the unit. Microscopes were typically used to check agglutination tests carried out in test tubes. 11 × 9 cm.
Wellcome Collection, London, PP/SAR/F/6/1. Copyright Medical Research Council. Reproduced with the kind permission of the Medical Research Council, as part of UK Research and Innovation.

Race and Sanger expressed effusive delight at being sent unusual samples by their US friends: "How the goodies pour in"; "What a splendid family"; "A pretty birdie"; "Never a dull moment at the Knickerbocker." Letters to and from the unit were addressed to "Aunty Ruth," "Queen of the Red Blood Groupers," "Very Honoured Herr Professor Dr. med. Amos."[60] Cultivating warm correspondence with the Knickerbocker and many other labs in North America brought vital resources to the Lister. It extended the unit's surveillance network beyond NBTS donors to rich new worlds of serological possibility. In his playful letters to Race and Sanger, Cahan frequently alluded to the perceived centrality of the Research Unit amid the extensive traffic of blood specimens, referring to the lab as the blood grouping "Mecca": "It is nearing sundown and we will shortly get out our rugs and kneel and face toward the Institute. Once again we supplicate your help. It's our antibodies. They need light. We have water, brine and pointed heads, but no light."[61]

As puzzling clinical samples were turned into research materials, they also functioned as gifts between researchers. Thanking Race for help with a difficult sample of blood in 1954, Cahan declared, "You have been so helpful to us in our past appeals for assistance that we are once again sending this token of our great esteem—a freshly drawn specimen from the patient with this curious serum."[62] The sharing of blood samples forged and consolidated relationships.[63] Letters to and from colleagues in blood banks, hospitals, and universities were densely patterned with dialogue about rare and unusual types of blood, the interpretation of data, the reliability of antisera, the integrity of cell panels, requests for blood and grouping expertise, exchanges of blood gifts, and personal news.

As well as consciously keeping doctors, scientists, and serologists abreast of their project, Race and Sanger often struck up relationships with donors. Their genetic research depended on gathering blood samples from the extended families of as many donors as possible. They recounted a "lovely day" spent visiting and taking blood from a family of farmers near London.[64] Establishing that warm relationship enabled Race and Sanger to follow up with visits to further members of the same family and ensured the possibility of repeat donations. In other instances the scientists reached families by recruiting general practitioners. One particularly helpful individual was a GP's wife in South London. Tongue in cheek, Race thanked her for her "marvellous list of potential victims." He continued, "Your patients seem to be an extraordinarily helpful lot of people. Every one of the families . . . has cooperated."[65]

Sometimes Race contacted NBTS donors themselves. In February 1950, he drafted a template letter to NBTS blood donors who had particularly curious blood. In persuading individuals to participate, Race explicitly linked the moral virtues of donation for transfusion with the importance of donation for research; "We realize that you have already helped the Blood Transfusion Service a great deal but hope that you will not mind being asked to help again in another way." He noted that he was making this request to people who had already freely given blood for transfusion, linking the benefits of these two kinds of donation. As he put it, "research into human blood groups is of the utmost importance to the problems of blood transfusion," and he emphasized that his science was "entirely dependent" on this kind of generous cooperation.[66] Thus, in their direct appeals to would-be subjects, Race and Sanger relied on the NBTS not just for its material infrastructure but also for the emotional commitments that donors made to the service. The researchers turned loyalty to community and nation into new knowledge about blood.

Social relationships—just like bottles, test tubes, fridges, letters, index cards, and phone calls—helped to constitute the remarkable system of serological surveillance that stretched across the country and to select labs overseas. The Research Unit used that distributed system to produce new knowledge about blood groups and their genetics. RTDs around Britain processed tens of thousands of blood samples per year and sometimes chanced upon the rare specimens that yielded new antibodies or blood group antigens. Relationships with blood banks in the United States extended that surveillance machine to American populations. As Race put it in a report to the MRC, each of the samples sent to the unit represented "the pick of thousands of other sera . . . tested and found to contain no antibody or antibodies that present no new problem."[67] Blood group genetics relied on a vast public health infrastructure that served as an instrument for detecting heritable serological difference.

Serological Surveillance

In probing the immunological specificity of blood, Race and Sanger counted on the broad geographical scope of the national transfusion service, and on depot workers willing to send perplexing samples to London. The RTDs were sentinels for unusual antigens and antibodies among constituencies of donors and patients far beyond the Research Unit. The Lister researchers thus positioned themselves at the center of a recursive system that expanded the range of antibodies and red cells in their fridges and freezers. They also carefully cultivated personal and professional relationships to extract blood and make samples move. The resources available for serological genetics depended on donor recruitment campaigns, correspondence between researchers and doctors, and relationships forged with doctors, patients, and donors. This carefully managed network strengthened the authority of the unit within the NBTS and overseas.

As a result of this system, new blood group loci were being discovered every couple of years.[68] To geneticists, each new locus was potentially a new waypoint for mapping a region of the human chromosomes. The Research Unit scientists remained closely engaged with the old problem of studying segregation between blood groups and other human characters. As historian Daniel Kevles vividly describes, postwar London was a lively place for the study of human heredity, and Race and colleagues maintained strong relationships within London's small but dynamic community of geneticists. Sanger later recalled how the community kept in touch via telephone, over lunches, in pubs, and during visits to one another's laboratories and homes.[69] Fisher maintained a warm correspondence with both Race and Sanger and regularly visited from Cambridge. Sanger recalled how Fisher "loved . . . seeing our results and playing with them and was always very, very interested in anything we did, any new blood groups . . . We saw a lot of him."[70]

Following Fisher's move to Cambridge, Penrose had been made professor of eugenics at University College, where he directed research into the inheritance of complex human traits and oversaw the *Annals of Eugenics*. The Research Unit worked closely with Penrose's lab, and Haldane regularly joined both groups for lunch at University College on Saturdays.[71] In 1948 Sylvia Lawler moved from the unit to the Galton Laboratory, where she further helped to foster a close working relationship between the two labs (figure 5.3).[72] Her studies relied on blood samples drawn in hospitals, and to obtain those Race and Lawler forged relationships with doctors across London, from the National Hospital in Bloomsbury to the Maudsley Hospital in Camberwell and the Fountain Hospital in Tooting.[73] Without knowing it, highly distributed groups of donors and patients were incrementally helping to delineate some of the earliest contours of a map of human chromosomes, through bureaucratic and social networks forged with blood.

As blood traveled longer distances, the scale and scope of its administration and collection increased. Donor registries became larger and the specificity of blood groups became ever sharper. Through the labor of donors, patients, RTDs and the Research Unit, more blood group

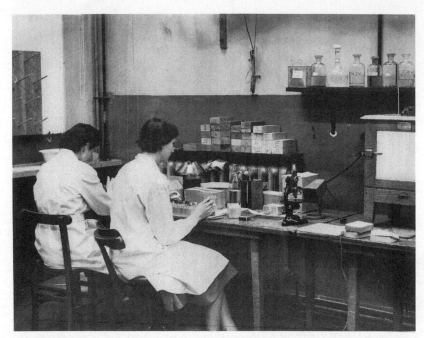

5.3 Photograph of serologists at work at Lionel Penrose's Galton Laboratory at University College London, ca. 1950. One of the two women is likely to be Sylvia Lawler, who worked at the Blood Group Research Unit before moving to the Galton Laboratory in 1948. The two serologists, wearing crisp white laboratory coats, work in front of wooden blocks filled with test tubes that presumably contain samples and reagents. Lawler later explained that she had worked in "the only room that had windows," and she commented on the simplicity of the equipment: "My requirements for apparatus were very, in those days, relatively simple . . . pipettes and a very small number of test tubes . . . and this sort of old microscope that Pasteur would have thrown out." Quotation from pages 5 and 20 of Daniel J. Kevles, interview with Sylvia Lawler, June 29, 1982, Daniel J. Kevles papers, Oral History Interview Transcripts, 1982–1984, box 1, folder 17, RAC.
Rockefeller Foundation records, photographs, series 100–1000 (FA003), series 401: England, subseries 401A: England—Medical Sciences, Galton Laboratory—Genetics, box 107, folder 2071, RAC. Reproduced with the permission of the Rockefeller Archives Center.

systems and variants were defined and incorporated into the system. Over the next few years, these connected features of postwar transfusion—that is, attention to blood group specificity, on the one hand, and a nationwide standardized infrastructure, on the other—would create the conditions for dramas, real and imagined, that centered on the "search for rare blood."

SIX

Valuable Bodies and Rare Blood

As new blood groups were defined, interest flourished in "rare" blood. This was the central plot device of the 1952 British film thriller *Emergency Call*, which told of a doctor's search for three donors able to provide a pint each of very rare blood for Penny, a gravely ill five-year-old girl. The film begins with doctors at a hospital hurriedly telephoning the local regional transfusion director (RTD) to find out whether he had any blood that would match Penny's (figure 6.1). Explaining that there is "none at the bank; they're going to check the register for suitable donors," the film's main protagonist, Dr. Carter, takes Penny's mother aside to reassure her: "There are bound to be donors on the register who belong to Penny's group. It's a nationwide organization. In a matter of hours we shall have all the blood we need."[1]

Despite the real-life connections between the NBTS and a nationwide network of donor panels and depots described above, film doctor's assurance is overly optimistic. To Penny's detriment, she has an exceptionally rare blood group; to the plot line's benefit, the only three people in the whole country with a match are, as one film review described them, a "coloured sea-man," a missing boxer, and a "murderer-on-the-run."[2] The sailor refuses to donate owing to a painful past experience when his blood was rejected because of the color of his skin. The boxer is preoccupied with evading the clutches of a criminal gang who attempted to fix one of his matches. The murderer is known to have the correct blood group because of forensic tests carried out at

6.1 "We're in urgent need of a supply of ORh-CDE. Can you help us?" A still image from *Emergency Call* (1952), by Nettlefold Films, directed by Gilbert Lewis. In the center is the film's principal protagonist, Dr. Carter (played by the well-known Anthony Steel); to his right is the hospital pathologist, who has just checked the girl's blood group; to his left the hospital consultant on the phone to the "local blood bank." The three are overlooked by a clock that reminds the audience of the urgency of their quest. Still image from 00:04:22.

Copyright Renown Pictures Ltd. Reproduced with permission from Renown Pictures Ltd.

the scene of an unsolved crime some years before but is now living under an assumed name and new identity. The doctors have only five days to find three pints of blood. The drama of *Emergency Call* is based on the ensuing pursuit, in which two figures of authority, a policeman and a doctor, negotiate the unruliness of peoples' lives and the limitations of the bureaucratic control of citizens. Combining themes of identity and civic duty, the film follows our heroic doctor as he sets out to find and persuade the would-be donors to provide the precious blood for the dying child.

The premise and plot of *Emergency Call* captured several features of postwar blood transfusion in Britain that came together to produce a bureaucratic, scientific, and dramatic preoccupation with "rare blood" during the 1950s. A highly distributed nationwide service with blood group testing in hospitals and depots had established an efficient system of serological surveillance that was also a machine for announcing new blood groups. New attention to blood specificities put pressure on procurement, which in peacetime was already more challenging now that donation could no longer be framed as part of the war effort. As donors were recruited to the service, they received a blood group identity—but some also learned that their blood was particularly rare and valuable. This charged transfusion with a new layer of drama. *Emergency Call* was not the only fictional work to pivot around rare blood; during the 1950s the BBC broadcast at least three radio plays with the same basic plot line.[3] Rare blood captured the attention of newspapers, which enthusiastically publicized the pursuit of exceptional donors. But "rare blood" was never self-evident. A great deal of bureaucratic and technical work went into defining a person or specimen as exceptional. Rare blood and highly prized donors were brought into existence through the bureaucratic technologies of the NBTS and the multiply transfused bodies of chronic patients. The special value of certain donors and specimens might also shift, as blood became marked in new ways.

Making and Managing Rarity

The war was over, but blood of all kinds was in greater demand than ever. Hospitals now relied on it not just to treat acute blood loss but also to manage anemia, in planned surgery, and in neonatal care. As the Cold War intensified, concerns about a new "national emergency" (that is, nuclear war) spurred plans to stockpile dried plasma. But voluntary blood donation was dwindling: it was hard to sustain the message of its crucial importance, especially as food rationing continued. From nearly a million registered donors at the end of the war, by 1948 only a third of that number were being bled, a situation that transfusion officers regarded as "desperate."[4] Donor recruitment took on a new urgency. As well as press advertising and sending reminder cards to registered donors, NBTS organizers made the most of the growing audiences for cinema and television, and of advances in film production. The successes of the wartime documentary films of Paul Rotha, one of which had been devoted to the EBTS, had convinced the Ministry of Health of film's potential to inform and persuade on matters of public health.[5] As a result, the MoH now spent considerable sums on documentaries and training and publicity films.[6] The films *Blood Is Life* (1957) and *Blood Can Work Miracles* (1961) were also edited to create TV "fillers" of under a minute, designed to be broadcast on BBC or the ITV network, and "filmlets" of about three minutes for the end-of-cinema newsreels.[7]

The NBTS desperately needed all kinds of blood. But the discovery of new blood groups now made some kinds more valuable than others. The likely inspiration for *Emergency Call* had been an episode in the late summer of 1950, when the *Times* reported the international search for a "rare blood" donor, a story that ran for several days. Kathleen Hall, a seriously ill patient slated for an operation at St. George's Hospital, London, needed blood of a rare type that could be matched by only "one person in 20,000."[8] The transfusion center in the county of Surrey (southwest of London) issued an appeal in national and local newspapers that resulted in a "continuous stream of volunteers" arriving at the transfusion center, and "hundreds of blood samples from all over the country . . . by express post and rail." The front page of the *Daily Express* declared, "Blood Hunt Goes on All Night."[9] The next day the search became international; the *Times* reported that "two pints of blood from seven people were flown in from Copenhagen," and more was expected from the Red Cross in Australia. But still no blood was of the correct group.[10]

Workers at the Surrey transfusion center were swamped with eager volunteers. Officials involved later complained that although the center had carefully explained to the press that the "blood type needed was mostly likely to be found among group O donors who are in sub-group R.H.—cdE," most of the volunteers lining up outside were not of that group.[11] One medical officer implored volunteers not to get in touch unless they already knew that they were of blood group O, and the *Times* urged readers to check this information on the "buff card issued from their transfusion centre."[12] The Surrey depot alone tested nearly a thousand samples before a blacksmith in another part of the country was found to have the correct blood group.[13] Newspapers announced the news: "He's One in 20,000"; "Rare Blood Woman Says 'Thank You.'"[14]

RTDs might have been gratified by the public enthusiasm, but they viewed the appetite for rare blood dramas as a mixed blessing. The NBTS saw that such high-profile cases could boost donor recruitment—it later staged a publicity photograph of *Emergency Call* star Freddie Mills sipping tea after donating blood in front of an *Emergency Call* poster.[15] But the RTDs were also concerned that such stories made the transfusion service look disorganized, and they believed that well-meaning but unsuitable volunteers had seriously hampered the day-to-day work of the Surrey center.[16] NBTS officers decided that specific searches for rare blood should be kept out of the public eye: "Directors should appeal first to their colleagues before resorting to widespread publicity and wireless appeals."[17] They also longed for more effective technologies for locating and tracking donors with the rarer kinds of blood. Above all, the NBTS needed a bureaucratic solution.

They decided to establish lists of people known to have the rarest blood groups who could reliably be called upon at short notice when a recipient needed blood of an exceptionally close match.[18] The plan was to create two complementary lists: regional panels would comprise "comparatively rare blood groups, which were yet sufficiently common for each regional transfusion center to build up its own register"; a nationwide panel would include people with "blood groups which were so rare that it was necessary to compile a central register."[19] The latter would be kept by Arthur Mourant's Reference Laboratory in London and would list 2,000 people with the very rarest groups nationwide. Mourant's team also carried out the tests to define whether a person's blood was sufficiently rare to be added to the registers.[20] The MoH had to hire extra clerical staff to help the Reference Laboratory "type out the National Register of blood donors."[21] The resulting panel was so important to the NBTS that a senior member of the Reference Laboratory had to be on call to cover emergencies at all times.[22]

The nationwide rare blood register took two years to complete, but it was finished by June 1952, when the *Lancet* announced it as "the first of its kind in the world."[23] This bureaucratic tool disciplined and formalized the search for rare blood groups that had seemed so unruly in the newspapers two years before and that had afforded such drama in *Emergency Call*. Mourant at the Reference Laboratory monitored and oversaw the upkeep of the register, and he entreated the regional directors to look out for donors who would be suitable for the list and make sure they were willing to act in emergencies.[24] When, occasionally, rare blood searches hit the headlines, Mourant reminded RTDs how important this bureaucratic technology was.[25] To keep such stories at bay, he regularly analyzed the list of blood groups to "find out whether any sections could be strengthened," that is, whether any blood groups could be better represented. If donors were removed from the list—which might happen if they were ill, had died, or had simply "resigned"—then RTDs were obliged to find replacements. More than once Mourant had to prompt RTDs to keep their lists up to date.[26] But by the middle of the decade, the Rare Blood Panel was a fully integrated component of the NBTS.

Certain donors now had special status, and the RTDs sought to reinforce their loyalty. Even if such rare donors could not hope for financial remuneration, the NBTS made particular efforts to make them appreciate their importance and value.[27] To help maintain the high quality of the register, Mourant drafted a letter that would be sent to those donors who had requested to resign from the national panel. It urged them to remain committed to the list, emphasizing how "exceptionally valuable" they were and what an important service they were performing. Mourant also reminded donors of the specific investments that the transfusion service had made in their blood, such as the "highly elaborate tests" needed to characterize it. So valuable were these individuals that the NBTS was willing to provide transfusion officers to bleed donors in their own homes.[28] This was far from the unruly, undisciplined, and thoroughly theatrical drama of *Emergency Call*. The regional and national rare blood panels consolidated a new regime of value for postwar blood, based on blood groups.

Living Archives

Blood was precious not only for transfusing into people but also for manufacturing the serum reagents for grouping tests. Thus another class of "rare" donor was those capable of producing large quantities of antibodies. Only a limited range of testing antibodies could be made using rabbits and guinea pigs, and humans were the main source of grouping antisera.[29] Anti-A and anti-B antibodies could be sourced from regular donors, who produced these "normal" antibodies naturally, without inoculation.[30] But people vary in the quantities of antibodies in their blood, and transfusion workers went to considerable lengths to find and bleed high-titer individuals—that is, people with high concentrations of antibodies.[31] Just as for donors of rare blood groups, MoH officials drafted letters to high-titer donors to convince them of just how valuable they were to the transfusion service: "Such strong anti-bodies are very uncommon and are invaluable for special Transfusion purposes"[32] In the early 1950s the NBTS consulted RTDs about new donor cards that would include information on whether a donor's blood had a high or low antibody titer, and which antibodies had been tested. Although it is not clear how widely they were used, such cards were marked with the note, "IMPORTANT: Your blood is more valuable than ordinary because it can be used for testing other donors."[33]

To underpin the supply of antibodies from the general donor population, the Lister Institute Reference Laboratory tested and monitored the antibody titers of Air Force personnel, a relationship that the NBTS had apparently inherited from Taylor's and Fisher's serum collections during the war. In 1950 the Air Ministry committed officially to keeping regional centers informed of the whereabouts of high-titer service people.[34] All in all, these efforts kept the transfusion service well supplied with ABO antisera.

Rh antisera were harder to come by. By the 1950s, the effects of Rh incompatibility on mothers and their babies was well known. NBTS officials urgently pushed to focus the attention of hospital doctors on the Rh status of pregnant women. Many Rh– women were still receiving transfusions of Rh+ blood, and this seriously compromised their chances of healthy future pregnancies. With new awareness of Rh, the demand for such tests was quickly becoming far higher than could be met with the sera available; in 1950 stocks were so low that more than half the pregnant women in England and Wales were giving birth without having had an Rh test.[35] This situation was further complicated by the remarkable number of Rh variants known by then. Although the anti-D antibody remained the most important for routine testing, anti-C and anti-E antibodies were also in high demand—and all of these were extremely rare and very valuable. Even at the Lister Research Unit, as Sylvia Lawler later recalled of working there, "You actually had to have a PhD in order to handle any anti-Rhesus serum and it was all very, very precious."[36]

The most reliable source of Rh antibodies was Rh-negative mothers who had already carried babies suffering from hemolytic disease: they had already been inoculated with Rh+ blood and so carried the antibodies that were needed to make testing antisera. An MoH pamphlet entitled *The Rh Factor: A Leaflet for Midwives, Nurses and Health Visitors* (1949) explained that it was important that mothers "co-operate" in donating their blood for this purpose, "even though it means some personal discomfort."[37] The MoH expected these women to be particularly obliging because they had been directly affected by the problem of Rh incompatibility.[38] In 1950, Rh-negative mothers accounted for almost the whole anti-D serum supply, but it was still scarce, so the transfusion services had to pursue other avenues. One option was to deliberately inoculate people to stimulate anti-Rh antibody titers. This could not be done on women who might have babies in the future because the antibodies would be dangerous to an Rh+ fetus, so some regional depots tried to restimulate the antibodies of immunized women who had reached menopause. Others tried to inoculate men, or women who had taken vows of celibacy. A *BMJ* article declared that inoculated nuns, men and menopausal women were "the greatest potential source of all."[39] Inoculation was risky, though: stimulating Rh antibodies would restrict the kinds of blood that an individual could receive in the event of transfusion. All in all, donors capable of providing Rh antibodies were few and far between and were very precious.

Since the mid-1940s, yet another class of donor had become extremely valuable to researchers. As we have seen, any transfusion inoculated a person's body against novel antigens, resulting in the production of antibodies, in much the same way as in response to an infection or vaccine (or, indeed, a Rh+ fetus). Therefore, repeated transfusions could result in the buildup of multiple antibodies against donor blood. Monitoring the kinds of antibodies carried by a multiply transfused patient could be a productive strategy in the search for novel antigens among donor constituencies. Those antigens might themselves be very common but previously unknown—so blood from such people had the potential to reveal important new blood groups. The bodies of multiply transfused patients became highly prized by Race, Sanger, and their colleagues.

The value of such patients was first proved through the high-profile discovery of the "Lutheran" blood group antigen, reported in 1945 during a collaboration between Race, then still at the Galton Serum Unit, and Sheila Callender, doctor and reader in medicine at the Nuffield Department of Clinical Medicine in Oxford.[40] Callender and her colleague, the Turkish medical student Zafer Paykoç, were carrying out a systematic study of the antibodies contained in the blood of transfusion patients.[41] Now that blood was becoming a routine therapy to treat some anemias, the buildup of novel antibodies was becoming a significant problem for chronic patients. The more antibodies that accumulated in a person's blood, the more hazardous further transfusions would be. The principal subject of Callender's and Paykoç's study was a twenty-five-year-old female patient under observation at the city's Radcliffe Infirmary, referred to in the published work as "F.M." This patient suffered from lupus erythematosus, resulting in a persistent anemia that had been treated with nine transfusions from eight donors. These inoculations of blood had resulted in the appearance of a "remarkable succession of antibodies" in her serum.[42] To find out just how remarkable, the researchers took samples of blood from F.M. after each successive transfusion, and tested those against panels of red cells with well-characterized arrays of antigens.

Some of the serum samples from F.M. agglutinated a pattern of red cells that had never before been observed—the implication being that one or more antibodies in those samples had reacted with hitherto unknown red-cell antigens. The bureaucratic systems in place at the hospital and the local blood depots meant that the doctors were able to trace those antigens back to particular donors with the surnames Willis, Levay, and Lutheran. Race helped Callender and Paykoç to locate and test the families of these donors and ascertain the inheritance of these groups. The first antigen (Willis) was found to belong to the Rh class of antigens (as a variant on the Rh allele C, it was named C^W); the second (Levay) was found to be exceptionally rare and unrelated to any existing group.

In this study, the Lutheran antigen was the real prize, as it, too, turned out to be unrelated to any known blood group system yet was rather common in Britain. Researchers tested the anti-Lutheran antibody extracted from F.M. against samples from unrelated donors, probing them for the presence of the Lutheran antigen. They traced the families of several additional donors, laboratory workers, and students found to be Lutheran-positive. Following the parents and offspring of seventeen families, the researchers concluded that a Mendelian dominant allele was the genetic basis of the Lutheran antigen.[43]

The NBTS donors, Willis, Levay, and Lutheran, all gave permission for their names to be given to the new antigens and blood groups.[44] Other major new blood groups defined during this period—such as Duffy and Kell—were also named after donors. This postwar naming practice, which fit with the valorization of the altruistic blood donor, diverged from traditions of eponymous naming in medicine. Many diseases were named for the physicians who first described them, but new blood groups were named for the donor in whom the antigen was identified.[45] Unlike a disease—for which multiple patients were typically needed to identify and define it—a new blood group antigen could be pinned to a single individual.[46] Moreover, for a donor's blood to lead to the full characterization of a new blood group system (that is, a blood group locus), researchers had to have an ongoing relationship with that individual and his or her family—we have seen how Race and Sanger followed up with donors' families, sometimes returning again and again for blood. This, to the researchers at least, likely underlined further the significance of the donor's family name. Testifying to this, later in life Mourant would recall that "Kell" had been an abbreviation of the name of "a very cooperative donor" with "a remarkable family" of (conveniently for the researchers) twelve siblings.[47]

F.M. had been suffused with the blood of multiple donors during successive transfusions and as a consequence held an array of detectable antibodies to novel proteins. F.M.'s antibodies divulged for the first time antigens that existed "out there" among Britain's donor population but that had not before been visible. This made F.M. a special kind of research subject—someone who, primed by the transfusion services, was capable of yielding new serological knowledge. The NBTS, then, forged two new kinds of population. First, it identified willing blood donors who harbored a wide range of antigens that had been invisible until their blood was moved into new bodies. Second, in multiply transfusing certain patients, it created highly valued living archives of antibodies that could disclose novel aspects of human serology and genetics.

As multiply transfused patients were recruited by Britain's hospitals and NBTS to the search for new blood groups, they helped to specify rare blood in increasingly subtle ways. At the same time, every transfusion narrowed the range of blood groups that could be transfused into that individual, leaving multiply transfused patients themselves in greater and greater need of that rare blood. Repeated transfusions saved the lives of individuals suffering from conditions such as anemia and hemophilia; they also modified their bodies in ways that were both dangerous and useful.

"White" and "Colored" Blood

During the early 1950s certain people became understood as having especially precious blood, and the identities of particular donors became attached to new antigens. Meanwhile, another social marker of human identity became linked to select blood samples when a new kind of blood arrived at the Research Unit. Neither the EBTS nor the NBTS had ever before used racial categories in its bureaucratic practices, and racial categories had apparently not so far entered into the work of the unit. But in the 1950s some intriguing cell samples from the Knickerbocker Blood Bank began arriving at the Research Unit with the labels "white," "negro," and "colored," and these shifted the ways that the London scientists ordered and investigated their samples. These racial labels brought a new dimension to the alignment, ordering, specification, and value of donors and their blood.

Since the unit had been established, Race and his colleagues had cultivated close relationships with colleagues in the United States, resulting in the routine movement of serum samples across the Atlantic. At least until the early 1950s, the exchange of such specimens was largely restricted to sera—that is, the antibody-containing component of blood. Sera, which are cell-free, could be frozen and so were easy to preserve and transport over long distances. Blood group tests, though, had to be carried out on red blood cells, as their antigens defined a sample's blood group. Freezing blood had the effect of rupturing cells (hemolysis), which precluded doing any grouping tests; so for red cells to be useful to the serologist, they had to be refrigerated instead, which limited their maximum storage time and travel distance. Using the transatlantic airplane postal service, cell samples sometimes arrived intact—especially if the sample was clotted and still with its serum—but they would often succumb to bacterial infection en route.[48] The London unit often reported regretfully to US colleagues that a batch of cells had not been testable, or had totally hemolyzed.[49] Thus, the fragility of red cells limited the distances that antigens could travel and still be interpreted by serologists.

This restriction was relaxed in the mid-1950s when Amos Cahan of the Knickerbocker Blood Bank implemented a new technique for preserving red cells en route to London. Adding a small quantity of penicillin to tubes containing whole blood made the samples more stable.[50] Suddenly, potentially fascinating red-cell antigens could reliably travel across the Atlantic. For the Research Unit, this innovation was announced by the arrival from Cahan of beautifully preserved complete red-cell panels—that is, arrays of red-cell samples that the Knickerbocker hoped to use for routine assays on sera.[51] In fact, now that red cells could be sent dependably through the mail, the Knickerbocker was marketing a panel of red cells to US hospital pathologists for carrying out their own complex grouping tests. Cahan marketed the panel under the name "Panocell," and this became a popular resource for blood banks. The Knickerbocker had to keep the panel up to date and fully characterize its red cells for all known antigens. In order to check those characterizations, Cahan sent the cells to the Research Unit for further testing, to Race's and Sanger's delight.[52] This new preservation technique had altered the material properties of testable, heritable red-cell antigens.

From the perspective of the Research Unit, these new Knickerbocker materials were novel in another way, too. Throughout the war, and for several years after, the American Red Cross had segregated blood according to the categories of "white" and "negro," a practice that became a major civil rights cause.[53] Only in 1950 did the American Red Cross cease marking the records of donors with racial designations. But private blood banks evidently used racial markers for far longer: in the mid-1950s, intriguing cell samples from the Knickerbocker began arriving at the Research Unit, labeled by racial group.[54] In 1950s Britain, racial prejudice powerfully affected experiences of health care, immigration, and work, but the NBTS had not used racial categories in its bureaucratic practices.[55] The array of specimens that accompanied the correspondence from the Cahan lab sharply piqued the interests of the London-based researchers, who eagerly began using these categories to further recover the meaning and relevance of blood group variation.

To the London researchers, these US red-cell antigens were a new kind of heritable material. In due course the unit's twice-yearly reports to the MRC changed noticeably. In the 1953–1955 report, the first section, entitled "Work on the Blood Groups of Negroes," detailed new blood group antigens that were apparently characteristic of that racial group.[56] It focused attention on an antibody that had appeared in the blood of a "Mr. V" after a long succession of transfusions.[57] One of Cahan's correspondents had isolated the antibody—later known as "anti-V." Cahan was investigating the antibody's properties and was apparently startled to find that it reacted with large numbers of blood samples labeled as "negro." He sent it on to the London Research Unit for further investigation, where Race threw himself into the investigation with great enthusiasm. Following up Cahan's tests, Race wrote to explain that he was now searching in London for a new kind of donor, "This past week I wrote to the Medical Officer of Health for Lambeth [local to the Lister Institute] asking him where the babies of the Jamaican negroes in London are being born." Using terminology that suggests racial stereotyping, he went on, "One fecund spot is Dulwich Hospital"[58] There, Race hoped to gather samples of red cells and genetic data from mothers and their children. Fired up by this apparent racial link, Race sought to acquire what he called "colored" samples from serologist colleagues traveling in Nigeria and Ghana.[59] A supposed connection between race and blood group had become a research question in its own right.[60]

In the film *Emergency Call*, the paternalistic doctor could educate the "colored" sailor (and the film-viewing public) in the lesson that blood transfusion had the power to dissolve racial categories: "White, black, brown, yellow: human blood's the same the world over."[61] To the Research Unit experts, though, blood the world over was not the same. Race and Sanger were well aware that human populations varied in their frequencies of certain blood group alleles—as we will see, their close colleague Mourant was a world expert on this kind of variation.[62] But they also believed that some blood groups could be diagnostic of race, perhaps drawing on recent claims that hemoglobin—another blood constituent—could offer "proof" of blackness.[63] In a lecture about their ongoing work, Sanger explained that anti-V was "the best single antibody yet known for distinguishing negro blood from that of whites."[64] To the MRC, Race claimed that "the diagnostic power [of anti-V] was well illustrated when the two New York whites who had V . . . were found, on enquiry, to be Puerto Ricans."[65] Race was implying that new antigens might offer a direct link to racial identity (and moreover that "negro" blood in New York shared fundamental features with blood from "Puerto Ricans," London-based Jamaicans, and donors from Nigeria and Ghana). Commenting on yet another blood group system, Race reported to the MRC, "If we assume that a third allele Fy is the explanation [for these antigens] then this is the biggest known single gene difference between negroes and whites. Skin colour, shape of face and so on do not count for they must depend on many and unidentified genes."[66] The way that Race framed it, certain blood groups were more direct indicators of race "even" than skin color.

The London researchers took the Knickerbocker's "negro" and "colored" categories for granted. And these labels apparently also cast "British" blood in a new light. Race and Sanger began to use the term "white" to characterize certain samples.[67] Writing to Cahan to update him on recent findings, Race and Sanger recounted taking blood from a friendly family of "very white . . . farmers," and they included in their MRC report a section entitled "Work on the Blood Groups of Whites."[68] The terms "negro," "white," and "colored" had been created in the context of New York City by a blood transfusion system that had a long history of blood segregation. The London unit now imported and applied these terms to individuals and families in Britain, where different histories of colonization and migration had forged a strikingly different social landscape.[69]

We have already seen that the British researchers' pursuit of serological and genetic variation depended on geographically dispersed networks. Specimens from highly distributed groups of donors brought new antigens and alleles into the purview of the Research Unit. Now, blood from New York marked with racial labels seemed particularly rich with serological promise. These novel samples had the potential to yield new properties of blood groups, new antigens and antibodies, and new waypoints on the human chromosomes. Sanger herself used a geographical metaphor when she commented that their interest in "negro" blood "was not so much in the anthropological significance of the findings, but in the less parochial view that they gave us of human genes." With wry self-deprecation, Sanger went on: "We had previously tended to feel that only British genes, and of course Australian, deserved serious study"—presumably referring to "white" Britons and Australians.[70] Race made the point more directly when he told the MRC, "By studying negro blood we learnt something about a locus that we could not possibly have done had we confined ourselves to white samples."[71] Moreover, analogous to the value placed on so-called primitive populations in studies of blood group diversity, "negro blood" at this time and place was judged valuable because it could reveal genetic and serological secrets that were not just specific to certain races but relevant to *all* human blood.

Panels and Peoples

Rare blood was identified in different ways. The work of the NBTS to mark individuals according to blood group not only made variation visible but also defined certain individuals as rarer (and more valuable) than others. Some donors were particularly prized for their rare combinations of blood groups; others, for their high-titer antibodies. Inscribed racial categories could mark donors and samples in a way that elevated their value to serological and genetic research. The NBTS also made certain people exceptional in a more literal way. The transfusion services changed patients' bodies, transforming them into devices for detecting blood group and genetic novelty. The highly distributed communication systems of the NBTS, and the bodies of donors and patients, functioned together as an instrument for defining blood groups. In turn, the identification of new groups recursively shaped the identities of blood and its donors.

As blood traveled further, the "rare blood panels" that had begun as national projects gradually expanded internationally. In 1956, the National Blood Transfusion Association of Eire (Republic of Ireland) formally requested to be brought together with the UK National Rare Blood Panel.[72] In 1960, by which time whole blood could routinely be sent across the Atlantic, the American Association of Blood Banks National Cell Register applied to work out a scheme of formally liaising with the UK National Panel.[73] In 1964 the International Society for Blood Transfusion sought to create a single International Rare Blood Panel—an initiative that was eventually established in 1968 in collaboration with the World Health Organization.[74] Under that scheme, national blood transfusion centers were responsible for testing candidates for the panel, who would eventually have their blood checked again at Mourant's Reference Laboratory. Thus, as blood became more mobile and more finely differentiated, donors were brought into more expansive bureaucracies.

As rare blood became familiar to newspaper readers, it also began making regular appearances in reports of another kind. In the courtroom, blood groups were still cited in paternity disputes; now they also became part of the drama of murder trials. London's Scotland Yard established a blood grouping laboratory to trace the identities of criminals over space and time.[75] Newspaper reports of trials began to cite courtroom evidence from serologists, and readers were learning that "rare" blood could offer particularly damming evidence of culpability. In 1949, a "rare blood group" was cited as key evidence by the *Times* in its report of a murder charge in North London.[76] During the same year, the headline on the front page of the Gloucester *Citizen* declared, "'Rare' Blood on Accused Man's Clothing," and explained that bloodstains from the alleged murderer were of the same group as the victim's, "a rare one found in only 2 ½ per cent of the population."[77] In 1953, the *New York Times* cited a Scotland Yard detective in claiming that human blood was "becoming almost as valuable as fingerprints" for the identification of individuals.[78] The forensic truth of rare blood was also woven into the plot of *Emergency Call*, where the final would-be donor (the "murderer-on-the-run") is identified from the (rare) blood left at the scene of his crime several years earlier. Although the criminal is by then living under a new name, his blood betrays his true identity to the doctor and police chief. In the end, the heroes of the film are almost too late: as they catch up with him, the murderer is shot. But before he dies, he consents to giving a last pint of rare blood, and the little girl's life is saved.

In the United States, "rare blood" took on an even more vivid role in the imagination of the nation. *Emergency Call* was apparently such a successful film that it was rereleased in the United States under the title *The Hundred Hour Hunt*, where it caught a wave of interest in rare blood that was fueled by concerns about a future atomic war. In the 1950s, many US cities and states implemented programs for mass blood grouping. Knowing one's blood group was a civic duty, and ideally, that information was intended to be visibly attached to a person's body. There were several efforts to issue citizens with "dog tags" that were "atomic radiation–resistant" and color-coded by blood group.[79] Remarkably, Indiana initiated "Operation Tat-Type" to tattoo blood groups onto the arms or chests of residents.[80] And as knowledge about transfusion expanded, Americans also learned that some blood groups were particularly unusual and precious. Newspapers and novels intensified interest in rare blood, and "rare blood clubs" merged a culture of fraternal secret societies with medical preparation for atomic war.[81]

The London-based Research Unit relied on and helped orchestrate an interplay between serological genetics, the national transfusion infrastructure, and international networks. Next door, Arthur Mourant's lab was also devoted to blood groups, but in ways that used transfusion networks to chart genetic diversity. Like Race and Sanger, Mourant would make productive use of British blood and also of the expanding international movement of specimens and data. But instead of studying genetic inheritance per se, he would use the circulation of blood and paper to map the genetic relationships of the peoples of the world.

SEVEN

Postwar Blood Grouping 2: Arthur Mourant's National and International Networks

On January 1, 1952, the Royal Anthropological Institute (RAI) established the first institution devoted to worldwide data on human genetic diversity. The Nuffield Blood Group Centre was to bring into a single place all knowledge about the blood group frequencies of human populations. The center was located in a modest cottage at the back of the RAI in Bedford Square in North Central London.[1] But it had lofty ambitions. As the US magazine *Science News-Letter* put it, the result would be a scientific picture of human history, revealing "the genetic relationships" of people across the world and making visible the "past nomadic wanderings and migrations of early human tribes over the face of the earth."[2] The project was certainly effective. Within a decade, the center was famous for its work to geographically map the world's blood groups, and its data represented tests carried out on several million people.[3]

The project cohered with a postwar inclination toward totalizing scientific archives.[4] Practically speaking, its collection activities were made possible in part by the center's director, Arthur Mourant, who was also in charge of another London-based institution, the Blood Group Reference Laboratory. As its name implied, the Reference Laboratory was a central "reference" point for puzzling samples of donor or patient blood. Based in Chelsea, a few miles from the Nuffield Blood Group Centre, the Reference Laboratory made and distributed antisera to the postwar National Blood Transfusion Service (NBTS); it also made blood-based standards that other laboratories used as metrics for assessing local blood grouping reagents. It would soon be recognized by the World Health Organization (WHO) as the official International Blood Group Reference Laboratory. In that capacity the lab would distribute standardized antisera and check the purity and specificity of reagents from WHO-accredited labs around the world.

These laboratory practices made possible Mourant's paper-based blood group archive. Mourant rarely collected blood himself; rather, he leveraged his position at the heart of the postwar transfusion service to routinely collect data via depot directors on all new NBTS donors. On the international stage, he supplied standard reagents and serological advice to doctors, anthropologists, and missionaries; to blood banks, hospitals, depots, clinics; and to researchers on fieldwork expeditions, and he asked for local blood group data in return. Beyond those networks, he corresponded with researchers intent on collecting blood group data on expeditions, often using British colonial institutions and contacts.

As a result, the Reference Laboratory became the most important international hub for testing reagents and blood specimens, and the Nuffield Blood Group Centre (henceforth "Blood Group Centre") amassed paper records from across Britain and beyond. This chapter examines the relationships between Mourant's exchanges of antisera and his accumulation of data, and their dependency on networks built on wartime blood transfusion, long-established colonial structures, and new postwar infrastructures of international health. It describes how Mourant's paper-based "anthropological" activities helped him to position the Reference Laboratory as an international authority capable of adjudicating standards, and how his standardization work allowed him to amplify his anthropological collections. Yoking these enterprises meant that within a few years Mourant could claim to be in charge of the most authoritative collection of blood group data in the world.

Antiserum and Authority

Mourant's earlier life and career prefigured his interests in laboratory medicine, anthropology, and mapping.[5] Born on Jersey—one of the Channel Islands between southern Britain and northern France—he earned an undergraduate degree in chemistry at Oxford, where he developed interests in archaeology, human prehistory, and, above all, geology. After doctoral research on the Precambrian rocks of Jersey, and a couple of years of laborious mapping work for the British Geological Survey, he returned to Jersey in the middle of the Depression, unsure where to turn next. On the suggestion of his family's doctor, he established the island's first private chemical pathology laboratory, where contact with local doctors and medical problems inspired him to apply to medical school in London. Once he gained his medical license, he decided on a career in clinical medicine, and in 1944, he was posted as a junior medical officer to the North East London Blood Supply Depot in Luton.

Now in the midst of the wartime EBTS, Mourant bled donors, drove vans, and carried out grouping tests. He also became intensely interested in the developing work on the Rh blood groups; he later claimed that during a feverish attack of the flu he had experienced a lucid delirium in which he had been able to integrate "all the facts and theories" of the increasingly complex system.[6] At around this time Mourant encountered a perplexing case of blood transfusion involving anemia at his depot and, through a series of tests, he discovered one of the Rh antisera that Fisher had predicted. Delighted by this discovery, the Cambridge-based Galton Serum Unit invited him to collaborate further with George Taylor, and after Taylor died unexpectedly in 1945, Mourant joined the unit.

At Cambridge, Mourant worked with Race and doctoral student Robin Coombs on the development of testing reagents to detect Rh antibodies, which were difficult to identify using conventional antisera.[7] The Galton Serum Unit was integrated into wartime transfusion infrastructure, and it made, tested, and distributed high-titer testing reagents to depots and labs around the country. Mourant took part in these activities enthusiastically, and when the war ended, the MRC assigned these serum-based responsibilities to the Reference Laboratory in London, under Mourant's direction.

Mourant quickly built up a sizable laboratory. The principal sources of antisera in the 1940s were high-titer human donors (for anti-A and anti-B sera) and laboratory rabbits (for anti-M and anti-N). To make Rh reagents, Mourant and his colleagues bled women who had become immunized against the Rh factor during pregnancy.[8] In this, Mourant was helped by Joan Woodward, who had been in charge of serum production at the Galton Serum Unit, and former unit member Elizabeth Ikin, who by then had ten years' experience of specialized blood group work. A great deal of the laboratory's work involved "serological investigations," that is, the routine testing of blood specimens sent by depots around the country—work that, as we have seen, provided the raw materials for serological genetics. Mourant considered Ikin "the most experienced blood grouper in Britain," and would later describe her as the "mainstay" of the laboratory she largely managed for the next thirty years.[9] Other technical staff at Mourant's laboratory came to work there from high school. In line with the increasingly professionalized lab technician workforce, and as was usual with blood grouping work specifically, almost all of the employees in the Reference Laboratory were women.[10] Mourant soon added a "medical officer" to his staff, to give practical advice on transfusion problems and hemolytic disease. By 1950 the lab comprised twenty workers, including scientists, technical staff, animal attendants, and people washing and packing glassware. (Mourant estimated that "nearly a million tubes, bottles and other pieces of glassware are washed in this room annually.")[11] Moreover, Mourant's staff provided a massive increase in blood grouping training in postwar Britain.[12]

From his new position at the Reference Laboratory, Mourant cultivated an international reputation through the manufacture and distribution of reference standards for blood grouping antisera—that is, the antibody-containing reagents used to test blood groups. Thanks to wartime developments, by the late 1940s, serum and plasma could be remarkably long lived.[13] They were stored frozen at −10°C or −20°C, or freeze-dried by evaporating off the liquid at low temperature and pressure. Freeze-drying resulted in a white powder that could keep its original titer indefinitely. Whereas the fragility of red cells limited the distances that antigens could travel and still be interpreted by serologists, freeze-dried antisera could travel internationally with ease, permitting new connections between institutions.

How were standard antisera made? In the late nineteenth century, immunologist Paul Ehrlich had first established the principles of standards for biological agents. Like humans, laboratory animals vary in the titers of antibodies in their serum. Ehrlich established a method of comparing the biological potency of serum doses to a reference standard, that is, fractions of a single large batch of the same product.[14] The expansion of serum therapy in the early twentieth century led to the establishment of several institutes for serum research and standardization, including the Statens Serum Institut in Copenhagen and the Division of Biological Standards within the National Institute for Medical Research (NIMR) in North London.[15] To Mourant, the supreme stability of freeze-dried blood grouping antisera raised the possibility of producing a reference standard that could be distributed widely and used to measure the concentrations of antibodies in local reagents.

The idea was that any lab could titrate the standard antiserum in parallel with their own local supply so that the potency of the latter could be expressed in terms of the standard.[16] Mourant's Reference Laboratory col-

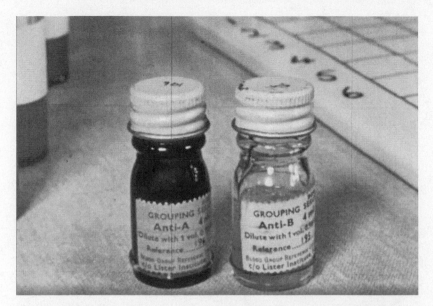

7.1   Still from the film *Blood Grouping* (1955), the purpose of which was to show students
and house officers some of the techniques used in routine blood grouping in a hospital
laboratory. This image shows standard antisera supplied by Arthur Mourant's Blood Group
Reference Laboratory at the Lister Institute. Anti-A serum is provided in a brown glass
bottle with a brown label, anti-B in clear bottle with a white label. Those labels explain the
required dilution ratios and give the reference batch from which the sample was derived (on
the right, anti-B is a sample from reference batch 195). Filmed at the Group Laboratories,
Mile End Hospital, London. Cyril Jenkins Productions Ltd., *Blood Grouping* (Imperial
Chemical Industries Limited, 1955), 20:33 min, sound, color. Still image from 00:02:22.
Wellcome Collection, London, https://wellcomelibrary.org/item/b17505963.

laborated with the NIMR to set up standards for blood grouping antisera
for hospitals and depots around the country (figure 7.1). Mourant viewed
this standardization and supply work as being in keeping with Britain's
new national model of health provision. Just as procurement of donated
blood for transfusion conformed "particularly well with the spirit of the
National Health Service," so he believed that the "free supply of testing
sera of human origin was a logical extension of the system."[17] To Mourant, the donation, standardization, and supply of blood grouping antisera helped to forge a publicly spirited form of health care.

In 1948, Mourant and the Reference Laboratory began extending those
anti-A and anti-B standards internationally. Between the wars, the League
of Nations Health Organisation had coordinated standards for various bacterial antitoxins, antidysentery serum, insulin, and vitamins. Postwar, the

139

WHO continued this work, with the NIMR and Statens Serum Institut in Denmark continuing to serve as centers for the preparation, maintenance, and distribution of international standards.[18] The methods used by the NIMR to make anti-A and anti-B standards are a remarkable demonstration of how blood grouping serum could be considered metrically and politically "international."[19] The project was overseen by Ashley Miles of the NIMR Department of Biological Standards, and he recruited Mourant's lab as central to the task.[20] Miles also elicited the collaboration of researchers at the US National Institutes of Health (NIH) to collect blood from donors there, freeze it, and fly it to London.[21] At the NIMR, samples were defrosted and blended with sera collected locally, producing a "final pool" of (US- and UK-derived) serum, which was then dried in a centrifugal freeze-drying machine at the Lister Institute.[22] To determine the antibody concentration of the pooled sera, Miles and colleagues carried out careful titration tests, making serial dilutions in successive tubes, adding red cells, and determining the maximum dilution at which agglutination could be observed. For further tests on the pooled sera, they sent aliquots (portions) to eight laboratories in Canada and Europe (Czechoslovakia, Denmark, France, Holland, Italy, Norway, and Sweden), as well as to the NIH and the Biologic Products Laboratory in Washington, DC—all of which used their local titration methods to test the robustness of the standard.[23] Thus, these "international" standards were really North Atlantic standards. The antisera were made from blood from UK and US donors, and their titers had been established by consensus between European and North American labs.

The Turn to Anthropology

While Mourant was establishing the Reference Laboratory in West London, several London-based researchers were seriously discussing a "clearinghouse" for blood group data. Enthusiasm for an institution devoted to their collation and analysis was motivated in part by the accumulating records of the NBTS. After Fisher's efforts early in the Second World War, blood donor records looked to many like a fabulously valuable genetic resource. Fisher's friend and colleague John Fraser Roberts, for example, actively exploited the opportunities afforded by the records. Having served on the MRC Human Genetics Committee in the 1930s, he shared many of Fisher's interests in blood group genetics, and he used the wartime transfusion-center records in Bristol and Wales to publish several papers in *Nature* and the *Annals of Eugenics*.[24] In 1948, the *Times* quoted him extolling "the great practical importance" of blood groups, which "had led to the testing of millions of donors, thus permitting applications to geographical and racial variations without parallel in living forms."[25] Cyril Darlington was also enthusiastic about blood group diversity studies. An eminent plant cytogeneticist, enthusiastic eugenicist, and public advocate for genetics, Darlington had a strong interest in the relationship of blood groups to language distribution.[26] Fraser Roberts and Darlington were both concerned that with the steep decline in active donors since the war, NBTS records might be selectively deleted, potentially destroying whole collections of valuable genetic data. In 1948, both men appealed to the Nuffield Foundation for money to support a project to mine and evaluate this wealth of information.[27] They proposed that the records might form the basis of a comprehensive survey of the genetic diversity of the whole of the British Isles.[28]

Fraser Roberts and Darlington had a keenly interested audience. The country was struggling to come to terms with the prospective dismantling of its empire. It was also grappling with a demographic crisis and with a severe labor shortage that was producing new patterns of emigration and immigration.[29] Perhaps responding to perceived challenges to British identity, the RAI established in 1948 a British Ethnography Committee under the chairmanship of distinguished anthropologist and geographer Herbert Fleure, the remit of which was to "consider means of promoting the ethnological study of Great Britain." The committee incorporated the blood group survey into its program and included Fraser Roberts in a series of lectures in the early 1950s. He lectured on Britain as a singular site for studying genetic diversity, owing to its "long and stable history, well authenticated records, high racial diversity and recognized genetic gradients."[30] He elaborated further in a 1951 BBC radio Home Service broadcast called *History in Your Blood*, in which he wryly quoted the comic book *1066 and All That* (1930) on the history and national identity of the British Isles:

The Scots (originally Irish, but by now Scotch) were at this time inhabiting Ireland, having driven the Irish (Picts) out of Scotland; while the Picts (originally Scots) were now Irish (living in Scotland) and vice versa. It is essential to keep these distinctions in mind.[31]

Fraser Roberts was elevating genetically "mixed" populations in a way that echoed the rhetoric of the 1935 book *We Europeans*.

In the same broadcast, Fraser Roberts announced that a "big count" of blood groups had been carried out in the northern counties of England, "thanks to the help of the officers of the National Blood Transfusion Service at Newcastle upon Tyne," a large and densely industrialized city in the north of England.[32] Fraser Roberts was referring to a recent pilot study he had carried out to evaluate whether transfusion records could be used to map blood group diversity—a larger, much more systematic version of Fisher's wartime enterprise. Fraser Roberts had good reasons for choosing to study the transfusion records in that particular part of the country. He was concerned that donors with common blood groups (who were regarded as less valuable) might resign disproportionately from panels and that this kind of selection might threaten the integrity of the data of an entire regional set of transfusion records. Speaking about this, he worried: "If there is any possibility that some of the cards of resigned donors may have been destroyed, the whole record is unusable for anthropological purposes."[33] So he needed a set of records in which "resigned" donors had not been discarded, and the clerical practices at Newcastle suited him perfectly in this respect.

This pilot study of British blood groups sowed the idea of a larger-scale "anthropological" center. Other researchers besides Fraser Roberts were keen to talk about this possibility. Wales-based physician Morgan Watkin was in the midst of a large-scale study of Welsh blood groups.[34] Mourant was involved in population diversity studies in England and Denmark and was closely engaged with research on the Rh groups of the Basques.[35] Under his direction, the Reference Laboratory was already carrying out a significant quantity of "blood group anthropological work"—that is, testing specimens for the purpose of elucidating population blood group frequencies.[36] Darlington had recently published on the distributions of blood groups and language sounds in Europe.[37]

With all of this support, the RAI put the idea of an anthropological blood group center on a more formal footing in 1951 when it convened a one-day meeting, held in the Eugenics Theatre at University College London. It brought together geneticists, anthropologists, and transfusion workers to "survey the functions and need of blood group studies in anthropology."[38] Chaired by Fleure, the delegates agreed that they needed to address the question of "how to collect, assess, and make available to anthropologists the vast and rapidly growing mass of data on blood groups," both published and unpublished. The proposed center would "tabulate and analyse results and data, . . . act as a clearing house and information bureau," and "assist fieldwork and publication."[39] An RAI Blood Group

Committee, which included Fleure, Fisher, Fraser Roberts, Darlington, and Mourant, was appointed to oversee the planning.

For Fisher and Fraser Roberts, the formal inclusion of blood group studies in the program of the RAI helped to fulfill a vision of genetics that they had articulated more than a decade before.[40] For the RAI, it pointed to a new agenda for anthropology. After the war, some of the institute's members had worried about how the atrocities of National Socialism would impact studies of race. In 1946, Fleure had used his inaugural speech as RAI president to announce that it had been "a mistake to divide mankind into groups termed 'races,'" and to recommend that anthropologists focus their efforts on how "drifts of people in different directions carried ancient characters far and wide."[41] Fleure, who had spent much of his career studying the racial geography and history of Wales, felt that population genetics offered a cogent way of reforming the questions and methods of race science. In line with arguments developed by Julian Huxley and others in the 1930s, and soon to be deployed publicly by UNESCO in its campaign against racial prejudice, Fleure urged anthropologists to "welcome increased co-operation from researchers in genetics."[42] Fleure believed that the Blood Group Committee was essential for promoting such cooperation.

Applying again to the Nuffield Foundation, the Blood Group Committee obtained the £14,000 needed to run the center for its first five years.[43] Mourant was already enthusiastic about anthropological collections. By 1950 his Reference Laboratory was regularly testing blood specimens for the purposes of determining population blood group frequencies, and as a result he and Ikin had already coauthored three papers in the *American Journal of Physical Anthropology*.[44] Mourant was appointed "honorary advisor" to the center—but was invariably referred to as "de facto director" or "director." Statistician Ada Kopeć—who had a mathematics doctorate—was in charge of its day-to-day running, and Kazimiera Domaniewska-Sobczak was its clerical assistant. Another Polish colleague, Janina Wasung, joined them soon after as librarian. The three women were multilingual and scoured the published literature on blood groups.[45] They maintained a card-index bibliography of publications, assembled an offprint collection, and correlated and tabulated all relevant blood group research in preparation for computation.[46]

Work at the center was broadly divided into the collection of "overseas" blood groups (of which more in the next section) and British blood groups. The British Blood Group Survey was established by Mourant and Kopeć and operated along similar lines as the Newcastle pilot study by

Fraser Roberts. Like the latter, it was built on data extracted from donor registrations and was made possible by Mourant's position within the NBTS. At regular meetings of regional transfusion directors (RTDs) at the Ministry of Health, Mourant persuaded depot directors to cooperate with the researchers, despite the potential administrative burden it would place on the transfusion service. By the time the project began in earnest, Mourant reported that he had been "assured of the friendly cooperation" of his RTD colleagues.[47]

That "cooperation" became institutionally embedded in the NBTS when officials modified its donor record cards to "incorporate information on places of birth, occupations and maiden names of married women."[48] These inscriptions gave a new and additional meaning to such records: they became anthropological documents as well as medical ones. Postal codes and birth names made donors and their donated samples amenable to anthropological analysis. Just as during the war, donors themselves were still apparently unaware that their blood was being used for this purpose: there is no evidence that the transfusion services worked to describe or explain the British Blood Group Survey to its recruits. Rather, it was RTDs whom Mourant had to recruit and maintain good relationships with. At their regular meetings at the MoH, Mourant repeatedly reminded delegates to send their records to the center.[49] At one, Mourant invited Fraser Roberts to give a lecture on the aims and objectives of the survey.[50] Kopeć displayed maps and gave demonstrations at conferences in London and Paris, and Mourant hosted a meeting of RTDs from across the country to show them the product of their anthropological endeavors.[51] By 1955, all the Scottish transfusion centers and most of the English ones regularly cooperated with the Nuffield Blood Group Centre. Mourant eventually persuaded the NBTS to adopt a formal inscription protocol whereby depot workers would copy the registrations of new donors and send them to the center as a matter of routine.[52] Over the next fifteen years, Kopeć received, sorted, and analyzed blood groups from England, Wales, Scotland, and Northern Ireland.

International Networks

By the early 1950s, Mourant was ideally placed to collect blood group data not only from the British Isles but also from overseas. Many countries now had national transfusion services, often run by the Red Cross.[53] Beyond national borders, the infrastructures for transfusion were expanding internationally. European colonies promoted the transfer of practices and materials between countries. The British Red Cross ran transfusion schemes in Kenya, Uganda, Southern Rhodesia, Northern Rhodesia, Nigeria, and the Gold Coast.[54] The MoH attempted to transfer transfusion technologies to and from Canada, New Zealand, and India.[55] For a short time before the independence of French West African countries, several cities in the region were connected by the movement of blood and plasma from the Dakar transfusion service in Senegal.[56] Exchanges were also offered by new international societies. The International Society of Haematologists was founded in 1946 and held annual congresses for its rapidly expanding membership. In 1951 the International Society for Blood Transfusion launched *Vox Sanguinis*, which remained a premier journal for blood group research for many years. The increasingly international movement of labor, blood, and sera opened new possibilities for collaboration, standardization, and exchange.

In 1950, the International Society of Haematologists held its congress in Cambridge, where Mourant's collaborator Norwegian serologist Otto Hartmann formally proposed that the WHO designate (and fund) Mourant's Reference Laboratory as the organization's "International Laboratory."[57] Established in 1948, the WHO was an agency of the United Nations, with headquarters in Geneva and regional officers around the world. This was a body that was designed and created by the industrialized nations— especially the United States, the United Kingdom, and Western Europe. Within that framework it brought together into a single organization the work of centralized epidemiological surveillance, campaigns against epidemics, disease control, and health system reform.[58] Blood transfusion was part of the last of these.

Essential to the machinery of the WHO were expert advisory panels and committees, which provided the organization with technical advice on particular subjects and disseminated their findings through "technical documents."[59] The WHO's Expert Committee on Biological Standardization was already coordinating efforts to produce international reference standards for vaccines, antibiotics, and other therapies, and the International Society of Haematologists proposed that standardized blood grouping reagents should be regulated in a similar way. The increasing significance of rare blood for transfusion made this particularly pressing. By now most laboratories found it straightforward to prepare antisera containing anti-A, anti-B, anti-C, anti-D, and anti-E antibodies. But many rare blood groups could be detected only using equally rare antisera, which were hard for many labs to make or obtain.[60]

The WHO was not just concerned with antiserum but also sought to make and control contact between institutions. Guided by European scientists (Mourant included), the WHO designated certain preferred laboratories around the world as "national blood grouping laboratories," which would serve as points of contact for regional blood banks, transfusion centers, hospitals, and local health clinics. In creating a single "international" reference laboratory, the WHO produced a node for the exchange of testing reagents and advice among its member countries.[61] This node was not an obligatory passage point, but it did elevate Mourant and his laboratory to a visible and highly respected position within a large network of institutions. By designating the "international" and "national" laboratories, the WHO selected and yoked together key sites that served as centers of authority and expertise for those working with blood.[62]

Ashley Miles of the NIMR was appointed head of the WHO Expert Committee on Standardization, and his prior collaborations with the Reference Laboratory likely helped the latter's elevation to the WHO role. Several other features also perfectly positioned Mourant's lab for the part. The Reference Laboratory already gave regular courses in blood grouping technique to pathologists and technicians in Europe, the Commonwealth, and many other countries, which extended its authority far beyond Britain. Owing to its extensive international contacts, the lab already had an unparalleled collection of frozen antibodies, which could be used to carry out detailed testing of red-cell panels.[63] Conversely, lab workers in Britain and overseas routinely sent the Reference Laboratory samples of rare antibodies for verification. As a result Mourant could claim that "almost certainly no other laboratory in the world . . . holds such a wide range of sera." Just as with the Research Unit next door, this was a consolidation of power based on the recursive relationship between antibody-containing sera and antigen-associated red cells. Another key feature in favor of Mourant's lab was the National Rare Blood Panel: the register of donors with the most unusual blood groups who could be called upon to donate in an emergency. Mourant oversaw the panel and envisaged its international expansion: "It would be a relatively straightforward matter to include persons from other countries in such a register." Such a move would not only help hospitals locate rare blood for transfusion but also enable a far greater range of laboratories to access unusual blood specimens for research.[64]

Thus, in 1953 the WHO officially appointed Mourant's lab as its International Blood Group Reference Laboratory, a move that brought extra funding for WHO-related work.[65] With this appointment, the lab was required to perform practical tasks for the national blood grouping laboratories of other countries: distributing the rarer kinds of antisera to determine the purity of sera, and testing panels of red cells submitted by those labs. As a result, Mourant could project the Reference Laboratory as a place of unparalleled expertise. Indeed, Mourant himself was a key source of advice when the WHO was establishing its list of "national" reference laboratories, thus helping to define the shape of the WHO blood transfusion network.[66] This expanded Mourant's range of postal exchanges and intensified the flow of blood group results to the Nuffield Blood Group Centre in Bedford Square.[67]

"Worldwide" Data

Now that it was WHO-affiliated, the Reference Laboratory could broaden its network, and in so doing it mobilized a brisk traffic of reagents, blood samples, and data to and from labs worldwide. Mourant himself characterized his standardization work and his anthropological interests as reciprocal and mutually beneficial. To the WHO, he claimed that his anthropological endeavors strengthened the authority of the Reference Laboratory among the international transfusion community.[68] To the MRC, he explained that he was uniquely placed to elicit "collaboration of local workers [collecting] specimens of blood for anthropological research." These activities were apparently so happily interdependent that Mourant was able to describe his anthropological studies in routine reports to the MRC without needing to explain how they cohered with his clinically related duties.[69]

Mourant's extensive archive of correspondence testifies to the varied exchanges that drove the movement of sera, blood, and data.[70] Sometimes Mourant appended requests for blood grouping results to letters that accompanied reagents to hospitals, blood banks, clinics, or transfusion centers. In other cases, scientists or clinicians shared data when they asked Mourant for advice or hard-to-obtain antisera. Sometimes Mourant wrote to authors of anthropological studies to ask for points of clarification or to inquire after new results. On occasion, he brokered contacts between distant researchers and clinics or blood banks that could serve as local sources of reagents or equipment. Sporadically, Mourant himself requested samples of rare sera, such as when he contacted the US company Spectra Biologicals (which supplied reagents to "the complete blood bank"), in order to carry out what he called "special 'African' work."[71] Beyond simple lists of data, Mourant encouraged certain correspondents to send chilled blood samples to the Reference Laboratory, where they would be tested by Ikin or other laboratory staff.[72]

In describing the movement of anthropological blood and data, I use the term "exchange" to acknowledge that in sharing their materials, Mourant's correspondents at least expected something in return—whether that be serum, contacts, scientific credit, help and advice in the future, or the published acknowledgment of an internationally known blood specialist. These interactions established bonds of reciprocity between collectors operating in overlapping social and professional worlds, often at some geographical distance from one another.[73] The exchanges between Mourant and his correspondents were heterogeneous and informal: no single correspondent was obliged to cooperate with either the Reference Laboratory or the Nuffield Blood Group Centre. But Mourant's WHO-approved position carried a great deal of weight, and he had considerable influence.

Mourant's letters of the 1950s give a sense of the geographical and political variety of his correspondents: for example, the University College of the West Indies in Trinidad; the Armed Forces Medical College in Pune, India; Shanghai's Second Medical College; and the Connaught Hospital in Freetown, Sierra Leone.[74] Many of his correspondents had trained in Britain or the United States and now worked in European colonial hospitals and universities; others learned about Mourant's services through the WHO or related professional networks.

Mourant had a tendency to instrumentalize "local" scientists as collectors or informants who could serve his own research projects. Historian Elise Burton draws attention to a notable case in which one researcher, Israeli anthropologist Batsheva Bonné, resisted Mourant's paternalistic formulation of their relationship, leading to a professional conflict over priority rights to publish and interpret data.[75] This not only makes visible the asymmetrical power relationships between Mourant and some of his scientific correspondents but also highlights Mourant's absolute reliance on scientists like Bonné to negotiate, collect, and analyze anthropologically and genetically meaningful data. Mourant traveled regularly to attend international conferences, but he rarely collected blood himself, either at home or abroad. He was a London-based, MRC-funded, WHO-accredited technocrat, who accumulated data by making use of international networks of public health and transfusion and European colonial investments in medicine and research.

While a great deal of data and blood arrived in London from medical institutions, Mourant also supported European researchers on "expeditions." For example, he supplied Paul Julien, president of the Netherlands Anthropological Society, with rare antisera for a study of "pygmies" in Gabon, central Africa.[76] In 1962, he arranged serological training for Cambridge students embarking on an expedition to Cambodia and Indonesia.[77] The Reference Laboratory tested chilled samples sent from Oxford anthropologist Derek Roberts on an expedition in Sudan.[78] And Mourant's colleagues taught blood grouping and supplied antisera to another Oxford student, Anthony Allison, who orchestrated two blood-collecting expeditions in the late 1940s and early 1950s, one to Kenya, another to northern Scandinavia.

Allison's activities offer a snapshot of some of the institutions and professional and personal connections that supported research by Mourant's correspondents in the 1950s. Allison embarked on his first student expedition to British-controlled Kenya under the auspices of the Oxford University Exploration Club. Preparing for his trip, Allison visited the Lister Institute to gain serological training.[79] Mourant also provided him with advice on how to collect and transport blood, and offered written introductions to Kenyan laboratories and medical institutions with relevant equipment and expertise. British colonial Kenya seems to have been a favorite destination of the Oxford University Exploration Club, which by then had already organized two expeditions to the country since the end of the war.[80] Affirming the significance of Kenya as a scientific destination for UK researchers, over half the money for Allison's expedition was granted by the Colonial Office, which had established a fund in 1940 for expanding scientific research in Britain's colonies.[81] Despite the early signs of a violent resistance movement against the British colonial government, a Kenyan blood group survey apparently fitted neatly with the stated goal of the Colonial Office to support the scientific study of colonial environments and societies.[82]

British colonial institutions and administrative links helped to make certain populations accessible and amenable to bloodletting. To reach the people that Allison made into his research subjects—as he put it in a report of his first trip, "Kikuyu, Masai, Luo and Girama tribesmen and Arabs"—Allison leveraged contacts with his family and friends and relied upon local networks of medics and public health workers.[83] He claimed that the "semi-official" status of the Oxford Exploration Club helped to persuade doctors and scientists to provide local introductions and access to labs.[84] Testifying to some of the local institutional contexts in which he collected blood on that initial trip to Kenya, Allison referred to his Luo and Kikuyu research subjects as "hospital patients" and "groups of labourers" in districts in or near Nairobi. Without further explanation, Allison also reported blood grouping tests "carried out in the field . . . on 233 Masai tribesmen" further to the south.[85] He recruited professionals to help secure blood using a branching network of contacts: the director of the Medical Research Laboratory in Nairobi was a family friend, and through him Allison elicited introductions to district medical officers and, in turn, to people he termed "local medical assistants," whom he recruited to accompany him on his collecting forays.[86] Allison evidently trusted those professionals with the important job of identifying potential research subjects, and to readers he offered no further details of how they chose individuals to bleed.[87]

Allison's expeditions targeted human populations that were becoming typical of postwar genetic collecting projects: marked as "isolated," "ancestral," "unmixed"—these terms often referenced those groups subjected to the greatest administrative control (and contrasted with the "mixed" British populations promoted by Fraser Roberts).[88] Allison's approach was similar to other blood-collecting expeditions of the 1950s, in Africa and elsewhere. For example, in 1955, Ikin and Mourant coauthored a paper with a researcher from the South Africa Institute for Medical Research entitled "The Blood Groups of the Hottentots," in which they explained that blood from 200 people had been collected "in a number of their Reserves in South-West Africa" by a local Senior Health Inspector.[89] In South Africa, "reserves" were areas of land that had been demarcated by the Native Land Act in 1913, in which "native" people were under the administrative control of the colonial government.[90] A study of San people in Southern Africa by the same authors, entitled "The Blood Groups of the Bushmen," explained how difficult it had been to collect blood specimens "from such a primitive, nomadic race, living in remote and inaccessible regions," and noted that sampling had been made possible by the "Regional Health Officer."[91] In Sudan, Mourant's colleague Derek Roberts sampled blood from Shilluk, Nuer, Dinka, and Burun people with help from the "Province Medical Inspector" and the "Public Works Department."[92] These published papers do not recount how donors were approached and blood drawn, but they do show that colonial administrators helped to make those encounters happen. States reified and routinized racial categories through census practices, patterns of employment and land appropriation, the establishment of "native reserves," and educational and medical services.[93] These administrative constructs helped to isolate certain population groups, while white medical officers and health inspectors made the bodies of apparently isolated, anthropologically distinctive people permeable to blood collectors.

Bleeding

Donors and research subjects themselves were at the very periphery of Mourant's networks, almost out of his sight. Mourant rarely encountered them personally; almost all of his negotiations around blood and data were with doctors, depot directors, and researchers.[94] But his correspondents did broker blood, usually with a great deal of coaxing, coercion, persuasion, and (sometimes) payment. Blood is abundant and renewable, but it is also messy and dangerous. Human tissues have often been regarded as something that cannot and should not be given away except during specific ritual practices—so scientific collectors were obliged to attend to the settings and exchanges of bleeding encounters.[95]

Many of Mourant's correspondents, therefore, extracted and tested samples in hospitals and clinics as part of routine medical practices. By the 1950s, drawing blood for testing was relatively uncomplicated in these settings, even if a needle was daunting.[96] Withdrawals outside hospitals and clinics were more complex: we have already seen that Mourant's Lister Institute colleagues Robert Race and Ruth Sanger carefully cultivated warm, cheerful relationships with donors and families, but field studies in countries foreign to the collectors had the potential to be far more fraught. An important reason for recruiting "local" health workers was to mitigate and smooth encounters.[97] The status and credibility of those interlocutors was crucial, although in published work authors seldom mentioned these individuals by name.[98]

By this time Mourant rarely drew blood himself, but he did recommend bleeding methods in the RAI's much-used handbook *Notes and Queries in Anthropology* (1951), which gave fieldwork advice to anthropologists. He noted that the easiest and cheapest method was to prick an earlobe or finger, though his phrasing testifies to the intimidating character of the equipment: "either a triangular surgical needle or a 'blood-gun.'"[99] Mourant preferred the more sterile (though potentially more formidable) "venipuncture" technique. Applying a tourniquet to the arm, the collector would use a purpose-designed Bayer's glass "venule"—an all-in-one device for withdrawing and sealing two to five milliliters of blood.[100] Blood collections often took place in community buildings or schools. And although lining up to meet a medic with a needle would have been familiar to people subject to vaccination regimes, participants would have been right to be wary of blood collectors. Mourant explained to his anthropologist readers the risk of transmitting hepatitis even with alcohol-sterilized needles: the "blood-gun" was "best kept in a tube of spirit and rubbed with spirit on cotton wool before and after use and between successive subjects. . . . *Ideally* . . . needles should be boiled or flamed each time they are used" (my emphasis).[101] Regarding the venule, Mourant warned fieldworkers that venipuncture should only be performed by medically qualified professionals who were "fully aware of the dangers of sepsis."[102]

Mourant was careful to encourage sterile procedure, but his "ideally" suggests that in some circumstances even proper sterilization might be forgone in the pursuit of blood group data.

Every exchange could present opportunities for both parties: some research subjects leveraged blankets, knives, and other goods in exchange for their blood.[103] Researchers sometimes offered basic medical supplies in return for specimens—with the additional effect of framing blood withdrawal as medical. One of Mourant's correspondents reported frankly that to persuade participants, he gave them "a rough clinical examination followed by prescription and . . . the distribution of some medicine."[104] In another letter to a colleague, Mourant bluntly referred to such exchanges as "bait," a word that acknowledged the nature of the exchange.[105]

Thus, drawing blood required needles, cotton wool, bottles, sterilizing apparatus, and training, as well as persuasion, bribes, and (sometimes) a comfortable atmosphere. In Britain, even the most enthusiastic "village bleeds" during the Second World War had required the careful cultivation of a morality of giving blood, and transfusion-service officials had worried constantly about how to maintain donor loyalty.[106] In 1950s East Africa, narratives of "blood stealing" and "blood sucking" associated with vampire stories made some communities and individuals deeply suspicious of biomedical researchers.[107] Willingness to give blood, and the terms on which donations happened, were extremely variable.[108]

The acquisition of blood and tissues is rarely neutral, whether in transfusion centers, hospitals, schools, or donors' homes, and whether in Kenya, the United Kingdom, Israel, or Oregon. Social, professional, and political identities shaped who could do what to whose bodies. The moment of encounter between donor and collector was conditioned by the authority of collectors and the settings in which they subjected donors to bleeding.[109] This affected how often collectors could call on donors, how much blood they could take, what data could be aggregated, and what kinds of sampling strategies were possible.

Making Exchanges

Blood and data did not flow of their own accord. Mourant was institutionally positioned both to intervene in the tight administrative infrastructure of the NBTS and to make use of looser, WHO-structured international networks. Britain and its allies had won the war and had forged alliances during it, all of which added to the cultural capital of a London-based health official. Mourant was a consummate administrator and was devoted to persistent persuasion and negotiation. Former colleagues recall a small, neat, dapper man ("like Hercule Poirot," one interviewee commented) who was very kind, especially to students (figure 7.2). Most of his letters were polite and efficient—those of a postwar London technocrat using an expansive range of postal correspondents. Mourant cultivated these relationships over many years: some archived folders of letters span three decades. By the mid-1960s, many wrote to Mourant with new data as a matter of course, also asking his views on statistical analysis, notifying him of new publications, and requesting comments on manuscripts. Mourant's collecting project was one perfectly suited to a fine lab manager and administrator, who enthusiastically made it cohere with postwar institutions like the NHS and the WHO. Mourant neatly coupled his pursuit of blood group genetic variation to his activities within expanding infrastructures of public health, which were reciprocally shaped by his anthropologically inflected collecting.

Mourant's twin enterprises—of standardizing blood grouping reagents and collecting anthropological blood group data—were inextricably linked. The making of standards, the detection and verification of rare sera and unusual antigens, the cultivation of authority, the forging of professional contacts, and the collection of diversity data: all of these were parts of a cohesive system for making and mapping genetic variation. In the history of genetics, the creation (and control) of chromosome maps often relied on the cultivation of informal correspondence and exchange networks—for example, in the fruit fly genetics laboratory of Thomas Hunt Morgan at Columbia University in the 1910s.[110] Mourant's geographic maps also relied on exchange networks, although these were far more varied. He was engaged in sending and receiving sera, blood, data, advice, contacts, and published acknowledgments; he sometimes collaborated closely with correspondents; he sometimes merely exchanged offprints.

Mourant could claim access to blood and data because he occupied a political position (in a WHO-sanctioned reference laboratory) of a very specific place and time, namely, postwar London. From Mourant's perspective, his laboratory's position as an MRC-funded, WHO-accredited center made it unquestionably suitable for an elite, large-scale collecting project, in a way that would not have been the case for a hospital in the north of England or a transfusion center in Australia. As we have seen, Mourant traveled extensively to attend conferences, but he

7.2   Arthur Mourant at an academic conference in the United States, wearing a smart brown striped suit, tie, leather shoulder bag, and name label. Color photograph taken and printed in the United States, dated 1954, probably sent to Mourant by a colleague. Image cropped from a larger photograph with another unknown individual. Whole photograph 11 × 8 cm.
Wellcome Collection, London, SA/BGU/L.10/4. Copyright Medical Research Council. Reproduced with the kind permission of the Medical Research Council, as part of UK Research and Innovation.

almost never chose to go abroad on collecting trips himself: he perceived his role as managerial and a step above the work of collection. Mourant saw himself as an internationalist—he was a longtime pacifist, an active supporter of the League of Nations, and proud of his fluent French. Maintaining his managerial role and persona made him a superior narrator of worldwide data and reinforced his frame of reference as a metropolitan "view from everywhere."[111] His distribution of standard antisera helped Mourant to view his role as part of a liberal, technocratic elite that believed it could help to build a progressive postwar world. To do this, he linked a national public health system (the NBTS) with the WHO and leveraged long-established imperial medical connections.

Lists of blood group results did not simply represent natural objects, waiting to be discovered and mapped. They were brought into being through deliberate, negotiated exchanges among donors, doctors, nurses, depot directors, colonial administrators, WHO officials, technical laboratory workers, clerks, secretaries, and MRC scientists, all of whom lived and worked in configurations defined by a postwar world order. Those connections—forged through sera, standards, and postwar politics—shaped the very kind and quality of the anthropological data that accumulated at the Nuffield Blood Group Centre. There, its sorting, indexing, and mapping would give the data new life and meaning.

EIGHT

Organizing and Mapping Global Blood Groups

By 1954 the Nuffield Blood Group Centre had been running for two years, and Mourant had just completed his first major compilation: *The Distribution of the Human Blood Groups*. It was the third of three volumes on blood published by Blackwell Scientific Publications, after *Blood Groups in Man* (1950) by Race and Sanger and *Blood Transfusion in Clinical Medicine* (1951) by Patrick Mollison.[1] Mourant's book represented as "anthropological" data the frequencies of the different blood groups (A, B, O, Rh, M, N, and many others) in populations across the world. It contained extensive tables that ordered data according to continental, religious, national, racial, and "tribal" categories. It offered more than 150 pages of text describing the varying blood group frequencies across (in this order) Europe, the "Mediterranean area," sub-Saharan Africa, "Asia," "Indonesia and Australia," and "the Aborigines of America." It included nine fold-out maps that dissolved population categories altogether, offering the impression of a smooth diffusion of blood group alleles across geographical space (figure 8.1). In these maps, social and political groupings melted away, leaving apparently purified, objective population genetic data. The tables, prose and maps apparently created a picture of the historical migration of human populations. In this way, Mourant's book was a realization of the proposal made by Haldane more than twenty years earlier (chapter 2). Genetically inclined anthropology journals published glowing reviews. The *American Journal of Physical Anthropology* described the book as

APPROXIMATE DISTRIBUTION OF THE Rh BLOOD GROUP GENE C IN THE ABORIGINAL POPULATIONS OF THE WORLD

MAP 7

PERCENTAGE FREQUENCIES:

over 90   80·90   70·80   60·70   50·60   45·50   40·45

30·40   20·30   10·20   under 10

8.1   One of nine fold-out maps in *The Distribution of the Human Blood Groups*. It shows the percentage of individuals that carry blood group allele C within select "aboriginal" popula-tions in different geographical regions of the world. The map uses isolines and shading to indicate threshold frequencies across space, which obscures evidence of the patchiness of sampling, the circumscription of geographical boundaries, and the political borders that structured collections. This map seeks to convey a dense concentration of the Rh allele C in Papua New Guinea, Indonesia, and the Philippines, radiating out across Asia into Europe and Africa. Mourant's decisions regarding threshold frequencies, and the judgments that he made with his cartographer about isolines and their tinting, had a profound effect on the information conveyed by this and similar maps. By altering these, while keeping the same underlying data, Mourant could draw together whole populations or push them apart. From Mourant, *The Distribution of the Human Blood Groups* (1954). 25 × 17 cm. Reproduced with the kind permission of the Mourant family.

"indispensable"; *American Anthropologist* called it "brilliant"; and the RAI journal *Man* considered it "the most important" contribution to "the anthropology of blood groups" to date.[2]

How did Mourant and his colleagues at Bedford Square collate, order, and represent blood group data as credibly "anthropological" and authentically "genetic"? The institutions and networks that channeled blood and blood group results to London are part of the answer, as are the ways in which Mourant and his colleagues collected, sorted, analyzed, and represented them. Ada Kopeć, Kazimiera Domaniewska-Sobczak, and Janina Wasung of the Nuffield Blood Group Centre amassed data in the form of blood group frequencies, which are properties of population groups. They classified and labeled specimens and grouping results according to donors' nationalities, geographic location, "tribe," caste, race, or religion. As a rich scholarship attests, human population categories were often forged through territorial claims, the assertion of political power, and the contours of settler or internal colonial administrations.[3] But these dynamics were not always visible to Mourant or his colleagues at Bedford Square, who preserved in card indexes and publications the categories chosen and defined by fieldwork collectors. Particularly striking are the ways that the ordering and representation of British blood group data differed from those collected overseas. This chapter explores how Mourant and his colleagues made blood group population data "genetic" and anthropological.

Constructing "Foreign" Populations

The ordering and representation of blood group frequencies in publications like Mourant's *The Distribution of the Human Blood Groups* often began before and during collection. Population categories were almost always defined before a study began.[4] The kinds of blood and data that traveled to the Nuffield Blood Group Centre were shaped by the political conditions of their collection, the institutions in which donors were bled, and the administrative networks that made donors accessible. These elements came together during encounters around blood.

A striking case of how this operated comes from one of Mourant's very first "anthropological" blood group collaborations, which dealt with the blood groups of the Basques. The Basques were (and are) a population with a distinctive cultural identity, located geographically in an area in southwest France and northeast Spain. By the 1940s, the category "Basque" had served a multiplicity of historical narratives about human history. The population seemed particularly striking to anthropologists, historians, and linguists for their physical features (modern Basques apparently resembled "human remains in the Neolithic dolmens of the Basque country") and their local language ("many Basques still speak the same language as in the Stone Age"). The Basques were "interesting relic[s] of Iberian times" and a "perfectly definite ethnic group, both in their racial characters and in their traditional culture."[5] They were so captivating to scientists that one sarcastically pointed out that at one time or another the Basques had been supposed akin to the "ancient Egyptians, Guanches, Berbers, Etruscans, Phoenicians, Lapps, Finns, Bulgarians, or to Asiatic races," even "the sole survivors of Atlantis!"[6] Nevertheless, Mourant and his colleagues wondered whether an idiosyncratic pattern of blood groups might confirm the Basques as a distinct population with an unusual history.

The Basques were also uniquely accessible to Mourant.[7] Severely oppressed and threatened by Francisco Franco's dictatorship, large numbers of Basque people were in exile by the 1940s, and some lived in London. There, a prominent Basque ethnographer and Catholic priest, José Miguel de Barandiarán, publicly pointed to the scientific importance of the group's cultural identity in an appeal in the *Journal of the Royal Anthropological Institute*, in which he called for scientific research on this "special race without analogy with any other known group." Barandiarán suggested that through their research, anthropologists might help persuade governments to "respect and protect the ethnic elements of the Basque people, not only for [their] antiquity, but [also] for [their] scientific interest."[8] Blood grouping studies, he proposed, might help to affirm and elevate the status of this special group.

When Mourant collected Basque blood, his earliest samples were drawn from exiles in London and Paris, later augmented with samples sent by a doctor from San Sebastián, in the Basque region of Spain, who was known to Barandiarán. At all three sites (London, Paris, and San Sebastián), Mourant had asked collectors to select participants based on their "personal names," which he emphasized incorporated "the names of several generations of ancestors."[9] Mourant was unhappy with the first round of collection—as it did not yield the blood group frequencies he had been expecting—and he resolved to obtain what he called a more "representative" sample. A hematologist colleague, Marshall Chalmers, traveled to southwest France to collect "a larger number of specimens under expert anthropological guidance." This "guidance" came once again from Barandiarán, who accompanied Chalmers to assure him of the "family relationships" and "racial purity" of each person tested. Even after collection, Chalmers further sifted the specimens by "eliminating the small number of persons who were believed to be of mixed race, and the few others who were blood relations of other persons tested."[10] In this case, "blood relations" likely meant people connected through mutual grandparents (including too many cousins in the survey would distort population frequencies). The researchers wanted to affirm the racial purity of their sample, although the authors did not elaborate on what guided Chalmers in his estimates of who was "mixed" and who was "Basque."

Based on the calculated Rh frequencies, the researchers judged that this second round of collecting had produced 383 samples from "the 'purest' available Basques." (Mourant and his coauthors themselves added the scare quotes around "purest.") These samples yielded a Rh-negative frequency of 29 percent, which was satisfactorily higher than the corresponding frequency in the rest of Europe.[11] The authors believed that the results supported the hypothesis that the Basques had originally been a pure, isolated, Rh-negative population and had slowly been mixing with the Rh-positive peoples of the rest of Europe: "The Basques, while they may be akin to the Celtic and other peoples of the fringes of Europe, have retained a racial purity" not found elsewhere.[12] Reporting the story, *Science News-Letter* made much of the conclusion that the Basques were an "almost pure representative race of ancient Europe."[13]

That the collectors insisted on individuals' "racial purity" before taking their blood points to one way in which collection protocols shaped published results. Another was the decision to put sampling choices in the hands of a trusted local expert: in this case, Juan Miguel de Barandiarán. We saw in chapter 7 that "local assistants" were sometimes used by blood collectors to mediate bloodletting encounters. Such individuals were also apparently important for producing credibly homogeneous samples. In Kenya, Allison's "assistants" apparently "knew their fellow tribespeople from those belonging to other tribes." As they collected blood, they checked the identities of subjects' "parents and grandparents."[14] When Allison went on to Scandinavia to test the blood groups of Sami people, he remarked that his interlocutors knew "who were purebred Lapps and who were half-breeds from their names, the languages they spoke, and enquiries about parents and grandparents."[15] In the US state of Oregon, anthropologist and serologist William Laughlin, a correspondent of Mourant's, adopted a similar strategy when he trusted a local Basque lawyer to help him with an intensive survey of the blood groups of the local Basque community. Anthony Yturri, a "Basque university graduate," as Laughlin described him to Mourant, "offered to contact two hundred Basque speaking Basques, with intact lineages."[16]

In all of these cases, researchers relied on the expertise of individuals who were projected as "local," a move that itself seemed to guarantee the authenticity of the data. Whether made by "local assistants," medical officers, health inspectors, or hematologists, population identifiers established in the field remained stuck to blood and data as they traveled to London to be incorporated into the collection at Bedford Square.

Making British Populations

Back in London, the methods used to construct population groups for the blood group data of the British Isles were strikingly different from those applied to data from overseas. NBTS donor records were not inscribed with a donor's ancestry, names of grandparents, or racial identity. Rather, they were structured by the organization of the blood depots and "regions" of the service. The only "anthropological" clues to donors' identity were their names and where they lived. When Fraser Roberts carried out his pilot survey of Britain's blood groups using NBTS cards, he made judgments about the "relevant particulars" for making genetic data—which at the time included surname, sex, and blood group.[17] Fraser Roberts worked at the London School of Hygiene and Tropical Medicine, where he had a group of assistants copy that information from donor cards to punch cards for computation. Those workers also gave those donor records a geographical identity—that is, they attached to the punch card information about where the donor lived at the time of registration. They could not simply arrange the cards according to transfusion center, because most people donated blood closer to their place of work than to their home. Instead they scrutinized the addresses on every record and used those to locate the donors on a wall-mounted Ordnance Survey map. These clerical methods, shaped by the NBTS donor cards, gave a fine-grained structure to this population data.

Not only were the British blood group surveys defined by NBTS procedures, but NBTS administration was apparently shaped by the surveys. Fraser Roberts noted that relying on the postal address had been "the most difficult and time-consuming" aspect of processing the population data.[18] Reflecting on these practical challenges, he explained, "What is required is a code made once and for all, so that any address can be immediately changed into a code number."[19] Negotiating this, Fraser Roberts met with members of the Post Office to discuss the coding of postal areas for research purposes. What are called "postcodes" in the United Kingdom today were not systematically introduced until the late 1950s when the Post Office began using electromechanical sorting machines.[20]

Nevertheless, by the time that Fraser Roberts embarked on his study, the Post Office was privately organizing its sorting on the basis of 1,700 independent areas and an additional 300 subdivisions of cities, "precisely delineated on maps." Although these were not available publicly, Fraser Roberts persuaded the Post Office to make them available for use in transfusion depots, together with periodic updates detailing changes to boundaries. He hoped that transfusion centers and donors could be properly disciplined, and he suggested that a "directive" be issued to the transfusion centers "pointing out mistakes and omissions and asking them to ensure that addresses are filled in as correctly as possible."[21] Anticipating a full-scale survey of the blood groups of the British Isles, Fraser Roberts and his colleagues persuaded the NBTS to incorporate into their record cards this coding system for home addresses.[22]

Fraser Roberts also grappled with another distinctive feature of the blood groups of the British Isles. In producing data on the frequency of a particular group, he was confronted with a problem of delineation not faced by researchers analyzing data collected overseas. The NBTS records gave Fraser Roberts no ready-made categories for classifying donors; rather, he attempted to derive population boundaries from the donor records themselves. Starting with the Ordnance Survey map showing donor addresses, he first aggregated the records into groups of twenty to seventy individuals, keeping separate "individual towns, parts of towns and villages" and adding together "rural areas . . . where it was practicable." In deciding where to draw boundaries, Fraser Roberts also took into account what he called "natural" aspects of the landscape, such as valleys, and features "that might be expected to facilitate communication," such as roads and railways.[23] Having established these small groups, he sequentially collated them to make larger and larger populations, at each step using the chi-squared test to discern whether the new groupings were significantly heterogeneous in blood group frequencies.[24] He reported that eventually he and his assistants were able to distribute 54,579 donor cards into 321 areas and to calculate a frequency for each area. The population categories from which Fraser Roberts determined these frequencies were derived from a combination of the internal heterogeneity of the data and the researchers' own judgments about landscape.

The pilot study was seen as a great success, and it resulted in a detailed map of blood group frequencies distributed across the middle part of Britain (figure 8.2). So compelling were these results that the British Blood Group Survey became a routine strand of research carried out at the Nuffield Blood Group Centre, where it was led by statistician Ada Kopeć.

The Post Office scheme of post towns and postal districts had been so useful to Fraser Roberts that Kopeć chose to use those to define the relevant "territorial units" at the start of her survey, so that cards could be sorted and coded as they arrived at the center.[25] The British Blood Group Survey bound together the bureaucracy of the blood transfusion service to the administrative topology of the Post Office.

In another coupling of postwar bureaucracies, researchers around the country extracted population data from the records of the NHS. This was the era of large-scale cohort studies, mass data gathering on health surveys, and epidemiology, projects that were facilitated by the NHS.[26] "Association studies" sought to document population-level correlations between blood groups and disease, studies that were later incorporated into a successful strand of international research on population genetics.[27] In 1950s Britain, such studies were made possible by the standardized documentation practices of the NHS and NBTS. Their authors typically collected blood group data from tens of thousands of hospital patients and (as controls) donors at blood transfusion centers. However, they did not survey patients directly. Instead, they chose people affected by diseases that were routinely treated with elective surgery—those patients had their blood groups recorded on their hospital records. Now that blood transfusion had become a routine part of surgical planning, compatible blood was prepared in advance of such procedures.[28]

One large-scale study collected data on blood groups from patients suffering from peptic ulcers and several kinds of carcinoma, all conditions treated with elective surgery.[29] Another dealt with the association between blood groups and hypertension and relied on the fact that patients with abnormally high blood pressure were routinely prescribed the therapeutic removal of blood at local transfusion centers.[30] These patients, then, had both their clinical diagnoses and their blood groups routinely incorporated into NHS patient records, along with addresses and fields such as "social class" and occupation. Like the NBTS donor cards, information could be lifted from these records and thus transformed into genetic data. In the peptic ulcer study, one member of the team visited multiple hospitals to extract information from 13,000 case notes, data that were then coded and transcribed onto punch cards. "Control" subjects for these studies were individuals from the same hospitals or local transfusion depots. Therapeutic transfusion had made it possible to construct a paper trail connecting blood group to patient to disease. Yet again, the bureaucracy of the NHS had made it possible to construct and study clinically significant genetic variation.

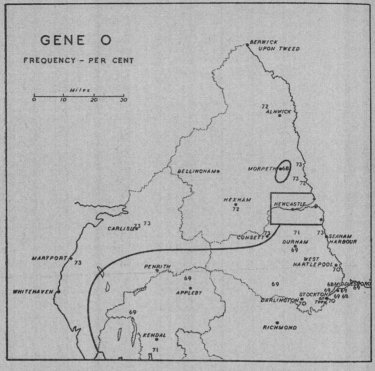

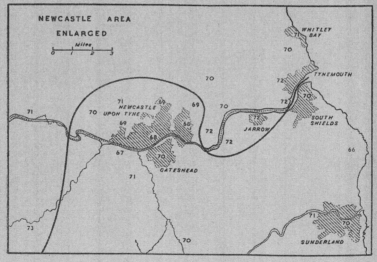

Sorting and Calibrating Data

At the Nuffield Blood Group Centre, blood group data was abstracted away from the messy complexities of blood collection and the difficulties of delimiting populations. It took on a new life as Kopeć, Domaniewska-Sobczak, and Wasung sorted, assembled, and represented "world" genetic diversity. Their principal tool was a card index, which held bibliographic information on each blood group publication. By the mid-1950s, the workers at the center had produced a bibliography of nearly 3,000 titles, indexed under authors' names, subjects, populations and countries.[31] Many of these cards were also cross-referenced to the center's offprint collection. A surviving collection of cards, which Mourant and his colleagues apparently used for the second edition of his book, shows that when they received unpublished data, the researchers handwrote a summary of the results into the card index on thin paper. When the relevant article was published, the information was transcribed with a typewriter onto a card. In this system, cheap paper and handwriting represented provisional knowledge; cards with typewritten print corresponded to published knowledge.[32] Data were typically (but not always) included in population calculations only once they had been published.[33] Mourant used the card index extensively as a reference tool and referred to it frequently with correspondents. The card bibliography served the analytic tasks of the center and was the basis for supplying specialist bibliographies to inquirers.[34] Letters and publications were mined for blood group and other genetic data: "data are copied, gene frequency and other calculations are performed and the results recorded in a special file."[35] Such work formed the basis for the compilation of thirty or so tables published in *The Distribution of the Human Blood Groups* (1954).

8.2   Two maps presented together in a paper by John Fraser Roberts, showing the frequencies of blood group O across the width of northern England (top) and an enlarged view of Tyneside (bottom). The uppermost dotted line denotes the Scotland-England border, the internal dotted lines apparently indicate rivers, and the shaded areas represent major settlements. Numbers indicate the frequencies of blood group O, and a single isoline bisects both maps. Fraser Roberts alludes to the judgments that he made in positioning this line: "A number of experiments were tried and it soon became clear that it is possible to draw a single line from east to west, dividing the whole region into two in such a way that the two areas are homogeneous within themselves and all the significant heterogeneity lies between them." From Fraser Roberts, "An Analysis of the ABO Blood-Group Records of the North of England" (1953). 24 × 14 cm.
Copyright Springer Nature. Reproduced under license.

The book mobilized the data contained in the card index. It comprises forty tables, which cluster together population groups perceived to have close geographical relationships. Larger geopolitical regions (e.g., "Africa" or "Asia") are subdivided by nation, religion, or caste ("Jews," "Koksnath Brahmans") and sometimes further divided into smaller areas (e.g., "Lucknow," "Bombay"). Data from the British Isles and elsewhere in Europe—collected principally via blood transfusion centers—tend to be categorized by nation ("English," "Scottish") and place ("Stornaway," "Inverness West"), whereas overseas blood groups are classified by country, race, and "tribe." Europe was far more densely covered than other places in the world. Mourant placed "the Jews" inside the chapter called "The Peoples of the Mediterranean Area," remarking that "the Jews of the Dispersal differ physically from the peoples among whom they live."[36] Judging by the population categories inscribed on index cards, the Nuffield Blood Group Centre appears to have preserved the racial judgments of the collectors, even such arbitrary categories as "Indians (British Columbia)," "Chippewa Indians (Full Blood)," "Chippewa Indians (>¾ Indian)," and "Chippewa Indians (<¾ Indian)" (figure 8.3). Horizontal lines delineating the rows in some instances denote national boundaries and sometimes more opaque divisions. Moreover, the ways that groups were ordered geographically and politically did not always correspond to collection sites; the category "Chinese" might be qualified by the name of the city in which the donors had lived. In other words, the clerical staff sometimes relocated blood group data to the places that either the authors of articles or Mourant and his colleagues judged to be donors' native lands.

The tables make visible some aspects of Kopeć's statistical work to transform blood group results into genetic population data and to assess the credibility of that data. Columns to the furthest right in figure 8.3 quote blood group results in terms of percentage—not of individuals with a given blood group but of the respective alleles.[37] This was routine in population genetics, but quoting allele frequencies also often had the effect of amplifying apparent differences between populations.[38] (For example, owing to the dominance behavior of different alleles, 50 percent of a given population might be blood group O while 71 percent carry the O allele.) Kopeć calculated those allele percentages from "observed phenotype," that is, blood group, listed in the second column from the right. To assess the credibility of the data, that column also listed "expected" phenotype values. These she calculated from allele frequencies using the Hardy-Weinberg equation, and therefore she gave

TABLE 35
KELL GROUPS
SAMPLES TESTED WITH ANTI-K AND -k

		Number Tested		Phenotypes %			Genes %	
				KK	Kk	kk	K	k
ENGLISH	Ikin et al. (1952) 658	1166	OBS	.09	7.63	92.28	3.94	96.06
			EXP	.15	7.57	92.28		
WELSH	,,	116	OBS	.00	8.62	91.38	4.41	95.59
			EXP	.19	8.43	91.38		
SCOTTISH	,,	527	OBS	.00	8.92	91.08	4.56	95.44
			EXP	.21	8.70	91.08		
IRISH (Northern Ireland)	,,	106	OBS	.00	7.55	92.45	3.85	96.15
			EXP	.15	7.40	92.45		
BRITISH (United Kingdom)	Parkin (1952) 1602	935	OBS	.10	6.74	93.16	3.48	96.52
			EXP	.12	6.72	93.16		
CHIPPEWA INDIANS (Full Blooded)	Matson and Levine (1953) 934	161	OBS	.00	14.91	85.09	7.76	92.24
			EXP	.60	14.31	85.09		
CHIPPEWA INDIANS (>⅔ Indian)	,, ,,	128	OBS	.00	10.16	89.84	5.21	94.79
			EXP	.27	9.88	89.84		
CHIPPEWA INDIANS (<⅔ Indian)	,, ,,	206	OBS	.48	7.28	92.23	3.96	96.04
			EXP	.16	7.61	92.23		

8.3   Table 35 of *The Distribution of the Human Blood Groups*, showing group phenotypes and allele frequencies of the Kell blood groups in several human populations. The table displays (from right to left) population categories, citations to the papers where the data is from, the number of people tested, the observed and expected phenotype percentage frequencies (with inferred genotypes KK, Kk, and kk), and the calculated blood group allele percentage frequencies (*K* and *k*). Racial categories are displayed alongside national groupings without further explanation. From Mourant, *The Distribution of the Human Blood Groups* (1954).
Reproduced with the kind permission of the Mourant family.

the expected numbers for a large, randomly mating population. Any discrepancies between "observed" and "expected" values were meant to give the reader an impression of the reliability of the gene frequency calculations. To Mourant, the most likely explanation for a discrepancy was that the grouping tests themselves had been inaccurate, owing to ineffective reagents.[39] Such a discrepancy might also have been due to inadequate sample sizes, although Mourant chose not to include chi-squared values, which would have indicated the statistical significance of the difference between observed and expected values.[40]

Mourant organized much of his book according to geopolitical, religious, and "tribal" groupings similar to those described in the tables. His chapters follow a geographical sequence from Northern Europe, to the Mediterranean, to "Africa South of the Sahara Desert," to "Asia," to

"Indonesia and Australasia," to "The Aborigines of America." Within each chapter, section headings vary from the geographical ("The Iberian Peninsula") and the geopolitical (using the names of countries or regions such as "Eastern Europe") to the religious ("The Jews"). In some sections, local categories include "tribal" ones, such as various populations in Africa, "the Eskimos" of North America, and the Maori of Aotearoa/New Zealand.[41] These chapters consist of long descriptions of the blood group frequencies of different populations, alongside information about the geography, anthropology, history, and languages of the populations under study. Despite the range of information that Mourant marshaled, he presented not a single reference to any paper that contributed historical, linguistic, or anthropological knowledge. By leaving out the sources for his information about the populations under study, Mourant took their identities for granted.

Blood group surveys rarely mentioned how individuals were recruited as research subjects, and where they did, it was often with reference to a "local" expert, for example, Barandiarán's "expert guidance" on "pure Basques" and Yturri's skill in identifying Basques with "intact lineages." As we have seen, Mourant and his colleagues presented "local" assistants as having authentic and reliable knowledge of subject populations. By introducing these "assistants" in published reports, the researchers obscured further details about how those choices were made and characterized the relationships between research subjects as self-evident. Through these strategies, Mourant stabilized assumptions about which populations were genetically interesting.

Mourant's presentation of blood group data and the fact that he and other researchers rarely concluded anything very new about the populations they studied suggests that blood groups were not living up to the lofty ambitions held for them.[42] But another way of viewing these representations of blood group results is that, for Mourant's audience, they served to calibrate blood group population genetic data in relation to existing knowledge. I use the term "calibration" to point to the work done to align the methods and judgments offered by blood group genetics in relation to established physical anthropology, geography, and history.[43] After all, this was a young field of research. Simple lists of blood group gene frequencies would never have been interesting enough, but aligning these with racial, historical, and geographical knowledge made that genetic data legible. This served to show that the nascent field of human genetics had the potential to say something meaningful about human life and its past.

Mapping

The final pages of Mourant's book comprise nine detailed maps, each showing the distribution of a single blood group allele across geographical space (figure 8.1, above). Five maps showed data from "the aboriginal populations of the world." Four were of Europe, the source of more detailed data, which could be represented with finer granularity.[44] Each spanning a separate fold-out page, the maps present an outline of land masses. Onto those are superimposed isolines that indicate threshold blood group frequencies and offer the impression of a smooth diffusion of blood group alleles across geographical space. The isolines efface the category boundaries, and sometimes cut countries in half, or in several pieces; some flow across oceans to connect distant continents. Shading density indicates the blood group frequencies, offering visual connections between countries. For the Rh allele maps, the Soviet Union is left completely blank. The Soviet Union left the WHO in 1949 and did not join again for over a decade; the absence of information from this part of the world testifies to Mourant's reliance on WHO networks for this class of data.

There is little evidence of Mourant's mapmaking processes for the first edition of his book, but archival materials survive from the second.[45] Mourant prepared rough drafts of his maps, using as a starting point copies of a base map of a single continent or country—in the example in figure 8.4, Mourant used a French colonial map of North Africa that fixed "des tribus du Maroc" (the tribes of Morocco) in geographical space. Onto this older map Mourant superimposed the blood group frequencies calculated from his correspondents' data and used those values to sketch rough isolines. In drawing the isolines, Mourant was guided by the cartographically fixed population groups, and he likely took into account geographical landmarks such as rivers and mountains.[46] His superimposition of data and isolines onto the map was another kind of calibration: that is, the alignment of blood group frequency data with prior population categories. Mourant employed a professional cartographer to transpose the rough graphical representations onto standardized maps and to tint areas defined by isolines using shading of varied densities. Mourant's decisions regarding threshold frequencies, and judgments about isolines and their tinting, had the potential to profoundly affect the information conveyed by the maps. By altering these, even while keeping the same underlying data, Mourant could draw together whole populations or push them apart.

169

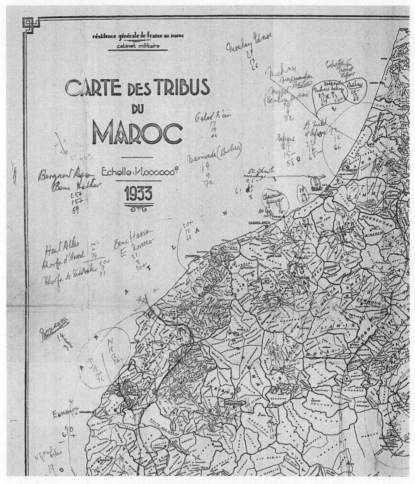

8.4 The top-left portion of a map showing the administrative boundaries of different "tribes" in Morocco, superimposed with Mourant's notes on blood group percentage frequencies. The underlying map facsimile had apparently been produced for the French government, copied in sections and pasted together. Probably in the late 1960s or 1970s, Mourant noted in red pen the blood group allele frequencies of local populations and sketched rough isolines indicating approximate percentage frequency thresholds. In preparing the second edition of *The Distribution of the Human Blood Groups*, Mourant sent drafts such as these to professional cartographer John Hunt. Whole map 60 × 48 cm.
Wellcome Collection, London, PP/AEM/E.37. Reproduced with the kind permission of the Mourant family.

The published maps offered an idealized picture of what blood group data could be. Population categories had dissolved, leaving apparently purified, objective information, which, as it accumulated, would presumably reveal detailed patterns of human diversity. Mourant's emphasis on large numbers was another way of abstracting and transforming human blood groups into something self-evident, distilled from the messy contexts in which blood had been extracted. Large numbers were crucial to the objectivity of the data. In a flourishing of the vision for human genetics that had been formulated by Fisher, Haldane, and colleagues in the 1930s, Mourant reasoned that, in the end, any errors would be swamped by the vast quantities of data that would be collected in the future. One of the reviewers of his book pointed out, more or less approvingly, that Mourant tended "to give those who report the data the benefit of any doubts which exist, apparently believing that tolerance is likely to produce more data to which time will apply the corrective."[47] Population genetics was a statistical science and Mourant judged that this approach could well be tolerated by the standards of the field. Large numbers were valuable in and of themselves: the more data that could be collected, the more accurate the estimates of population frequencies, and potentially the more fine-grained the map. When it came to applying for renewal of the Nuffield grant in 1955, Mourant emphasized that the need for the center was more urgent than ever, and moreover, "the compilation of all the available blood group data for the world . . . should, as far as can be foreseen, continue indefinitely."[48] To Mourant in the mid-1950s, collecting had no end in sight.

Making Humans Genetic

Mourant was a consummate collector, an enthusiastic collaborator, and an effective administrator, who framed his work in relation to the institutional and political exigencies of the time. He also had powerful confidence in his own ability to marshal and control his data. His project was remarkably free of either cutting-edge theoretical ambition or conclusive results: this was a descriptive, data-driven genetics built on infrastructures of blood transfusion, but with roots in interwar racial anthropology. It ordered and differentiated human bodies to link genetic data to specific human identities: place, society, nation, and race.

This was also a project to make humans knowable in genetic terms. Like Fisher before them, Mourant and Kopeć used the concept of "genetic equilibrium" to authenticate the data they were collecting as representative of "natural" or "breeding" populations. To them and many others, that this equilibrium could be found in racial, religious, and national groupings affirmed the merits of the project. As Mourant put it in the introduction to his book: "A scientific anthropology is coming into being and must in the long run prevail as a scientific zoology does."[49] He likened the genetic approach to human diversity to zoology, understood as an unquestionably natural science.

Mourant's large-scale collecting program involved the disciplined and expert labor of medical assistants, translators, statisticians, and clerks. Especially when dealing with "overseas" samples, he and his interlocutors relied on people with the "right" professional expertise and community or national affiliation to broker blood. These were epistemologically privileged observers who could credibly link the construction of genetic identity to kinships based on "tribe," "race," and nation. But their methods and their names rarely made it into print. In the late 1940s and early 1950s, Mourant and other authors took their identities for granted, and if they mentioned how research subjects were recruited, it was often with reference to a "local" expert.

By alluding to such interlocutors as local, the authors of scientific papers and books presented those individuals as having authentic and reliable knowledge about the populations in question. Their expertise was taken for granted, and this had the effect of making racial population categories (such as "full-blood Chippewa," "Lapps") seem natural. This move helped to align what was, after all, a new kind of human data (blood group allele frequencies) with existing knowledge (racial, geographic, and more subtle anthropological categories), even if that knowledge looks remarkably fragile today. Race, tribe, nation, and religion were accepted by Mourant and his readers as important frameworks through which genetic data could take shape and be made meaningful. Obscuring details about collecting practices and sampling choices was part of that process of calibration.

Continuing the theme of calibration, Mourant referred to *The Distribution of the Human Blood Groups* as an "instrument for research," a characterization echoed by geneticist Joseph Birdsell, who called the book an "important and finely polished research tool."[50] As Mourant put it, genetical anthropology was a young science, and the "study of mixed populations" was "only just beginning." At the present stage of research, he wrote, the main task "is the most complete possible genetical analysis of the *parent* populations" (my emphasis).[51] To Mourant, there was powerful appeal to collecting more and more data. Apparently, only when enough data had been collected on these "parent," or "aboriginal," populations, would blood group genetics be a fit instrument for probing more subtly mixed human groups, though for Mourant this end point was ever further deferred. His book's purpose was not *yet* to probe migrationary origins but to calibrate a new tool. Mourant's mapping offered a cogent way of making his genetic data authentically relevant to human difference.

Mourant used the (partial) identities of human specimens to make meaning from blood samples and elaborate new genetic knowledge. Historians Veronika Lipphardt and Jonathan Marks (among others) rightly observe the continued use of older racial categories in postwar genetic diversity research, even as researchers claimed that their work offered insights into human life and history, unencumbered by racial prejudice.[52] The practices described here underline the centrality of racial and other human identities to the making of meaning. If blood group genetic diversity data had *not* correlated with commonplace racial categories, what kind of interest would they have attracted? The work that went into making populations and human genetics "biological" operated as a set of cultural practices that depended on human characteristics and identities forged far beyond the scientific laboratory.

We have seen that where overseas data collection at that time tended to assume population categories as given, populations constructed from Britain's own NBTS donor cards required a fiddly statistical protocol: there were different ways of achieving authenticity for "abroad" and "home." From the early 1950s, however, standards changed, and genetic anthropologists began to be more explicit about their sampling procedures, sometimes recruiting linguists and social anthropologists to coauthor papers with them.[53] When geneticist Leslie Dunn traveled to Rome in 1954 to sample blood from individuals of an apparently ancient Jewish community, for example, he took with him his cultural anthropologist son.[54] In 1955, anthropologist Derek Roberts felt able to state that in studying the blood groups in the Nile Valley, he had given considerable attention to "devising satisfactory sampling procedures, an important aspect too frequently neglected in other studies whose value has thereby been lessened."[55] In 1956, US anthropologists Bertram Kraus and Charles White tackled head-on the question of who was best equipped to evaluate population groupings. They argued in *American Anthropologist* that social anthropologists were the right professionals for the job of defining "breeding groups" necessary for collecting human blood group data. "Only a careful study of social institutions, particularly such social groups as the clan, the local community, and the band, can possibly determine the true breeding population." To apply "modern" population genetics to humans, it was necessary to take into consideration "a form of environment not encountered with other organisms; namely, culture."[56] Without social anthropological expertise, Kraus and White warned, the extrapolation of genetic characteristics from nonrandom samples was fraught with the potential for serious error.

Their assertion provoked other professionals to vaunt their own disciplinary expertise. The well-known serologist Bruce Chown, of the Rh blood grouping laboratory in Winnipeg, countered that the genotype interpretations of Kraus and White were wrong: "An anthropological team studying population genetics should include a competent erythroserologist" (i.e., a blood group expert).[57] But Kraus and White's claims were in keeping with the direction in which genetic fieldwork was heading.[58] In a review of Mourant's updated ABO tables in 1960, Birdsell noted that for understanding microevolution, it would be necessary to be explicit about the "factors, both cultural and environmental, which structure the populations."[59] The methods of genetic population sampling were finally coming into view, although which were best suited to the task was still up for debate. These disciplinary clashes would continue to reverberate all the way into the 1990s during the Human Genome Diversity Project.[60]

Back in the early 1950s, sampling methods were not explicit. Mourant's maps presented a "natural," geographical, apolitical genetic topography that promised a cool-headed, newly objective understanding of the deep history of human migration.[61] Published during bloody episodes of British decolonization, and a highly visible phase of migration to Britain, Mourant's book deliberately overlooked the recent past. Genetics (the maps argued) was not about the politics of the British Empire, or the awkward complexities of race relations, but deep history. The transnational movement of serum—a substance made from human blood in the service of humanity—had helped genetics affirm the unity of all humankind. That notion, of genetics as modern, democratic, progressive, and redemptive, would soon be projected on a much larger stage.

NINE

Blood Groups and the Reform of Race Science in the 1950s

On November 10, 1952, television-owning households across Britain witnessed the country's first on-air demonstration of blood grouping. Arthur Mourant himself was appearing on a program called *Race and Colour*, the aim of which was to present "the scientists' view of race," as a preamble to a longer British Broadcasting Corporation (BBC) series about race relations. *Race and Colour* (broadcast in black and white) featured Mourant, Julian Huxley, and several other "experts," who seized the opportunity to explain how science undermined racial prejudice. The program singled out blood group genetics as a particularly powerful science for refuting ideas of racial superiority. The significance of blood groups for transfusion showed (the argument went) that humans were genetically diverse but "all the same under the skin." The world was unified in its diversity. Huxley and colleagues framed "science" as a rational, universal, and unprejudiced endeavor, and *Race and Colour* positioned blood groups as model postwar genetic traits.

This television program was emblematic of a broader movement in the early 1950s to project genetics as a modern, universal, and progressive science; an image that remains familiar today.[1] We saw a similar rhetorical fashioning in the 1930s when Huxley, J. B. S. Haldane, and other biologists in Britain intervened in discussions about the scope and meaning of race science. Supported by colleagues in other European countries and the United States, they had argued that genetics offered unprejudiced and "scientific" methods for studying human diversity. They successfully moved those arguments onto an international stage after the Second World War through their involvement with the United Nations Educational, Scientific and Cultural Organization (UNESCO), a specialized agency of the UN.[2]

An overarching principle of UNESCO was that education, science, and culture were crucial to the promotion of peace and world order. To Huxley, its first director general, the dissemination of scientific facts was unquestionably crucial to the promotion of civilization and social equality.[3] Indeed, UNESCO soon launched a campaign to combat racial propaganda.[4] Some of the pamphlets and books of this campaign projected genetics as uniquely positioned to reveal fundamental insights into human origins and diversity. Blood groups played an important role in this positioning.

UNESCO's image of genetics was captured in a single-page spread of a 1953 issue of *Life* magazine, which depicted the figure of Gregor Mendel, accompanied by crucifix, habit, and pea plant, under the heading "U.N.'s scientific pictures show what races are, how they originated, and how they became intermingled" (figure 9.1). Continuing the religious theme, the article juxtaposed Mendel with Adam and Eve, whose identities were signaled by the "Tree of Knowledge." The work that led to this segment of UNESCO's race campaign involved several of the scientists introduced so far in this book, including Huxley, Mourant, and American geneticist Leslie C. Dunn, whose collaboration with UNESCO helped project genetics as a reformed, internationalist, and humanist science that had overcome its eugenic past.

From one angle, the appeal and apparent success of this image of genetics is surprising. So far, much heredity research had been part of— and made possible by—two world wars and the large-scale administration of colonized countries. Indeed, after 1945, some high-profile scientists were concerned that the association of genetics with eugenics and Nazi race science had discredited the field.[5] Yet genetics emerged in the postwar decade as a seemingly purified, universally applicable, and politically neutral science for understanding human difference and ancestry. Moreover, this image dovetailed with Mourant's totalizing collection project in Bedford Square. Here I turn to the aesthetics and rhetoric of blood and genetics as they unfolded in the early 1950s, under the auspices of UNESCO and the BBC.

EARLY MAN, symbolized here by Adam and Eve, was probably dark-skinned progenitor of world's races.

HOW THE RACES OF MAN DEVELOPED

U.N.'s scientific pictures show what races are, how they originated and how they became intermingled

For the past three years sociologists, anthropologists and geneticists from all over the world have been working together to clarify one of civilized man's most confused and contentious concepts: the concept of race. Reviewing evidence from many fields, they finally reached general agreement on what is true, and not true, about the races of man. The results appeared recently in a booklet called *What Is Race?* whose illustrations by Jean Eakin Kleiman appear on these pages. Because racial myths and misunderstandings frequently lead to serious conflicts, the United Nations Educational, Scientific and Cultural Organization, which sponsored the project, is translating its booklet into several languages and giving it worldwide distribution.

Scientists agree that there are three main races in the world today—Caucasian, Negroid and Mongoloid—and that each is composed of many smaller racial groups (p. 106). To explain how races developed, the booklet reviews the history of man on earth. All races presumably originated from a common ancestor —early man. As primitive men increased in numbers they colonized the continents of the earth. But because the continents were relatively isolated, the population of each developed independently and grew perceptibly different from the others. Thus distinct races appeared. Later, as transportation developed, migrations led to the mingling of races, the production of mixed groups and even the isolation of new groups. These factors in turn have led to the variety of peoples in the world today.

It was the science of genetics which first explained how differences in populations arise (*following pages*). Though races differ genetically, UNESCO's scientists feel that major factors—intelligence and aptitude—show little or no variation. Environmental factors such as education and opportunity, they feel, are far more important in producing the differences among people today. But as the economy of backward areas advances environmental differences will diminish. And as races mingle more and more freely, most genetic differences may disappear.

GENETICS' FOUNDER was 19th Century abbot named Gregor Mendel. Growing peas in his monastery garden, Mendel discovered how characteristics are transmitted from generation to generation and explained for the first time the nature of heredity.

CONTINUED ON NEXT PAGE   101

9.1   First page of feature article "How the Races of Man Developed," from *Life*, May 18, 1953, 101. Its subheading declares: "U.N.'s scientific pictures show what races are, how they originated and how they became intermingled." The article explains that scientists had "finally reached general agreement on what is true, and not true, about the races of man." It is juxtaposed with Jane Eakin Kleiman's depictions of Gregor Mendel, with crucifix, habit, and pea plant, and Adam and Eve, whose identities were signified by an apple tree. The message conveyed by the images and text of this page was that the science of genetics held secrets for understanding race.

© 1953 The Picture Collection Inc. All rights reserved. Reprinted/translated from LIFE and published with permission of The Picture Collection Inc. Reproduction in any manner in any language in whole or in part without written permission is prohibited. LIFE and the LIFE logo are registered trademarks of TI Gotham Inc. Used under license.

Genetic Humanism

"Since wars begin in the minds of men, it is in the minds of men that the defences of peace must be constructed": so announced Britain's Minister of Education Ellen Wilkinson at the UNESCO founding conference in Paris in 1945.[6] Following lobbying from representatives from the Philippines, Brazil, Egypt, India, and the US National Association for the Advancement of Colored People, UNESCO incorporated into its constitution the objective of deploying science to fight "the doctrine of the inequality of men and races."[7] In 1948 the organization launched a program for ascertaining the "scientific facts" that would "remove what is commonly known as racial prejudice."[8]

The UNESCO race campaign was tied to the United Nation's defining agenda for a new world order, and it prescribed scientific expertise as a remedy for prejudice.[9] The supreme confidence in science as an instrument for the promotion of social cohesion harmonized broadly with the "scientific humanism" that had taken shape in interwar Europe and which had consolidated during wartime discussions about the future United Nations.[10] Huxley himself was a high-profile exponent of the view that science had potential as a diplomatic tool and social remedy, and that it offered a mode of knowledge production that would allow people in diverse parts of the world to understand one another.[11] To function effectively in the UN's program, science had to be elevated above politics.[12] Thus at its foundation, UNESCO projected a mode of scientific humanism that promoted the teaching of objective social and natural science for world citizenship.[13] In the words of director general Huxley: "science and the scientific way of thought is as yet the one human activity which is truly universal."[14]

Huxley's promotion of UNESCO's campaign on race was related to an intellectual project, promoted by a loose collective of scientists, to integrate a range of biological perspectives into a single evolutionary framework, a project that later became known as the "evolutionary synthesis."[15] Its advocates sought to knit together perspectives from paleontology, taxonomy, botany, and genetics into a single coherent evolutionary narrative.[16] Several of those involved also worked to integrate human life into this framework, as articulated in books such as Huxley's *Evolution: The Modern Synthesis* (1942) and the collective volume *Genetics, Paleontology and Evolution* (1949), as well as at meetings such as "Origin and Evolution of Man," held at Cold Spring Harbor in 1950. There was a significant overlap between those biologists involved in UNESCO's campaign on race and those who saw themselves as cocreators of the evolutionary synthesis, such as Huxley, Haldane, Dunn, and Theodosius Dobzhansky. For them, an evolutionary worldview encompassed human as well as plant and animal life, and this was the basis of a moderate yet liberal eugenic ideology that combined a democratic social sensibility with a belief in human progress.[17]

UNESCO's race campaign was organized by its social sciences department in collaboration with its department of mass communication, which was responsible for promoting "peace and human welfare" through articles, films, and broadcasts.[18] Taking as its initial impetus the UN's Universal Declaration of Human Rights (1948), which stipulated rights and freedoms for all people regardless of race,[19] UNESCO was tasked with initiating the campaign to show how science could dispel racial prejudice.[20] An internal memo explained the campaign's significance:

The public believes that race differences are important, and race prejudice is widespread. Scientists generally regard race as unimportant, and see no scientific justification for race prejudice. It will be the task of this convention to reduce the gap between popular and scientific knowledge in this respect.[21]

To "reduce the gap" was UNESCO's mission. In this formulation, science was the route to truth, and UNESCO's job was to bring it to "the public."[22]

As part of that campaign, the organization convened a meeting of scientific experts to draw up a list of "scientific facts" about race.[23] The multiple controversies generated by the resulting "statements" on race, and the many respects in which they failed to live up to expectations, have been explored in rich accounts elsewhere.[24] Most relevant to this story is that UNESCO carefully defined "race" as a "biological" term that (in the wording of one formulation) "designates a group or population characterized by some concentrations, relative as to frequency and distribution, of hereditary particles (genes) or physical characters, which appear, fluctuate, and often disappear in the course of time by reason of geographic and/or cultural isolation." This biological concept was temporal and dynamic (genes "appear," "fluctuate," and "disappear"). Crucially, the statements also explained that the biological "fact" of race should not be confused with "national, religious, geographic, linguistic or cultural" groupings.[25] Scientists could be trusted to handle the term in an expert manner, but the layperson could not: "Outside the scientist's laboratory the word 'race' has frequently been misused to justify policies producing economic and social discrimination."[26] Echoing the internal memo from two years prior the campaign insisted that the term "race" was in danger of being misused in "popular language" and could *only* be handled properly by biologists.

The message was clear: "race" was a biological reality, but the term could only be used credibly by scientists. Going further, *geneticists* were the expert scientists best placed to handle the word, since genetics shifted race science away from the study of typologies and toward a deeper and more subtle understanding of human population dynamics. Human genetics promised access to deep commonalities that tied together the peoples of the world.

Blood groups were made crucial to this argument, in ways illustrated most vividly in a picture book published by the UNESCO Department of Mass Communication in 1952 called *What Is Race? Evidence from Scientists* (figure 9.2).[27] The book used simple text and pictures to make scientific concepts intelligible to "people of secondary school age . . . and on up to adult education classes." The designs were by artist Jane Eakin Kleiman. Printed with a restricted palette—using the blue of UNESCO's flag as a background color and red as an occasional spot color—their most striking characteristics are the bold, highly abstracted figurative forms and diagrammatic illustrations.[28] These, so the preface explains, were devised to make scientific concepts "more easily intelligible to the layman."[29]

The lavishly illustrated *What Is Race?* presented "in a popular way, certain essential facts about the biological aspects of race."[30] Its opening page asks, "Which of the following would you call races?" followed by a list of population categories that include "Aryans," "Semites," "Nordics," and "Negroes." Underneath are the "answers" (Aryans = no; Semites = no; Nordics = yes; Negroes = yes), reaffirming the notion that race is "real" and has a specific meaning. Cautioning the reader that "the word 'Race' is a difficult one to use correctly," the book advises that "even scientists, whose business it is to be precise, are obliged to use the word 'race' in different ways at different times." The book insists that "race" is an expert biological or "zoological" term and frequently refers to the "major races of mankind . . . Caucasian, Negroid and Mongoloid," the suffix "-oid" carrying technical scientific connotations.[31] Again, scientists ("whose business it is to be precise") were the only people to handle "race" properly, and geneticists were the scientists with ultimate authority over the term.

Having established race as part of specialist biological terminology, *What Is Race?* also explained the significance of genetics. Like the statements

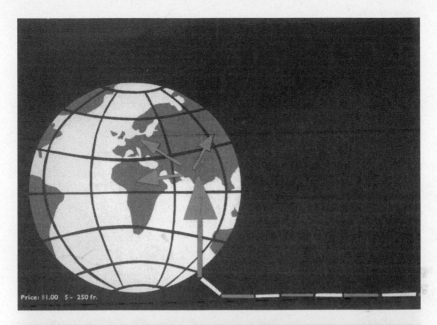

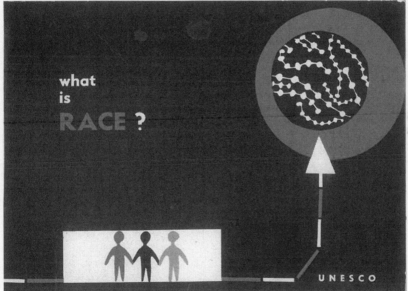

9.2 Back and front covers of *What Is Race? Evidence from Scientists* (Paris: UNESCO, 1952), in bold black, white, and red. Diana Tead, the UNESCO staff member who oversaw the production of the book, described the covers as featuring "three little men of each race holding hands, and all connected with one enlargement of a cell with chromosomes swimming around. . . . The back cover follows the cell around to India where presumably Man came into being." Quotation from Tead to Schaffner, October 29, 1951, 323.1 (094.4), UNESCO Archives, Paris. Image by Jane Eakin Kleiman, from UNESCO, *What Is Race? Evidence from Scientists* (1952). Front and back cover together 43 × 15.6 cm. Reproduced with permission from UNESCO.

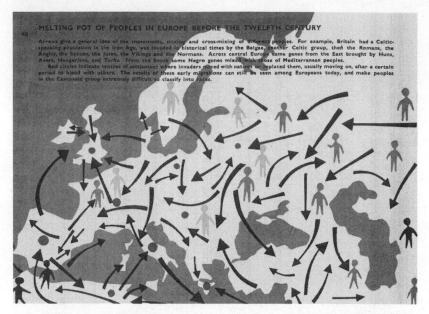

9.3 A single page from the UNESCO picture book *What Is Race?* arguing that genetics reveals Europe to be a "melting pot of peoples." Claiming to rebut Nazi notions of race purity, it claims that the continent has long been home to a mixture of races, represented here in different shades of gray. Its caption explains: "Arrows give a general idea of the movements, mixing and cross-mixing of different peoples. . . . Red circles indicate centres of settlement where invaders mixed with natives or replaced them, usually moving on, after a certain period to blend with others. The results of these early migrations can still be seen among Europeans today, and make the peoples in the Caucasoid group extremely difficult to classify into races." Image by Jane Eakin Kleiman, from UNESCO, *What Is Race? Evidence from Scientists* (1952), 40. 21.5 × 15.6 cm.
Reproduced with permission from UNESCO.

on race, the book emphasized that this science offered a dynamic understanding of racial difference. This it illustrated using the figure *Melting Pot of Peoples before the Twelfth Century* (figure 9.3), in which abstract human figures dance across an outline map, representing "successive waves" of migration across Europe. The illustration argued that the continent had long been home to a mixture of races, represented here in shades of gray. Although apparently chaotic, the illustration was in fact based on recent advances in mathematical techniques for modeling genetic change in populations. Its use of red circles to indicate centers of settlement graphically alluded to the available data on such genetic change—that of the blood groups.

Redemptive Blood

If the UNESCO biologists claimed that population genetics could offer deep insights into biological race, how did they think it should actually be studied? This is where the campaign made use of Mourant's blood-collecting project. *What Is Race?* devotes a whole section to the science of blood groups, again presenting what it calls a "popular" view of blood before explaining the "true" scientific one. A schematic demonstrates that "people of all races can be found in each blood group" (figure 9.4). The blood groups are emphatically unrelated to the superiority or inferiority of races. *What Is Race?* claims that the blood groups attest to the truth of the geneticists' view of race: "Nowhere can we show more clearly . . . that human groups, whether they be called races, tribes or peoples, seem to have the same basic assortment of hereditary characters."[32] On the next page, a two dimensional planar schematic describes the compatibility

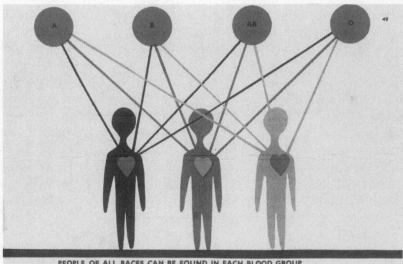

PEOPLE OF ALL RACES CAN BE FOUND IN EACH BLOOD GROUP

Individuals from any race can give blood to those from any other race—but only according to the Blood Transfusion Chart (page 50).

9.4   Schematic arguing that "individuals from any race can give blood to those from any other race." Four red circles denote blood labeled with the four major blood groups: A, B, AB, and O. Lines connect the circles to three human figures in different shades of gray, yet each with the same red heart, apparently representing "people of all races." Image by Jane Eakin Kleiman, from UNESCO, *What Is Race? Evidence from Scientists* (1952), 49. 21.5 × 15.6 cm. Reproduced with permission from UNESCO.

relationships between the different blood groups—thereby flattening and neutralizing racial hierarchies. Together, these diagrams framed blood group genes as the perfect mediators of racial difference. The message of both diagrams together was that as long as the blood groups matched, a member of any particular human group could donate blood to members of any other human group.

In using blood groups as emblematic of a scientifically enlightened genetics, the UNESCO book drew on a contemporary message about transfusion, which in the early 1950s was also being actively projected as proof of the fallacy of racism.[33] As we have seen (chapter 6), a key story line from the contemporary British feature film *Emergency Call* illustrates the notion that blood compatibility cut across racial lines. In it, the doctor and police chief cajole an urgently needed blood donation from a sailor, "Robinson," who is adamant that "white people don't want colored people's blood in them." The officials insist: "What do you mean, 'colored' blood? Your blood is red, isn't it?"[34] They eventually persuade Robinson to give his blood to Penny, the child dangerously close to death in the hospital. Viewers are asked to believe that Robinson, the "colored sailor," experiences this donation as redemption for the racist encounters he has suffered in the past. After donating, the humble, generous Robinson also gives to Penny's mother a photograph of his own small daughter: donor and mother realize together the kinship forged by his gift.

Two years after *Emergency Call* was released, the civil rights novel *Youngblood* (1954), by American writer John Oliver Killens, portrayed a doctor also being questioned about whether "white blood" could safely "mix with colored." In it, he replies: "There's no white blood and there's no black blood. . . . All blood is red-blood. The only difference is in the different types. Blood doesn't know any color line."[35] The novel narrated intense racial violence in the American South, during a period in which the American Red Cross Blood Transfusion Service had only recently ceased segregating the blood of people labeled "White" and "African American." *Youngblood* is ambivalent regarding the redemptive qualities of blood transfusion. In Killens's narrative, transfusion threatens but eventually fails to disrupt some deeply held racial kinships and prejudices. By contrast, *Emergency Call* narrated blood giving as a practice that promised to bring people together, unite communities, and supersede prejudice. The film underlined the midcentury characterization of blood transfusion as democratic, progressive, and antiracist.

Despite the pernicious blood segregations in the United States, which *Youngblood* captures so poignantly, *Emergency Call* could apparently credibly project the notion that giving and receiving therapeutic blood was a democratic, enlightened practice. Transfusion demonstrated that all humans were the same under the skin. As British sociologist Richard Titmuss would put it a few years later, "the life stream of blood that runs in the veins of every member of the human race proves that the family of man is a reality."[36] Transfusion was an international, humanist enterprise that could bring people together—a projection still used today.[37]

Thus, during the same period that genetics was being framed as a progressive science, blood transfusion was being characterized as a democratic and antiracist therapy, and blood groups linked the two. Scientific experts—such as Mourant and Robert Race—could quite safely study the relationships between blood groups and race.[38] The public was expected to learn, however, that blood groups and blood transfusion refuted the logic of racism. Blood groups mediated the complex terrain between transfusion as a redemptive public health enterprise and genetics as a "biological fact" of race.

Race on the BBC

Back in Britain, the BBC reframed UNESCO's message for the television program *Race and Colour*. The BBC had been established in the 1920s as a state-run broadcasting company whose aims were to "educate, inform and entertain." For the first thirty years of its existence, it did this primarily through the medium of radio. Television had first been introduced in the late 1930s, but, having been taken off the air during the Second World War, by 1950 it remained only a fledgling service, still trying to overcome its low status in relation to radio. Only in October 1950 did television gain its own department, a controller, and a seat on the BBC board of management, and it began to be seen as a medium with distinct possibilities. By then, an estimated 14 percent of British households owned a television, most in the Greater London area, a figure that rose dramatically after the 1953 coronation of Queen Elizabeth II. Watching television, for major events at least, was a public activity, with audiences gathering in homes and pubs.[39] During this time, BBC television was broadcast live from Alexandra Palace: it was an ephemeral performance, and not a permanent medium.

By this time, science was an established part of the BBC repertoire. Since the beginning of the war, several national science institutions had been lobbying for greater representation of scientists within the BBC and on its airwaves.[40] The Association of Scientific Workers had written to the director general of the BBC in 1941 to stress that it was "imperative that the general public should be infused with the knowledge that the body of science and its method of development—the scientific method—are instruments that can be controlled and utilized to whatever ends a community may choose."[41] So successful was this campaign that by the 1950s, people with scientific qualifications were occupying significant positions throughout the BBC. Eventually, the organization appointed a scientific advisor to ensure contact between producers and scientists. The person they appointed—Henry Dale, Nobel Prize–winning biologist, British pioneer of biological standardization, and director of the Wellcome Trust—held rather narrow views on what television could offer. He saw televised science as a popular science lecture but for a larger audience, with visual aids such as blackboard, wall diagrams, lantern slides, films, and the performance of experiments. Dale was out of tune with the contemporary ideas of producers, but his influence was widespread. Although he stayed at the BBC for only a short time, his model of science broadcasting was standard for the programs made, and *Race and Colour* perfectly fit that model.

Race and Colour belonged to the BBC's culture of "talks," originally derived from radio and now the foundation for nonfiction television. The Television Talks Department—which produced the show—was the main center for scientific programs for most of the 1950s, specializing in studio-based live television. Robert Barr, a documentary producer at the time, described the character of a typical talk program: "In a TV 'talk' expert opinion or information is conveyed *directly* from the *authority* to the viewer. It is Haldane talking on science, it is public men personally discussing events in the news."[42] *Race and Colour* had all of these elements: scientific men engaged in discussions and, with a variety of visual aids, conveying scientific information directly to the viewer.

The BBC's engagement with race also had a longer history. The corporation had from the beginning sought to produce programs that would promote communal identity among its audiences, whether at regional, national, or imperial levels.[43] In the 1940s, its output began to reflect new anxieties about race relations in Britain, and the BBC regarded part of its remit to reduce race prejudice.[44] The program *Race and Colour* was conceived in this context as a scientific prelude to a series of programs—called *International Commentary*—on "race relations." However, just as UNESCO in its race campaign deflected its message away from the racial politics of the United States and toward the politically safer ground of Germany's Nazi past, the BBC decided that to focus the program on race relations at home would be too inflammatory, choosing instead to focus on race relations on the African continent.[45]

Like UNESCO, the BBC was intent on shaping popular opinion among ordinary citizens, and it deployed the expertise of scientists to do this. The idea for *Race and Colour* had come from BBC producer George Noord-hof, who had read *What Is Race?* and decided that its material "could be made suitable for television, with very little difficulty." Noordhof saw the program as a "straightforward statement of the facts as contained in the book" and as an opportunity to redeploy several of the book's striking illustrations.[46]

Noordhof invited on to the program Mourant, Huxley, physical an-thropologist Jack Trevor, and cultural anthropologist Maurice Friedman. Mediating the discussion was well-known broadcaster Richie Calder, a newspaper science editor and public expert on issues relating to interna-tional peace. Precisely echoing the main statements of *What Is Race?*, *Race and Colour* emphatically conveyed that scientists were to be the main au-thorities on race. Nevertheless, the BBC already had a good idea of what they wanted their scientific experts to say. James Bredin, another of the program's producers, was concerned that the panel would be reluctant to commit themselves to "definite conclusions" about race. Bredin suggested that it would therefore be wise to hold an advance meeting with the guests before the program, "with the intention of persuading the scientists to fit what they want to say to the conclusions which we want them to reach."[47] Trevor was a particular concern in this regard, having what Bredin reported as a "retiring nature" that would "require a good deal of 'producing.'" Calder's job, Bredin added, was to bring the scientists to "as many definite conclusions as he [could]."[48] Like UNESCO officials, the BBC producers were instrumentalizing scientists to promote a distinct social agenda.

Race and Colour conveyed the UNESCO-mediated messages using an array of visual devices: diagrams from *What Is Race?*; a collection of four human skulls borrowed from the Haddon Library in Cambridge; and the display of seven living people, or "personalities," as they were described in the script. The subject turned to blood and its popular conceptions. Calder asked of Trevor: "We've been dealing with the bones, what about the blood[?] When people talk about mixed blood, a touch of the tar brush, blue blood, what does a scientist understand by blood in racial terms?" Trevor demurred, "That is really a matter for Dr. Mourant of the National Blood Group Reference Laboratory."[49]

Mourant had not been in the program's original lineup: in late Octo-ber the list of participants had instead included Isaac Schapera, profes-sor of social anthropology at the London School of Economics, so it is likely that blood grouping had not originally been conceived as a central feature. Surviving lists of graphics used suggest that blood groups were introduced to viewers first and foremost as having important functions in transfusion. Mourant's exposition was accompanied by an animated chart of the blood groups and a schematic showing possible transfusion combinations, presumably similar to the planar schematic of *What Is Race?*

Mourant then performed blood grouping tests on the seven "personalities," with the help of technicians from the Reference Laboratory. The first six people tested were a "Mr. Lo," Jacob Kisob, Yeh Ming, Hsu Long Chang, Ebenezer Bamghose, and Peter Keen. Remarkably, the seventh—representing the "Caucasian" race—was a young David Attenborough, later to become the United Kingdom's most celebrated wildlife presenter.[50] Following the demonstration, Mourant or another presenter made some general observations about "our blood groups [*sic*] differences and similarities."[51] Unfortunately, no fragments remain of the draft scripts and notes that related to the discussion of the blood groups. It is likely, however, that what Mourant had to say in *Race and Colour* was similar to what he said in 1951, when he participated in a program made in the UNESCO radio studios called *Science and Racial Barriers*, for which the script does survive. There he outlined what blood groups could say about race, arguing that "blood groups have no connection whatever with any idea of purity of blood in the sense of racial superiority." He went on to explain that although human populations around the world had different blood group frequencies, these were differences of degree, not kind.[52]

There is no evidence that the BBC presenters concluded anything further from the grouping performance in *Race and Colour*. Indeed, in dealing with blood group frequencies, what could have been concluded from seven blood tests on seven individuals? Rather, it was a performance to show what the genetic race science of blood groups looked like. The message from juxtaposing blood grouping tests with the various visual representations of the "major races" was that as genetic traits, blood groups *promised* to shed light on human diversity. The combined impression from these publications and broadcasts—the UNESCO statements, the *What Is Race?* book, and the BBC program *Race and Colour*—is that blood group genetics offered a much-needed potential route toward a "scientific" understanding of race.

"Much Too Technical"

Insofar as the BBC sought to make clear that race was the province of scientists, they perhaps succeeded too well. Optimistically, *Race and Colour* was broadcast at peak viewing time—between 7:45 and 8:25 p.m.—on Monday, November 10. As was routine for the BBC, especially for programs that might be contentious, the corporation conducted a viewer research report to garner responses among the general public. The reviewers were dismayed by the results: the program received a "disappointing" reaction index of 54, which was not only "well below the current average (62) for television talks" but also in "striking contrast to the figures . . . for the twelve programs under the title 'International Commentary.'"[53]

Although viewers complimented Huxley for his contribution, several felt that the technical scientific descriptions by the other presenters had obscured what they called the "real issues of the racial problem." One respondent—listed as a Linotype machine operator—claimed to see little point "in worrying whether man, 3,000 or 4,000 years ago, had a square head, a round head, or no head at all."[54] The report went on to say:

Most [respondents] were clearly not averse to a program that promised to enlighten them on a subject which struck many as being of paramount importance. This initial interest was, however, rarely maintained, largely because of disappointment with the angle of approach. The most emphatic criticism was that it was "much too technical."

Viewers found the section on blood groups particularly difficult to follow. The report explained that many people "felt like the Postman who wrote: 'I *do* think the experts could have simplified the discussion by cutting-out their phrase-codes when referring to blood-categories and so on. It was like listening to a lot of doctors at a medical confab.'"[55]

The BBC, like UNESCO, was attempting to strike a delicate balance. On the one hand, the construction of genetic race science as "much too technical" aligned perfectly with the image of elite scientific expertise that the two organizations were trying to cultivate. On the other, this portrayal risked undermining the hoped-for political function of science as a remedy for social tensions. The television-viewing public were not ignorant of the reality of race relations in postwar Britain, yet as inhabitants of a country that struggled to come to terms with changing relationships within its empire, a severe labor shortage, and a demographic crisis, they were skeptical of what the biological study of race had to offer. The program's reception made visible the tensions inherent to the UNESCO race campaign in its attempt both to elevate race to a new standard of scientific objectivity and to make the scientific understanding of race transparent and relevant.

Although disappointing, the complaints about this part of the program in fact cohered with the image of blood groups that Mourant and his colleagues were seeking to cultivate. Mourant's assertion in the UNESCO draft that "the term 'race' . . . should never be used when speaking of such human groups," is a reminder that blood groups as objects of research were *not* accessible to the general public, unlike visual characters such as skin color or facial features. Rather, blood groups could only be revealed by people with scientific or medical training. As Mourant wrote two years later in the introduction to *The Distribution of the Human Blood Groups*: "Though a non-scientific racialism is by no means dead, a scientific anthropology is coming into being and must in the long term prevail."[56] It was to Mourant's advantage that blood groups could be framed as traits that needed expert interpretation. Blood groups, and genetics more generally, offered a methodology for studying race that could be contained within the scientific realm.

A Model Genetics

Blood groups were technical entities. They could not be seen or felt; they could only be divined by trained technicians, scientists, or medical doctors. They were crucial to the progressive, cutting-edge, lifesaving enterprise of transfusion, and they could apparently trace deep patterns of human history. Mourant and his colleagues at UNESCO and the BBC saw blood group genetics as a model for what human genetics could be: readily abstracted and amenable to mapping, and apparently possible to uncouple from the political contexts in which blood was collected. Blood groups were seen as racial but not racist; they were social but not political. In short, blood groups, and by extension, genetics, projected a new race science fit for the postwar era.

UNESCO's account of human genetics was in keeping with the field's postwar disciplinary fortunes. The redemptive public framing of genetics included the renaming of its major journals in the postwar period—the *American Journal of Eugenics* became the *American Journal of Human Genetics*, the *Annals of Eugenics* became the *Annals of Human Genetics*—and the first international congress of the new field in 1956. The study of human genetics was also bolstered by widespread concerns about the effects of atomic radiation on mutation rates.[57] In the 1960s, genetics achieved a new kind of respectability when its findings became regularly reported in medical journals, and in 1964, the British Medical Association published the inaugural issue of the *Journal of Medical Genetics*, the first "exclusively medical" periodical on human genetics.[58]

UNESCO's race campaign was not just the story of an expert struggle about the right ways to think about human difference; it was also an early expression of the identity, objectives, and power of a new postwar institution: the United Nations. *What Is Race?* and other elements of the campaign were more than attempts to ameliorate social tensions and to reach a consensus on race; they were also performances to demonstrate the power of science, and more specifically, the authority and ambitions of UNESCO. In this context, science was presented as a neutral palliative for the racial and Cold War tensions of the postwar world, which could be framed as rational and uncomplicated by politics. Among a select but influential group of UNESCO and BBC experts and advocates, blood group genetics was a modern, international, unprejudiced science which could reach all corners of the world; it was the perfect tool to perform scientific internationalism. These themes would be expanded as blood continued to flow into genetics labs over the next decade.

TEN

Decoupling Transfusion and Genetics: Blood in the New Human Biology

In 1961, Arthur Mourant commented that "the study of the blood groups in human populations in general is becoming less and less a series of accidental off-shoots of blood transfusion work, and more and more a planned and integrated investigation."[1] In the 1940s and 1950s, Mourant's collecting program had largely depended on the infrastructures, social relations, and materials of blood transfusion; now human population genetics called for a more sustained international effort. In drawing attention to the planning and collaboration needed for large-scale collecting, Mourant captured the spirit of the age. International, large-scale, administration-heavy scientific collecting projects were becoming a feature of Cold War science.[2] In 1957, the International Geophysical Year (IGY)—a cooperative effort to make geophysical observations around the world—had already seized the imaginations of biologists. Inspired by the IGY, in 1958, the International Council of Scientific Unions began planning the International Biological Programme, an interdisciplinary project that would seek to accumulate data sets on living phenomena on a global scale.[3] These phenomena included human genetic diversity, and blood would take center stage.[4]

The Second World War was now receding into history, and the study of human genetics had been bolstered by widespread concerns about the effects of atomic radiation on human germplasm.[5] New funding was facilitating collaborations between geneticists and clinicians, and these were yielding connections between complex bodily conditions, protein sequence variation, and chromosome morphology.[6] Human blood was yielding new polymorphisms—that is, not just blood groups, but new arrays of other heritable traits with several possible variants. These included variable white-cell antigens, enzyme levels, and different forms of hemoglobin. Blood was diversifying genetically, but it was also taking new material forms. To define and reveal those novel variable traits, researchers were processing blood in new ways: by carefully culturing white cells, by separating out serum, by washing and hemolyzing red cells. Most striking, researchers could now freeze serum indefinitely, archiving it for future analysis.[7] Human blood was becoming fragmented both materially and genetically, under very specific technical conditions.

These changes had consequences for the meaning and organization of blood group genetics. When Mourant and Robert Race had founded their labs in the mid-1940s, they had efficiently hitched the study of human genetic variation to the materials, social practices, and procedures of blood transfusion. The work they did for and with the transfusion services positioned them as authorities, and their labs as passage points for reagents and data. By the mid-1960s, this configuration was changing. Race's work continued apace and remained crucially important to the work of the National Blood Transfusion Service, although the fast-paced discovery of new blood groups in his lab and elsewhere was beginning to quiet down. In 1962, the MRC had taken over funding of the RAI's Nuffield Blood Group Centre (now known as the "Anthropological Blood Group Centre"), testifying to the perceived significance of population studies to medicine. Within a couple of years, Mourant had left behind his work at the Reference Laboratory to focus wholly on his blood group anthropology. In 1965 he founded a new Medical Research Council institution devoted to blood group diversity research, called the Serological Population Genetics Laboratory. There, during the last fifteen years of his career, Mourant tethered his collecting endeavor to new international programs of research on "human biology" and "human adaptability."

But where Mourant had previously been able to claim a central position amid the international movement of blood and data, now the ways of studying human genetic variation were too varied and too large-scale to maintain a single obvious center of expertise. The kinds of blood, and the kinds of tests involved in such collections, now meant that samples were flowing into new labs, and expertise needed to interpret them was far more highly distributed. This chapter is in part about how the genetic and material diversification of human blood disrupted the alignment between wartime regimes of blood donation and research on human population genetics. It is also about endings. During the final years of Mourant's career he faced decisions about what to do with his unwieldy but alluring collections of blood and paper. Both presented profound storage and management challenges in a rapidly internationalizing research program.[8] Genetic methods were shifting, techniques of data management were changing, and retirement beckoned, but Mourant wanted to keep alive and usable the blood and data that had resulted from millions of bloodletting encounters. He faced practical and diplomatic challenges as he tried to maintain the vitality of the research materials lying in freezers and filing cabinets in London.[9]

Blood Diversifies

Through UNESCO, on the BBC, in many journals, and at conferences, the early 1950s saw visually and rhetorically striking claims for the study of genetics. The rest of the decade witnessed an extraordinary proliferation of studies on the genetics of human populations, which went well beyond research on blood groups. In 1949, two papers in *Science* had showed that human hemoglobin existed in two biochemically distinct forms and that those forms were heritable.[10] Biochemist Linus Pauling demonstrated these biochemical differences using the Tiselius apparatus, a huge instrument that separated proteins along a glass tube and used optical refraction to distinguish biochemical variation.[11] Just a couple of years later, geneticist Tony Allison published studies based on sickling tests he had carried out in East Africa, which seemed to correlate incidence of sickle cell disease with the density of malaria. This became the first apparently clear-cut example in humans of a polymorphism under selection.[12] Soon, researchers began separating hemoglobin variants using paper electrophoresis and a simple system of batteries, bringing into view an array of new hemoglobin variants.[13] British-based researchers Elspeth Smith and Hermann Lehmann demonstrated the efficacy of this cheap and mobile paper-based technique, which soon proved its remarkable suitability to field studies.[14] Throughout the rest of the 1950s, Lehmann and many others carried this apparatus on blood-collecting expeditions, cataloguing arrays of different hemoglobin polymorphisms and their geographical distributions.[15]

Meanwhile, blood serum was yielding an array of new genetically inherited proteins. The systematic survey of blood serum proteins was already used as a method for tracking contagious diseases in populations, a form of blood-based surveillance that became a WHO-sanctioned epidemiological practice.[16] The serum proteins being tracked in this way were acquired antibodies, produced in response to microbial infections. But from the mid-1950s, some varying serum proteins were found to be inherited polymorphisms. Toronto-based biochemist Oliver Smithies used starch-gel electrophoresis to define and characterize two main classes of inherited serum proteins: haptoglobins and transferrins.[17] These serum proteins joined hemoglobin proteins and blood groups as polymorphisms studied in the course of diversity research.

Another new source of blood-based genetic variation came in the form of enzymes, several of which were discovered owing to variable responses to certain foods and drugs. A deficiency in the enzyme glucose-6-phosphate dehydrogenase (G6PD) had been identified in racial studies of responses to both fava bean ingestion and the malarial treatment primaquine.[18] Another enzyme polymorphism—varying levels of cholinesterase in serum—was found as a result of unusual responses to the muscle relaxant succinyldicholin.[19] The development of simple assays for the ready detection of these enzyme deficiencies enabled population surveys of these polymorphisms.[20]

By the early 1960s, blood could also be used to visualize human chromosome karyotypes—that is, standardized prepared arrays of chromosomes. Human chromosome studies, known as cytogenetics, were beginning to flourish. The first cytogenetic studies on humans had required intrusive and painful bone marrow and skin biopsies, but now visual spreads of chromosomes could be made from venous blood.[21] Now that chromosome morphology could be surveyed from simple blood samples, there flourished a range of epidemiological and anthropological studies.[22]

White blood cells, too, were brought into view. During the late 1950s several researchers observed that serum from one person could react with the leukocytes (white blood cells) of another. The year 1958 saw the publication of three papers describing antibodies that reacted with a subset of leukocytes.[23] These sera had detected heritable leukocyte antigens that were quickly understood to have implications for tissue transplantation.[24] In the third edition of *Blood Groups in Man* (1958), Robert Race and Ruth Sanger could comment, in passing, on the "splendidly complicated antigens of the white cells."[25] The first international meeting devoted to "leukocyte antigens" was hosted by Duke University in 1964, and by the following year, many researchers were presenting work on the role of these antigens in skin grafts. By the late 1960s, the fiendish genetics of these antigens was coming into view, and it became clear that the variable white-cell antigens could be ascribed to a single locus with lots of tightly linked genes—analogous to, but far more intricate than, the Rh system.[26] Over the next fifteen years, the genes and alleles that corresponded to the human leukocyte antigen (HLA) complex turned out to be responsible for the antibody-based immune system in animals.

By the 1960s, then, individual genetic differences could be resolved from hemoglobin, from serum proteins, from blood enzymes, from chromosome preparations, and from white blood cells, as well as from the blood groups. Human blood was an increasingly complex substance that could reveal a huge array of heritable genetic variation. That meant, reciprocally, that blood could be used to define individuals in extremely specific ways. For example, in 1961 the medical journal *Lancet* published an editorial that explained the dazzling biochemical specificity of human blood. Not only were the various blood group combinations that could be identified in individuals now numerous enough to "define several million different classes of people," but new kinds of blood-based biochemical variation had amplified this specificity by several orders of magnitude. The *Lancet*'s angle was that such "biochemical individuality" could have important ramifications for medicine, a claim that presaged later hopes for studies of genomic variation. Simultaneously, for anthropologically inclined biologists, this rich variation became a resource for learning about human evolution, past and present.[27] Human blood was now widely understood to have the potential to reveal hitherto unknown aspects of human diversity, history, and health.

Blood was not only diversifying genetically. It was also taking on new material forms. Over four decades, preservatives, refrigeration, and packaging had profoundly changed the mobility of human blood for transfusion and research. From the use of anticlotting agents in the 1920s, to refrigeration in the 1930s, to the freeze-drying of plasma and antiserum during the Second World War: all of these technologies meant that blood could be stored longer and travel further, from the hot tropics to the cold Arctic. The separation of blood into fractions made it more mobile, and more economical—for example, a remarkable feature of plasma is that it can be recovered from "expired" whole blood. These practices also meant that blood was now available for a range of specialist purposes.

From the late 1940s, researchers in the United States began developing serum albumin for restoring circulation, fibrin "foam" for preventing hemorrhage during surgery, and fibrin "film" for stemming bleeding.[28] One blood fraction, however, was largely resistant to long-term storage: red cells were arguably the most valuable component of whole blood, but freezing caused their membranes to rupture. From the early 1950s, the NBTS began using glycerol to make small quantities of frozen concentrated red cells—useful for transfusing into small babies—part of the organization's arsenal of treatments, but the quantities produced remained very small in relative terms.[29] Nevertheless, by the mid-1950s, novel preservation technologies and new blood products meant that disembodied human blood was traveling further than ever before, and into new kinds of spaces.

The new hemoglobin, chromosome, enzyme, and leukocyte polymorphisms resolved from blood depended on treating blood in new ways. This had consequences for where and how genetic variation could be studied. When blood groups had been the only kind of heritable variation yielded by blood, tests had been carried out on whole blood or clotted cells. Now, cultured white cells were used to prepare chromosome karyotypes, and serum could resolve variable proteins and enzymes. Moreover, because serum proteins could reveal genetic variation of their own, geneticists could archive blood itself.[30] Techniques to define that variation had also become more numerous; they included not only serology but electrophoresis, enzyme assays, and cytogenetics. The tools and materials of human genetics were decoupling from the pursuit of therapeutic blood.

As we have seen, when data on human genetic difference had been generated by depots—involving bottles, registries, antisera, postal correspondence, donation, and exchange practices—expertise was found in labs closely associated with blood transfusion, such as the Galton Serum Unit in Cambridge, the Lister Institute in London, and Alexander Wiener's labs in New York. But now the materials and expertise of blood-based genetics were diversifying, and human genetic variation could be studied in more ways. Expertise in defining the newer genetic polymorphisms was distributed differently, among university departments and medical schools. By the early 1960s, the study of human genetic variation was a thriving subfield that was raising new research questions and supporting new kinds of collaboration. This included a new totalizing, administration-heavy blood-based project under the auspices of the International Biological Programme (IBP). What changed, and how did this affect Mourant's formidable collections?

"Human Biology" and the IBP

The Lister Institute blood groupers observed with enthusiasm and wonder as blood diversified into new polymorphisms and fractions.[31] Race and Sanger continued their research into the new blood groups and avidly pursued linkage with human geneticist colleagues. By 1969, the authors could claim that of the eight instances in which linkage had been established between loci on human chromosomes, six included a blood group locus—a fact that felt like a magnificent vindication of an extremely productive research program.[32] Meanwhile, Mourant expanded his "anthropological" blood group work—in which he was supported by a new intellectual framework.

A small group of physical anthropologists in Britain—led by Joseph Weiner at Oxford—were inspired by efforts in the United States to reform and reframe their field into an integrated study of functional, behavioral, and population aspects of humans.[33] Since the 1930s, physical anthropologists in Britain had been shifting attention from the study of bones to measurement of living people, and more recently especially to their physiological and genetic characteristics. Joseph Weiner himself was doing research on bone growth, primate dentition, hemoglobins and haptoglobins, and environmental physiology. He was inspired by attempts by US physical anthropologist Sherwood Washburn to bring anthropology into the consolidating framework of what was becoming known as the "evolutionary synthesis," that is, bringing research on population genetics together with studies of behavior and physiology and the paleontological analysis of fossils.[34] Weiner and his colleagues sought to import this new intellectual framework to Britain. With colleagues Derek Roberts in physical anthropology at Oxford, Nigel Barnicot in the Department of Anthropology at University College, and Mourant, he established the new Society for the Study of Human Biology.[35] To assert the relevance of this field to the postwar world, Weiner believed that a central theme of human biology should be "adaptability," that is, the physiological, physical, and genetic adaptation of people to varied climates and terrains. This committed Weiner to a formidable international administrative project that would provide a concrete basis for Mourant's continued collecting work: the Human Adaptability section of the IBP.[36]

The IBP was formulated by the International Council of Scientific Unions in 1959 as a large, centralized, collaborative data collection and sharing project that sought to facilitate collaborative links between institutions. Organizers had hoped that its administrative structure would emulate the remarkably successful International Geophysical Year (IGY), which had run from 1956–57 and had produced a number of data centers for collecting and disseminating worldwide geophysical data. As it turned out, the IBP proved to be a far harder project to manage than the IGY, and there were difficulties in settling on what kinds of collection were suitable for a world survey.[37]

The Royal Society set up a committee to manage all IBP activities in Britain.[38] They established an IBP central office on Marylebone Road in London, which issued the *IBP News* every three months.[39] Other large countries involved in the project established their own committees, and the international conveners of the IBP's sections set up networks of correspondents and international symposia.[40] Organizers settled on seven distinct sections, one of which was Biology of Human Adaptability (HA), comprising studies in population genetics, environmental physiology, and growth and fitness.[41] "Human adaptability" was a theme that chimed with scientific humanists committed to international collaboration, as well as being practically relevant in light of the dangerous possibilities of the Cold War. Weiner himself was appointed international convener of the IBP HA section. Human biology now had a visible footing on an international arena.

If Mourant's project to collect all available data on blood group polymorphisms had posed administrative challenges, then the HA section elevated these to a new level. The IBP was science administration writ large; its purpose was to coordinate collaborations and make varied collecting projects commensurate, and it faced formidable difficulties in standardizing its objects of collection.[42] Its communication and recording infrastructure had to be flexible enough to accommodate different sciences but sufficiently stable to produce legible, cohesive, commensurate results. As well as standardizing methods and instruments, it sought to designate laboratories for training and data analysis. Rather like Mourant's collections of the early 1950s, the IBP assessed and sanctioned labs that could be relied upon to provide credible data. It lent its name to affirm the credibility of existing institutions that would commit to the collection, processing, and storage of data.[43] The IBP was not a direct source of research funding; instead, it sanctioned certain labs as IBP centers, and those institutions were expected to apply to other funding bodies for support.

Weiner convened the first HA planning meeting in London in December 1962. Participants agreed that a central aim would be to coordinate surveys of all known gene polymorphisms in human populations throughout the world.[44] In practice, the focus here was on collecting data from what the meeting called "primitive groups," that is, as they put it, the "closest approximation one can find to the conditions under which man has lived for the greater part of his existence."[45] The IBP planners were formalizing the "unmixed," "isolated," and "ancestral" groupings used by the blood collectors of the 1940s and 1950s. Mourant himself explained in the *Eugenics Review* that the IBP project was urgent, before "practically all human communities become obscured in a world-wide uniform culture."[46] This was a scaled-up, systematic version of Mourant's blood group collections and his gene maps of "the aboriginal populations of the world" (see figure 8.1 on p. 157).

Likely because of its convenient proximity to Mourant's genetic collecting project, Weiner established HA section headquarters at the Royal Anthropological Institute (RAI) in London.[47] Over the next five years, he would coordinate plans for around 250 research projects on "human population biology," to be carried out in fifty countries.[48] The IBP offered a new institutional context for human population genetics, as well as new communities of researchers. This presented new opportunities for Mourant.

The Serological Population Genetics Laboratory

With IBP planning underway, Mourant hoped to continue his collecting work detached from the blood transfusion service. By the end of the 1950s, space at the Reference Laboratory was tight. Mourant's eagerness to help correspondents generate blood group results on anthropologically distinctive populations (and thereby contribute to his collection at the RAI) meant that the lab had started to groan under the weight of "anthropological" testing (that is, tests on blood from distinctive human populations). Meanwhile, the lab's serological activities were expanding to include organ transplantation antigens, and Elizabeth Ikin was working toward the first international WHO standards for anti-Rh testing sera.[49] Mourant frequently worried that requests for anthropological testing were more "than we can possibly cope with," and on one occasion he remarked with trepidation, "The amount of blood that is going to pour into London for anthropological testing in the next two months is unprecedented."[50] He anticipated that the IBP would only increase demand for tests and put pressure on serologists to produce rare antisera.[51]

Joseph Weiner's IBP project, and the new framework of human biology in Britain, gave Mourant the opportunity to cleave the anthropological testing from routine duties at the Reference Laboratory. Mourant's network of contacts was now so substantial that he no longer needed serum distribution to leverage data, and in any case the IBP was likely to provide ample networks for extensive collecting work. He appealed to the MRC to fund a new lab devoted to anthropological work. He explained that there was an urgent need to expand both blood group collection and testing. The complexity of blood grouping tests had increased severalfold, and the rate of publication had increased many times over in the last decade.[52] Moreover, new synthetic tables of data were "urgently needed" by those planning population genetical surveys for the IBP.[53]

The MRC needed a fair amount of persuading, and it had particular reservations about establishing a new unit for the already sixty-year-old Mourant.[54] There was also some uncertainty about the name for the new lab, in part because the MRC had recently established a "Population Genetics Research Unit" at Oxford, devoted to clinical genetics at a population level. But the MRC was eventually convinced, and after some wrangling, in 1965 it finally sanctioned Mourant's new Serological Population Genetics Laboratory (SPGL).[55] Mourant resigned from the Reference Laboratory, leaving its directorship to his capable deputy, Kenneth Goldsmith, under whose direction the lab's reference work would continue unabated, as would its close collaborations with laboratories in different countries.[56] Negotiating premises owned by St. Bartholomew's (St. Bart's) Hospital, Mourant found a home for his new lab in East Central London, next to Smithfield's meat market. The SPGL would combine anthropological testing with the collation and analysis work of the Anthropological (formerly Nuffield) Blood Group Centre.

At that last institution, Janina Wasung, Kazimiera Domaniewska-Sobczak, and Ada Kopeć had for fifteen years continued their clerical and computational practices using anthropological blood group data. Domaniewska-Sobczak and Kopeć, now "clerical officer" and "scientific officer," respectively, were named coauthors on synthetic compilations with Mourant.[57] As they helped Mourant to move the filing cabinets and index cards out of the RAI and into the SPGL, the three women became the "statistical section" of the new laboratory. Testifying to how extensive and bulky the paper was, the statistical section was housed a ten-minute walk away from the lab's testing activities, in an office in the heart of corporate London, in the shadow of St. Paul's Cathedral.[58]

By framing his new laboratory as one devoted to "population genetics," Mourant made it fit particularly well with the themes of the IBP. And, reciprocally, Mourant played a significant role in the IBP HA section. Blood testing and data collection were now part of an interdisciplinary endeavor to understand the relationship of humankind to its environment.

Weiner recognized Mourant's extensive experience in leveraging and managing blood and data and appointed him IBP consultant for blood collections and genetical surveys. Mourant recommended to Weiner those labs he considered capable of high-quality testing, and he brokered contacts between Weiner and researchers in Iraq, Pakistan, Malaysia, the Soviet Union, Hungary, and Iceland, among others.[59]

Although the SPGL was wholly dedicated to anthropological testing and analysis—and not, as the Reference Laboratory had been, engaged in the production of antisera—it was still often overwhelmed with blood. Much of Mourant's correspondence dealt with the scheduling of large consignments of cold samples arriving in his lab, and he regularly applied to the MRC for extra funding to hire temporary blood grouping staff, often making use of his own experienced former technicians. As well as its extensive testing work, the SPGL became something of a conduit and sorting hub for blood destined for analysis in a variety of other laboratories.[60] The SPGL sent aliquots for electrophoretic analyses to physical anthropologist Nigel Barnicot and his team at University College London, and to Hermann Lehmann's MRC Abnormal Haemoglobin Research Unit in Cambridge.[61] As the IBP got underway, Weiner reasoned that for the sake of efficiency and commensurability, the HA section needed common standards for collecting blood so that it could be subdivided and circulated for different kinds of testing.[62] It was Mourant who pulled together those standards and outlined them in the IBP handbook *Human Biology: A Guide to Field Methods* (1969).[63]

As the SPGL continued Mourant's testing work, it focused principally on blood groups, but, in line with IBP activities, it also began testing samples for certain serum proteins and some red-cell enzymes. SPGL members carried out these newer tests on serum (not the whole blood or clotted cells needed to test blood groups), which meant that Mourant could freeze certain samples indefinitely and ease test scheduling challenges. Entirely in keeping with other IBP laboratories engaged in such collections, Mourant took measures to keep some samples deep-frozen, so that they might be retested for new factors in the future, with around 100,000 frozen specimens accumulating over five years.[64]

Blood presented Mourant with management problems, but so did data. Under the auspices of the SPGL, Mourant was determined to publish a second edition of his magnum opus, now with a longer title: *Distribution of the Human Blood Groups and Other Polymorphisms*. Like the first edition, it ordered and tabulated all published blood group population data, but it also now included some of the newer blood-based genetic polymorphisms. The task was huge, and Mourant's main difficulty was managing vast quantities of data.

Exploring new technological solutions, he met a young geneticist-mathematician called Anthony Edwards, who in the early 1960s had worked on computerized methods of phylogenetic analyses with Italian geneticist Luigi Luca Cavalli-Sforza. Cavalli-Sforza and Edwards had used Mourant's blood group population data to construct phylogenetic trees of human populations—that is, visual representations of the evolutionary relationships between diverse groups of people.[65] When, in 1963, Edwards was offered a fellowship at the Atlas Computer Laboratory in Oxford, Mourant discussed with him and the MRC the possibility that "all the existing and very bulky records of the centre [be] transferred to punched cards" and be computed by Edwards.[66]

Edwards soon moved to Aberdeen, Scotland, to establish his own MRC-funded Human Genetics Computer Project, but from there he enthusiastically took up the problem of managing proliferating blood group data. Edwards' solution was a "Blood Group Tabulation System," written specifically for Mourant and designed to tabulate available data on blood group phenotype and gene frequencies.[67] With the data in computable form, his proposed system would, Edwards hoped, offer a straightforward and automatic system for producing "isogene" maps and constructing evolutionary trees.[68]

However, Mourant unexpectedly pulled out of the computing project. Edwards believed he had gotten cold feet owing to worries about the effect that the new methods would have on his collaboration with Kopeć and Domaniewska-Sobczak.[69] Mourant wanted to leave the job of tabulating blood group frequency data in the hands of the two women, although they themselves ultimately sought help from digital computers for calculating some of the more complicated phenotype-to-genotype transformations.[70]

Mourant's struggles with paper-based blood group results, and his resistance to digital computing, were two reasons why it took so long to finish the second edition of his book, eventually published in 1976 as the massive, 1,000-page *The Distribution of the Human Blood Groups and Other Polymorphisms*. Mourant's choice of a book as the medium for synthesizing and representing his collection was in part conditioned by the kind of data involved. For him, there was no possibility of periodically publishing updates in a journal. His unit of analysis was the population—therefore any single data point would need to be entirely

10.1 Example of an index card used for *The Distribution of the Human Blood Groups and Other Polymorphisms* (1976). The typed text is the official published information about a single published article (in this case, in the *American Journal of Physical Anthropology*); pencil handwriting next to the author's name ("1006") fixes the article in the bibliography of the book. Other pencil markings summarize the blood groups tested (top right) and the population categories studied (bottom right). In the top left corner is a code apparently indicating the geographic and "population" site of collection ("A" = North America; "Am i" = American Indian). 15.2 × 10 cm.
Photograph by the author.

recalculated as new information came in. For this reason, Mourant and his colleagues had to choose a cutoff date for the data they intended to include; after that date they closed the possibility of further refining their maps. So although his book was mobile physically, it was static with regard to what it contained. Even the bibliographic card index, kept locally by Kopeć and colleagues, became closed to further additions: having established their cutoff date of 1969, the researchers sequentially numbered each card in their index, fixing a unique identifier for each published paper (figure 10.1).[71]

This last management problem—that is, the recalculation of population-level metrics—might have been helped by a computer. Edwards's tabulation system would have allowed new data to be added at any later date and new population totals to be quickly calculated and (potentially) mapped.[72] At the time, Edwards himself commented on the accelerating accumulation of genetic data and the promise of electronic computers:

All this information would be indigestible were it not for the fact that mathematical biologists now see their way to interpreting it by automatic, objective, methods. Using electronic digital computers, their appetite for readily-available information has suddenly become insatiable: they can now use all the data that have ever been recorded.[73]

But even using computers to order, tabulate, and analyze the data, Mourant would have been faced with increasingly formidable problems of *how* to collect it. Previously he had been able to leverage his Reference Laboratory–WHO connections to make the Nuffield Blood Group Centre into a passage point for all available polymorphism data. But the new diversity of blood-related data meant that no single institution was perfectly positioned to orchestrate such a vast and diverse collecting enterprise. Mourant was an efficient administrator with a passion for collecting and mapping, and an exceptionally patient attitude to letter writing. While he had been closely tied to the transfusion services, and while genetic data had been confined to blood groups, his authority had been unquestioned. But now that data had diversified, there was no clear route or end point for its travel and accumulation; it was no longer obvious where human genetic data was destined to reside. The singular, simple system that Mourant had developed, based on practical blood grouping, was coming apart.

Retirement

During the late 1960s and early 1970s, data continued to amass at Mourant's offices in St. Paul's Churchyard, and IBP blood samples continued to accumulate in the SPGL freezers. Mourant's retirement grew closer, and he struggled to finish the second edition of his massive compilation of population genetic data. He had hoped that he could hand over the reins of the SPGL to a new director, but in 1970 he learned that the MRC would cease its funding. To have any hope of finishing his book, he needed to look elsewhere for financial support. Mourant applied to the Royal Society's IBP fund to continue testing his frozen serum samples. But simply testing the samples was not enough: Mourant felt that that these precious specimens had to be kept secure for the future, and they needed both space and "a responsible curator."[74] In 1970, still unsure of the future of his lab, Mourant transferred 100,000 deep-frozen samples to the Blood Transfusion Laboratory of Bart's Hospital, in the custody of his trusted former technician Donald Tills. Mourant also implored the MRC for further funding to support the data work, pointing out that the institution had, in effect, invested £50,000 in the book, and only comparatively little more would be needed to finish it.[75] For the final three years of Mourant's career, he tenaciously continued to hunt for funds from a variety of organizations to support the final stages of several collecting projects, receiving grants from the WHO and Britain's Wolfson Foundation. Through a series of personal acquaintances in the field of medical mapping, he even received money from the European Research Office of the US Army to pay a professional cartographer to draw his isoline maps.[76]

The MRC agreed that long-term blood storage was an important problem that had to be solved.[77] But Mourant also kept returning to the question of what to do with the paper index cards, offprints, and books, all of which might remain useful resources for research. One class of material was the 600,000 "record cards" that had comprised the British Blood Survey.[78] Measuring five by eight inches in size, they occupied ten cubic yards of space (about 6.7 cubic meters) and in the mid-1960s had already suffered near catastrophe when they had been transferred from the Anthropological Blood Group Laboratory at the RAI to the offices in St. Paul's Churchyard. Several boxes had been dropped on the busy London street, and the cards had become dangerously disordered.[79] Nevertheless, when Kopeć's book was published in 1970, Mourant's staff believed that the cards should continue to be kept because of their potential to yield "a lot more information." But these and the mass of other paper documents were cumbersome and unwieldy, and it was not clear who would be a suitable custodian.

When Mourant's SPGL finally shut its doors in 1976, he transferred his survey documents, offprints, card index, and serological books to the subdepartment of anthropology of the Natural History Museum in Central West London.[80] Soon after that, Tills himself moved to the Natural History Museum—with several thousand of Mourant's frozen blood specimens.[81] Since the late 1950s, the Natural History Museum's anthropology section had included a small serological laboratory, which Tills now ran. In that capacity, Tills now took over some of the IBP testing work, and his first task on arrival was to process "the backlog of 3,000 bloods which have stored in liquid nitrogen."[82] Tills also brokered space at the museum for Mourant, Kopeć, and Domaniewska-Sobczak to spend their final working months there, putting the finishing touches on the computation, analysis, and representation of the *Distribution* data.[83] Tills made vigorous attempts to keep Mourant's data usable, later arranging for the British Blood Group Survey cards to be photographed and transferred onto microfilm by Her Majesty's Stationery Office. But the collection remained largely unused.[84] The cards were difficult to manage and were kept in the basement in an awkward series of metal filing cabinets, which became known among staff as the "Berlin Wall."[85]

Thus, while Mourant's data were subsumed into other projects, his physical collections of blood and paper entered into a kind of historical limbo. He retired back to his native Jersey in 1977, while the offprints, survey data, and index cards that constituted his paper archive remained in the basement of the Natural History Museum. The 100,000 frozen blood samples—collected by IBP- and WHO-affiliated researchers during the 1960s and 1970s—also remained in the museum, housed in its Serological and Biochemical Genetics Unit in –80°C freezers.[86]

The blood and the paper were rarely used. But neither quite lost their allure. Both were bulky and cumbersome and were hard to reorganize and tabulate. But these problems apparently never completely outweighed their material inertia or scientific potential. Although neither substance was understood to be tied to any particular institution—having moved twice during Mourant's postwar career—younger colleagues in the human biology community judged that the records might still form the basis for a revised survey of the blood groups of the world. And although the blood samples cost money to keep frozen, they might always be analyzed for new genetic markers, should novel techniques be developed. Frozen IBP blood samples have had distinctive afterlives, the enduring potential value of which have kept hundreds of thousands of them in a latent state in freezers across Australia, the United States, and Europe for the last half century.[87] For Mourant's collections, both his blood and paper archives outlived his retirement, in a semidormant state: unwieldy, quiescent, but with the tantalizing potential for reanimation.

Collections Continue

The combination of wet and dry scientific administration that Mourant made so productive in the early 1950s captured a particular moment of alignment between wartime blood donation, postwar internationalism, and a reformist population genetics. Then, the whole world had seemed tractable: vast amounts of material traveled to Mourant's labs, and the methods of resolving blood group variation were relatively simple. By the mid-1960s this was drawing to a close. Genetics and transfusion were decoupling: human genetic population data were developing within a rapidly internationalizing bureaucratic network that was oriented more toward the future of research than the present of therapeutics.

Nevertheless, researchers working under the auspices of the IBP inherited features from Mourant's program. Several scientists published global book-length compilations of specific kinds of genetic data in the model of Mourant's blood group volumes. These included *Abnormal Hemoglobins in Human Populations* (1967), by Michigan anthropologist Frank Livingstone, and *The Distribution of the Human Immunoglobulin Allotypes* (1981), by Ohio-based geneticist Arthur Steinberg.[88] After Mourant's 1976 second edition of *The Distribution of the Human Blood Groups*, the next attempt to compile population data on *all* genetic polymorphisms was the massive, 1,000-page *The History and Geography of Human Genes* (1994), by Cavalli-Sforza and his colleagues Paulo Menozzi and Alberto Piazza, based at Stanford. Like Mourant, the trio initially produced a local archive that sought to collect all available published data from books and journals, although unlike Mourant, Cavalli-Sforza and colleagues used a computer to order and tabulate it.

This last collection of polymorphism data itself quickly became outdated owing to another kind of change. In retrospect, 1994 was the last possible date that anyone could have hoped to publish a totalizing compilation of data on human polymorphisms. Until this point, human genetic variation had been collected in the form of phenotype data: blood groups, serum proteins, hemoglobins, and enzyme levels. These were all variable traits that could be revealed using serology, electrophoresis, or protein assays. Researchers used mathematical transformations to convert the population frequencies of those phenotypes into population frequencies of alleles. But this was the decade of the Human Genome Project, and DNA variation itself was coming into scientists' purview. Rather than using phenotypes as proxies for genetic variation, now (or so the idea went) the DNA found in blood and other tissues could be sequenced directly.[89] Cavalli-Sforza himself spearheaded a new blood collection endeavor—which he called the Human Genome Diversity Project (HGDP)—the aim of which was to map the genetic variation of the world's peoples by sequencing the DNA from the cultured white blood cells of samples collected from indigenous populations. White blood cells could, in theory, be cultured indefinitely, producing unlimited quantities of DNA.[90] The HGDP was formally launched in 1994, the same year *The History and Geography of Human Genes* was published. Thus, the totalizing ambitions of the latter were quickly overtaken by new technologies for refracting extracted human blood.

Blood and Promise

This book has followed blood as it circulated through people and institutions, as it connected eugenics to the local politics of giving blood, and as it defined power relations between donor, nurse, patient, doctor, and scientist. By centering blood, I have emphasized genetics as a practice with social and material dimensions. Serological genetics was enacted by donors, nurses, needle sharpeners, and clerks, as well as by scientists and doctors. I have argued that the quantity and quality of human genetic data that could be made from human blood were affected by the relationships between collectors and donors, and by the circumstances in which they encountered each other.[1] The prominent role of bureaucracy in my account has highlighted the ways in which disembodied blood was used to carve out social and professional networks. These, in turn, helped to make human "populations" into objects of medical surveillance and scientific research.[2]

That emphasis on bureaucracy has also turned the spotlight on the movement of blood-related documents, their transformation into credible data, and, more broadly, how genetics was dependent on inscription and paperwork. Transfusion lent human genetics a paper-based infrastructure that enabled the labeling and movement of blood and the repurposing of those documents to create and circulate data. That infrastructure also reinforced the racial ordering of people, in ways that some scientists found useful for building genetic science. Even as prominent spokespeople heralded the end of race in science, racial categories were used to calibrate and affirm blood group population genetics.

My account frames midcentury human genetics as a science enabled and shaped by cultures of bloodletting, a planned bureaucracy, demarcations based on race, and collectivities of family and community. It paints a largely congenial picture: donors eager to make a contribution to society, doctors keen to embrace a lifesaving technology, scientists making contributions to health care while also fulfilling their intellectual interests. In chapter 9 I have given an account of the ways in which this public presentation was stage-managed, through the circulation of the UNESCO pamphlets, and another kind of communication technology: television. What I have also emphasized is that all of these networks were heavily constructed and required continual maintenance. After all, the humanist and altruistic spirit of the British transfusion enterprise was created in response to the alienating nature of its industrial-scale technologies. And the movement of blood and data between donors and patients was heavily managed, socially and practically. None of these could be taken for granted, and a great deal of technical and social work went into their assembly.

From 1920 to nearly 1960, blood groups were used to provide a model for what human genetics could be: they were abundant, they had clearcut inheritance, and they permitted the mathematical analyses of human populations. They were useful politically: they promised local disciplinary reform and projected a science that was internationalist, redemptive, and value neutral. Blood groups also promised much intellectually— for understanding evolution, for recovering deep history, and for pinning down the genetics of complex, medically relevant traits. Blood groups were promissory. They offered a vision of what genetics could look like in the future.

In a sense, that future has been promise itself.[3] Although Arthur Mourant's closely circumscribed world of blood and paper came to an end, some cultural features of midcentury blood group genetics remained intact. One of those was the idea of an ever-deferred genetic future. In 1992, the Human Genome Diversity Project promised that the collection of abundant DNA variation from indigenous populations would result in "'enormous leaps' in our understanding of 'who we are as a species and how we came to be.'"[4] In 2000, newspapers announced that the draft human genome heralded "a new era of medicine," which would "revolutionize the diagnosis, prevention and treatment of most, if not all, human diseases."[5] In 2013, a new organization called Genomics England embarked on the 100,000 Genomes Project, an initiative (framed by reference to "English" nationhood) to collect a vastly amplified quantity of genomic variation and correlate that variation with rare diseases. Underlining the promise invested in this enterprise, the 100,000 Genomes

Project was announced as part of the National Health Service's sixty-fifth anniversary celebrations. In the postgenomic era, such promises have become ever more ambitious: they have offered not just powerful scientific and medical knowledge but also sovereignty, new democratic rights, and sources of wealth.[6]

While the scale of enterprises based on collecting human blood and DNA has been amplified, the perception of genetics as a universal way of understanding human difference has largely been sustained.[7] For many, it is a commonplace that genetics and genomics can offer socially neutral applicable methods for capturing human difference, for recovering geographic ancestry, or for understanding variable responses to drugs. Today, the interpretation of genomic variation is sold as a technology of self-understanding: one such company tells its consumers that it will reveal "what makes you, you."[8] Another highly successful marketing campaign promises consumers that such tests would take them on a "DNA journey"—not just of genetic discovery but of redemptive, personal moral discovery, by revealing unifying family connections to people from other nationalities.[9]

Some have fiercely challenged these claims. Especially since the 1990s, many have come to see such blood collections as exploitative.[10] Others are worried about the way that geneticists continue to use social and political categories as implicit biological realities.[11] One lesson from the blood collections of the 1940s and 1950s is that the construction of human populations is always a socially and politically specific practice, even if the data derived from these can be presented as neutral. Indeed, several philosophers, sociologists, and anthropologists have recently demonstrated how racial classifications forged in specific populations have become folded into supposedly neutral genetic classification technologies—such as "ancestry-informative markers" and "biogeographic ancestry."[12]

Despite all of these concerns, genetics has overwhelmingly maintained its image as an unbiased methodology for detecting human diversity in medical and forensic settings, and one that can reliably underpin stories about human identities, ancestry, and migratory history.[13] Mourant and the IBP scientists saw themselves as part of a Western, liberal, progressive, redemptive project.[14] Fifty years later, human population genetics continues to be framed as an enterprise that unifies and affirms humanity, even as it is built on population groupings shaped by sociopolitical forces.

Also persisting from Mourant's research are some of his collections. Their uncertain status helps to highlight what has been preserved and

11.1 A photograph of one of several glass-fronted cabinets of the tearoom of the Division of Biological Anthropology, University of Cambridge, containing a large collection of Mourant's papers as well as the papers of some other scientists. The Mourant collection included file boxes with flaking offprints, loose sheets of paper with numbers and calculations, and letters discussing meetings, hematological surveys, and the International Biological Programme. They contained cardboard folders with labels such as "Wales Donor Sheets," plans and reports of the 1970s Paris "Centre Royamont pour une science de l'homme," and a script for a 1957 World Service radio broadcast on hemoglobin by Mourant and Hermann Lehmann. Several folders were filled with loose two-sided questionnaires from an extensive survey of hospital patients in the United Kingdom, testing associations between blood group and mental health. The collection also included several card-index drawers, containing about 4,000 four-by-six-inch cards, each with a single typed bibliographic reference and annotated with blood group totals. The collection was assembled by Mourant and his colleagues at two principal institutions: many of the papers were stamped with "Anthropological Blood Group Centre" (the later name of the Nuffield Blood Group Centre, ca. 1957–1962) and "Serological Population Genetics Laboratory" (ca. 1963–1973). Some papers were apparently derived from the later Natural History Museum collection of Donald Tills.
Photograph by the author.

what has changed between the 1950s and today. I first encountered remnants of Mourant's collections entirely by accident in the University of Cambridge Division of Biological Anthropology tearoom, in unlocked glass-fronted wooden shelves (figure 11.1). These shelves were filled partly with paper coffee cups and partly with approximately 300 three-inch-thick cardboard file boxes labeled "Mourant." The files contained materials ranging from questionnaires used in an extensive survey of British hospital patients, to cardboard folders of offprints, to card-index drawers, to radio scripts. The collection was derived from the Nuffield (later, Anthropological) Blood Group Centre and the Serological Population Genetics Laboratory. This was the working paper archive that Mourant had built and maintained in the service of science.[15]

Mourant's collection of papers had traveled to Cambridge with nearly 20,000 samples of frozen blood, now kept in –20°C and –80°C freezers in the Division of Biological Anthropology and the separate but affiliated Cambridge Leverhulme Centre for Human Evolution (figure 11.2). Most of the samples were collected under the auspices of the WHO and International Biological Program and were presumably sent to Mourant's lab

11.2 Frosted metal racks containing plastic vials in a –20°C degree freezer at the Cambridge Centre for Human Evolutionary Studies, 2012. Each rack contains 200 vials—some apparently contain plasma, others packed red cells or hemolysates (ruptured red cells). These racks had been durably marked "WHO" using embossed hard adhesive plastic, a technology often known by the trade name "Dymo," marketed from the late 1950s. Racks are labeled with a country (e.g., "Egypt," "Nigeria"), a place or tribe (e.g., "Sinai," "Manchok"), a date (ranging from 1968 to 1985), and a code (e.g., "EAA 1–75," "ZWT 1–75").
Photograph by the author.

for testing. Other samples originated from blood transfusion donors in the United Kingdom and Ireland, or from hospital patients.[16] Each vial has attached to it a paper label with a code (some typed, some written by hand), and their caps are marked with symbols in marker pen. Their current custodians are unsure what the codes on the vials and racks meant—they might refer to the person who collected the blood or their institutional affiliation.

Both paper and blood had been transferred to Cambridge from London's Natural History Museum when Mourant's former technician Donald Tills had retired. He had been in poor health and had decided to leave the museum to continue his anthropological investigations in a lab in his garden. In the early 1980s, Tills had contacted his former PhD supervisor, James Garlick, who was a lecturer in biological anthropology in Cambridge. The (then-called) Department of Biological Anthropology took custody of the blood in the hope that it would provide a useful resource for studies on human genetic variation. Over the years, the samples have been used sporadically for mitochondrial studies of genetic ancestry.[17] In 2016, researchers local to the collection judged that the frozen blood remained "a unique and highly valuable resource of genetic material from a diverse set of populations."[18]

This collection of blood is by no means unique. Tens of thousands of serum samples collected during the IBP in the 1960s and 1970s reside today in freezers around the world, especially in the global North.[19] Since their initial creation, the status and meaning of these collections have been transformed by changing standards of consent and privacy, by genetic concerns about the safety of blood specimens, and by new genetic technologies—which have changed what can be learned from blood.[20] Physically, the maintenance of the Mourant frozen blood collection in Cambridge has not been straightforward. Over the years, some vials have leaked and been disposed of. The department's Health and Safety Committee has worried about potential health hazards of the samples, especially the possibility of hidden pathogens. In 2003, the WHO published a "global action plan" for the laboratory containment of wild polioviruses, and to comply, the department conducted a sustained search for potentially infectious materials. The collection's custodians worry that the samples remain dangerous.[21]

In the past two years, the frozen blood has become vulnerable in other ways. The building that houses the Division of Biological Anthropology is due to be extensively remodeled, with some parts knocked down or altered and others built. The United Kingdom's Human Tissue Act (2004) requires each site to pay for a license for any human tissues that it keeps. It is not entirely clear to its custodians that the frozen samples should be classed as "human tissue," but department members, mindful of the act's regulations, judged they could avoid complications and cut costs by incorporating all of the frozen blood into the "Duckworth Collection"—a historic anthropological collection of human remains that has been accumulating for over 200 years.[22]

There are several further barriers to the blood's future use. Neither Mourant nor the IBP researchers had procedures for obtaining written consent from donors.[23] Today, the extraction of human blood for research is carefully controlled and requires detailed written procedures of informed consent—protocols that are still undergoing reevaluation today. But it is unclear to the researchers in Cambridge how and whether they might use Mourant's historical samples. The regulatory frameworks around the use of historical human remains vary in different countries. In Australia, for example, researchers must seek fresh consent from sample donors or their descendants and communities, whereas in the United States there are no such requirements.[24] The United Kingdom resembles the United States in this respect: there are no nationwide standards for the reuse of old blood; rather, local ethics committees in individual university departments make decisions about the repurposing of materials. One researcher observed that because the samples already exist in the department, it might be unethical to return to the same populations and embark on a new round of repeat collections. It is not yet clear how applications to use the blood will be judged by the university's review board.

Many researchers in the Cambridge department believe that crucial information about the frozen blood is contained in the 350 cardboard folders in the tearoom, including the meanings of the vial codes. Indeed, the main barrier to using the blood is the fact that the vials have become decoupled from their paper documentation. Just as Mourant had warned his IBP colleagues, conventional labels risk being dislodged from frozen vials and only a limited amount of information can be inscribed using embossed plastic, a felt-tip marker, or (as Mourant stipulated) a diamond-point cutter. Frozen blood needs durable documentation. Those records—like much of the paper discussed throughout this book—do not just give the blood meaning, they *are* its meaning: if the blood and records become irredeemably separated, then samples will no longer be capable of revealing serological secrets of human history. Rather, they will be just anonymous, frozen blood: expensive, messy, and hazardous.

Even more than consent, the Cambridge biological anthropologists are worried about maintaining the anonymity of the tissues.[25] They wonder whether the identities of the donors are buried in the papers—if so, then these must be secured and coded before the samples are used. Such anonymity is regarded by the researchers as crucial: without it, in a heavily bureaucratized world, samples might be linked to living individuals, potentially compromising medical confidentiality and risking access to insurance and health care. In short, many researchers believe that the paper documents must be located and secured in order to reanimate the blood.[26] In fact, Mourant's paper archive seems to have survived three decades in the Division of Biological Anthropology tearoom in part *because* researchers hope it will allow the frozen blood to be defrosted and used again, although this has not yet become urgent enough for any researcher to embark on the tedious work of sifting through the folders to find that information.

The Division of Biological Anthropology tearoom records represent a different kind of bureaucracy from the highly distributed paper infrastructure that produced the movement of therapeutic blood in the 1940s and 1950s. The transfusion bureaucracy was essential, multi-institutional, and nationwide, and it had to be dynamic, efficient, and robust. Since human genetics outgrew the blood transfusion services during the 1960s, those overseeing such collecting projects have had to build their own infrastructures: not just networks and materials for the movement of blood and DNA, but also regulatory frameworks and stable meanings. Human population genetics and anthropological studies have had to establish their own paths and methods for extracting and moving blood and data.[27] Moreover, research funding is organized in cycles; agendas change, laboratories close, and regulations shift. It is possible that the value of the blood in this story is already lost, although the Cambridge researchers hope that it is not.[28]

The safety of the frozen blood, the expense of keeping it, the lack of consent, privacy issues, and the absence of clear information about how it was collected and by whom, are all barriers to using the blood for future research. Nevertheless, in 2016 the department purchased a new, double-sized –20°C degree freezer at a cost of £3,000. Despite its uncertain status and material vulnerability, members of the Department of Archaeology and Anthropology continue to feel some responsibility to conserve the frozen blood. Perhaps, just as for Tills at the Natural History Museum, the blood still seems irresistible: too rare, too difficult to obtain, too intriguing to waste. And it has a fine pedigree—the WHO is a highly respected organization, and Mourant is still known by many human biologists as a father of their field.

The future of Mourant's paper collection is not yet clear, but it may soon change identity from a working, scientific collection of genetic data into an archive for historians to use. As it stands, the papers in the tearoom represent a perfect counterpoint to the human genetics archives held in the Wellcome Library, which I discussed in the introduction to this book. Those archives, held in secure, acclimatized facilities at 183 Euston Road in London, are beautifully catalogued and supremely accessible, having been made freely available online. But there are restrictions on their use: they have been carefully scrutinized for personal sensitive information in accordance with the Data Protection Act (1998), and many of the folders were recently closed for the next several decades. By contrast, in the tearoom on Pembroke Street in Cambridge, there are no barriers to my opening the cabinets and leafing through the files.[29]

But although the tearoom papers are open in one sense, they are inaccessible in another. The number of boxes and lack of evident structure make it doubtful how easily any historian could use their contents. For the papers to be used for this new purpose, they will be need to be catalogued, assessed for sensitive personal information, repackaged, and rehoused. The archivists who came to look at the papers with a view to accessioning them noted that (as with the blood) there are serious data privacy worries about the paper collection: among the files are large numbers of records that couple the names and addresses of individuals to their blood groups and a range of mental health conditions. If the papers are accessioned in the future, they will be assessed, sifted, and enveloped by a new bureaucracy: that of a special collections department of a library. This new layer of documentation will transform the papers that constituted Mourant's extensive scientific archive into a set of historical records.

But that work cannot be carried out yet. For now, the papers have an uncertain future. Several people involved are trying to establish who owns them. Although Garlick judges the collection to belong to the university, no official written documentation regarding ownership has yet come to light. Moreover, as I inquired about the possibility that they could be accessioned by the Wellcome Library, one of the custodians of the Duckworth Collection urged that we wait, explaining that more needed to be done to negotiate the future of the paper records in line with plans for the frozen blood. Any records that are deemed relevant to the blood (if those papers indeed exist) may remain in Cambridge, even if the rest of the collection is taken to London. For now, the samples remain separated from their documentation, and both paper and blood are still waiting.

The serological relations in this book defined the possible meanings of blood and what could be done with it. Blood, paper, serum, clotted blood, donors, scientists, antigens, and antibodies; changes to any one of these had the potential to affect another. Today, parts of Mourant's collections continue to be cared for, but their possible uses have been shifted by new relationships, new actors, and new narratives.[30] Paper inscriptions in the tearoom wait for archivists to order and protect them. Historians hope to transform the archived material into new stories about the past. Mourant's frozen blood endures, but now ethics committees prepare to speak for the interests of donors and their kin. Scientists wait to reanimate the blood and data using new kinds of genetic analysis. Paper, scientist, historian, archivist, freezer, and tearoom: serological relations continue to define the meanings of human blood.

Acknowledgments

Blood Relations began during my time at the University of Cambridge, in the Department of History and Philosophy of Science (HPS)—an institution with a remarkably friendly, generous, and critically engaged faculty, postdoc, and graduate community. Nick Hopwood, in particular, was instrumental in defining the methods and questions of the project, and he guided me through every aspect of its research and writing. The archives with which I worked are extensive and complex, and Nick has an unfailing ability to help me spot the interesting nugget among the rubble. Nick Jardine is unusual in having both historiographic sensibility and experience of having handled blood group data in his previous life as a mathematical taxonomist. Our conversations gave me confidence that the problems I had observed about postwar race classification were both real and interesting. I thank another Cambridge colleague, distinguished geneticist and mathematician Anthony W. F. Edwards, for his generous recollections and insights, especially of the work of R. A. Fisher. I am grateful to friends in HPS for creating a congenial and creative environment in which to work: Salim Al-Gailani, Leah Astbury, Mary Brazelton, Becky Brown, Sarah Bull, Elise Burton, Andrew Buskell, Helen Curry, Sophia Davis, Rohan Deb Roy, Matthew Drage, Kassel Galaty, Petter Hellström, Yuliya Hilevych, Ruth Horry, Anne Katrine Kleberg Hansen, Dmitriy Myelnikov, Josh Nall, Jesse Olszynko-Gryn, Sadiah Qureshi, Carolin Schmitz, Kathryn Schoefert, Keren Turton, Daniel Wilson, and Caitlin Wylie. Friend and colleague

Nick Whitfield provided a crucial, rich, and engaging account of British blood transfusion in the 1930s and 1940s; I made liberal use of his detailed attention to the workings of the EBTS, and of the archival images that he pointed me to. I am indebted to Tamara Hug, Aga Lanucha, and the Whipple Library staff for their incredible support in all aspects of departmental life.

From the very earliest stages of this project, Nick Hopwood thoughtfully facilitated connections between me and scholars beyond Cambridge, thereby putting in place the roots of a network that has sustained my research and career ever since. Soraya de Chadarevian showed me how rigorous engagement with the technical details of biomedicine can be a powerful route to recovering its communities. Susan Lindee has a compelling ability to describe the practices of science and medicine in human and political terms. Staffan Müller-Wille has forged frameworks for thinking about the *longue durée* of genetics and its paper technologies. All three generously brought me into conversations with their colleagues and students, and consistently gave me critically engaged feedback. One connection with long-lasting consequences that profoundly shaped my work and life was with Veronika Lipphardt, who invited me to her research group at the Max Planck Institute for the History of Science (MPIWG) in Berlin. From Veronika I gained a wholly new perspective on what it means to study the history of race, and the implications of this study for contemporary society. Birgitta von Mallinckrodt sustained the research group through her efficient organization, her constantly entertaining presence and friendship, and generous advocacy. The group depended on a team of student assistants not just for their administrative help but also for their contributions to the intellectual and social environment: Ricky Heinitz, Katrin Kleemann, Leon Kokkoliadis, Eric Llaveria Caselles, and Nina Ludwig. I was fortunate to share the orbit of the Lipphardt group with scholars who have become treasured friends. Sarah Blacker, Judith Kaplan, Lara Keuck, Emma Kowal, Joanna Radin, and Jenny Reardon have all, in different ways, shown me new ways of being critically engaged, and of being supportive colleagues. I am lucky that they remain my intellectual traveling companions. The intellectual and social life of the research group was created by many wonderful group members and guests, who included Susanne Bauer, Soraya de Chadarevian, Iris Clever, Samuël Coghe, Rosanna Dent, Yulia Egorova, Johanna Gonçalves-Martín, Susan Lindee, Ernesto Schwartz Marin, Yuri Pascacio Montijo, João Rangel de Almeida, Ricardo Roque, Ricardo Santos, Helga Satzinger, Marianne Sommer, Edna Maria Suárez Díaz, Mihai Surdu, Kathryn Ticehurst, and Sandra Widmer.

Thanks to the generous hospitality of Lorraine Daston, my final year at the MPIWG was spent in the convivial atmosphere of Department II. Raine gave me new ways of thinking about the ethical, moral, epistemological dimensions of sciences of the archive, and precise guidance on specific aspects of narrative. Colleagues in Department II created a stimulating environment in colloquia, reading groups, and corridors: Elena Aronova, Etienne Benson, Dan Bouk, Mirjam Brusius, Teri Chettier, Angela Creager, Raine Daston, Sietske Fransen, Donatella Germanese, Ana María Gómez López, Claire Griffin, Whitney Laemmli, Philip Lehmann, Elaine Leong, Minakshi Menon, Erika Milam, Tamar Novick, Christine von Oertzen, Susanne Schmidt, David Sepkoski, Skúli Sigurdsson, Laura Stark, Hallam Stevens, Bruno Strasser, Steve Sturdy, Annette Vogt, Dora Vargha, and Simon Werrett. Department II also offered generous administrative support, especially allowing me time to work with student assistant Anna Wölk, who helped me translate aspects of Myriam Spörri's valuable book.

I am grateful to the Wellcome Trust for funding the first three years of research on this book (grant number 089652/Z/09/Z) and the Max Planck Gesellschaft for several years afterward. Those institutions also helped to fund workshops that were crucial to the development of this research, including the 2012 workshop "Making Human Heredity" (with Soraya de Chadarevian), the 2015 workshop "(In)visible Labour: Knowledge Production in the Human Sciences" (with Judith Kaplan, and later Xan Chacko), and the 2017 workshop "How Collections End" (with Emma Kowal and Boris Jardine). At many other workshops, conferences, and seminars the following people offered comments and insights that left their mark on the book: Jenny Beckman, Sanjoy Bhattacharya, Michell Chresfield, Lisa Gannett, Mathias Grote, Jieun Kim, Robert Kirk, Ilana Löwy, Amade M'charek, Robert Meunier, Kärin Nickelsen, Edmund Ramsden, Simon Schaffer, María Jesús Santesmases, Laura Stark, Edna Suárez Díaz, and Sven Widmalm.

I am grateful to my kin within the SSEA project—Miriam Austin, Matthew Drage, Paul Gwilliam, Boris Jardine, Lizzy Laurence, and Alexander Page, as well as Lochlann Jain and Ana María Gómez López—for inviting me to bring some of the stories in this book into conversations about blood in art, religion, and everyday ritual practice.

This project began with an archive: the Arthur Mourant papers accessioned by the Wellcome Library from the National Cataloguing Unit for the Archives of Contemporary Scientists (NCUACS), University of Bath. I am grateful to the archivists at NCUACS and Wellcome for meticulously cataloguing the material, smoothing the way for my very first archival encounters. From 2009, I was lucky to be in the orbit of the

Wellcome Library as it forged new methods for making its materials available, and as it brought historians into discussion about the scope and shape of genetic collections, thereby giving me valuable glimpses into the world of scientific archives and their preservation and management. The Wellcome Library staff have been unfailingly helpful, in the routine day-to-day access to papers, in explaining their processes, and in obtaining high-resolution images. I also had generous help from the staff at the Rockefeller Archive Center in Sleepy Hollow, New York; the American Philosophical Society Archives in Philadelphia; the UNESCO Archives in Paris; the World Health Organization Archives in Geneva; and the custodians of the R. A. Fisher Papers at the University of Adelaide. In interpreting these archives, I gained much from the generous recollections of Tony Allison, David Attenborough, Anthony Edwards, John Hunt, Jim Garlick, and Patrick Mollison. I was astonished when this project also ended with the discovery of another archive: as I finalized the first full draft of this manuscript, a chance discussion with Cambridge archaeologist Andrew Clarke led to the unexpected tranche of Mourant papers in the tearoom of the University of Cambridge Division of Biological Anthropology, and Joanna Osborn generously helped me to access those materials.

Several colleagues and friends generously read drafts, chapters and proposals: Salim Al-Gailani, Sarah Blacker, Mary Brazelton, Sarah Bull, Elise Burton, Iris Clever, Angela Creager, Anthony Edwards, Petter Hellström, Emma Kowal, Josh Nall, Jesse Olszynko-Gryn, Stephen Pierce, Joanna Radin, Jenny Reardon, and Dora Vargha. I published some of the material in this book in other forms—in *British Journal for the History of Science*, *Studies in History and Philosophy of Biological and Biomedical Sciences*, *History of the Human Sciences*, and the edited volume *Human Heredity in the Twentieth Century* (2013)—and benefited enormously from the feedback of their editors and anonymous reviewers. I am indebted to Tracy Teslow and Michelle Brattain, whom I've never met in person but who made valuable comments on early draft papers. I am extremely grateful to Nick Hopwood, Boris Jardine, Richard McKay, and two anonymous referees who in its later stages read the manuscript in its entirety. For the final stages of the book, my heartfelt thanks go to staff at the University of Chicago Press: to Karen Merikangas Darling for coaxing the manuscript toward publication, to Tristan Bates for very kind support and advice, and to Tamara Ghattas for her astute edits.

This book is dedicated to my father, Andrew Bangham, who was professor of computer science at the University of East Anglia. During my childhood he moved between biology and the emerging field of computing. As he did so, he made many insightful observations about disciplinarity and kindled my own interest in the history and cultures of science. He and my mother, Kate Bangham, were supportive and loving when I made an unexpected switch of careers from genetics to history. My grandfather, biophysicist Alec Bangham, was particularly excited about this project: he knew Mourant during the early stages of his own career as a staff research scientist at the Babraham Agricultural Research Institute. Alec met Mourant through his research on cattle hemoglobin, and it was through his visits to Jersey (to bleed the island's famous cattle) that he and my grandmother Rosalind became lifelong friends with the extended Mourant family—Jersey becoming a favorite summer holiday destination for Alec, Ros, and their children. Through continued friendship between the two families, I was able to contact Jane Mourant, who kindly made possible support from Arthur Mourant's wife, Jean, for the reproduction of some of Mourant's materials in this book.

Many other family and friends supported me and this book: Sarah Bangham, Adam Polnay, Elena and Nick Polnay, Rosemary Bangham, Susannah Bangham, Jenny and Peter Zinovieff, Nick Jardine, Marina Frasca-Spada, Romilly Jardine, Luke Nashaat, Jenny Carpenter, Maggie Loescher, and Joel Chalfen. In the last weeks of writing, Rosemary Bangham, Jenny Zinovieff, Kate Bangham, Sarah Bangham, and Seraphina Rekowski generously provided child care, and Boris Jardine kept our family's body and soul together. Boris has been involved in this project from the very first grant proposal to the very final manuscript, which he read in its entirety. The book is imprinted with the many spaces we have lived in, places we have traveled, and—in the last two years—our shared life with Avery.

Glossary

AGGLUTINATION The clumping together of red blood cells observed when a specific antiserum (containing specific antibodies) reacts with a blood sample containing antigens specific to that antibody. As antibody binds with antigen, red cells clump together.

ALLELE Two or more alternative forms of a gene. Diploid organisms, such as humans, carry two alleles corresponding to each gene, one allele on each of two homologous chromosomes. Sometimes one of a pair of alleles suppresses the expression of the other—a condition known as "dominance."

ANTIBODY A class of proteins, naturally present in the body, or produced in response to the introduction of an antigen, that react with specific antigens. The older term, used by many of the people in this book, was "agglutinin."

ANTIGEN A substance that, when introduced into a human or other living organism, stimulates the production of a specific antibody. The older term, used by many of the people in this book, was "agglutinogen."

ANTISERUM A fluid, prepared from blood serum, that contains specific antibodies and is used to carry out blood grouping tests. Historically sometimes also called "testing serum," "grouping serum," or simply "serum."

BLOOD GROUP — Any of various classificatory groups into which the blood of humans and some other animals can be divided.

BLOOD GROUP SYSTEM — One or more blood group antigens controlled at a single gene locus, or by two or more very closely linked genes.

GENE — The basic unit of heredity in living organisms, originally recognized as a physical factor associated with the inheritance of a particular trait, and later understood to be located at a specific site on a chromosome and to consist of a sequence of nucleotides.

GENOTYPE — The genetic constitution of a trait or organism, specifically, the alleles present at a given gene locus.

LINKAGE — The tendency of genes that are close together on a chromosome to be inherited together.

LOCUS — The position on a chromosome of a gene or allele.

MENDELIAN TRAIT — A character with a clear-cut pattern of inheritance as predicted by the pioneer of genetics, Gregor Mendel.

PHENOTYPE — The observable characteristics of a trait or organism.

SERUM — The fluid part of blood that separates when blood clots. Serum contains antibodies and other soluble blood proteins and is used to make antiserum.

Sources

Archives

American Philosophical Society, Philadelphia (APS)
Barr Smith Library, University of Adelaide
BBC Written Archives
Bodleian Library Special Collections, University of Oxford
Division of Biological Anthropology, University of Cambridge
  (uncatalogued)
A. W. F. Edwards, private collection
National Archives, London (NAL)
Natural History Museum, London (NHM)
Rockefeller Archive Center, Sleepy Hollow, New York (RAC)
Royal Anthropological Institute London (RAI)
Somerville College Library Archives, University of Oxford
UNESCO Archives, Paris
University of Sussex Special Collections
Wellcome Library, London (WCL)
World Health Organization, Geneva (WHO)

Interviews and Personal Correspondence

Anthony C. Allison
David Attenborough
Anthony W. F. Edwards
James P. Garlick
John Hunt
Patrick Mollison

Notes

INTRODUCTION

1. For the history of human chromosome research: de Chadarevian, "Putting Human Genetics on a Solid Basis" (2013); de Chadarevian, *Heredity under the Microscope* (2020); for the ways that blood group research was interwoven with many other strands of human and medical genetics from the 1930s onward: Kevles, *In the Name of Eugenics* (1995), esp. 113–237; de Chadarevian, *Designs for Life* (2002); Lindee, *Moments of Truth* (2005); Harper, *A Short History of Medical Genetics* (2008); Comfort, *The Science of Human Perfection* (2012); Hogan, *Life Histories of Genetic Disease* (2016).
2. On some of the recent meanings of human genetics: Nash, *Genetic Geographies* (2015); TallBear, *Native American DNA* (2013); Nelkin and Lindee, *The DNA Mystique* (2004).
3. Nadia Abu El-Haj and Elise Burton show how genetics was built on deeply political nationalist projects, in their case relating to the Middle East: Abu El-Haj, *The Genealogical Science* (2012); Burton, *Genetic Crossroads* (forthcoming). Soraya de Chadarevian and Susan Lindee analyze the movement of materials and knowledge between the lab and clinic: de Chadarevian, "Following Molecules" (1998); Lindee, *Moments of Truth* (2005).
4. Porter, *Genetics in the Madhouse* (2018), 1.
5. Hyam, *Britain's Declining Empire* (2008); Clarke, "A Technocratic Imperial State?" (2007); Bennett and Hodge, eds., *Science and Empire* (2011); Tilley, *Africa as a Living Laboratory* (2011).
6. Amrith and Sluga, "New Histories of the United Nations" (2008).
7. On voluntarism in Britain from the late nineteenth century to the Second World War: Snape, *Leisure, Voluntary Action*

and Social Change in Britain (2018). On the history of the scout and guide volunteer movements: Proctor, "(Uni)Forming Youth" (1998), Proctor, *On My Honour* (2002). On the contributions of these to interwar donor recruitment: Whitfield, "A Genealogy of the Gift" (2011).

8. Sayers, "Blood Sacrifice" (1970 [1939]), 171.

9. Sayers, "Blood Sacrifice" (1970 [1939]), 171–2.

10. "Message from an Anonymous Blood Donor," July 24, 1940, folder 13–1-C Blood Donation, topic collection 13: Health 1939–47, Mass Observation Archive, University of Sussex Special Collections, http://www.massobserva tion.amdigital.co.uk.ezp.lib.cam.ac.uk/Documents/Images/TopicCollection -13/86, accessed February 21, 2020.

11. Strasser, "Laboratories, Museums, and the Comparative Perspective" (2010).

12. At the time, the mechanisms by which antibodies and antigens interacted were far from clear; Landsteiner himself joined fierce debates over the structural and electrochemical interactions that produced specificity. Those debates are the central problem in the following rich account of the history of antibody and species specificity during the twentieth century: Mazumdar, *Species and Specificity* (1995).

13. Bristow et al., "Standardization of Biological Medicines" (2006); Keating, "Holistic Bacteriology" (1998); Simon, "Emil Behring's Medical Culture" (2007); Mazumdar, *Species and Specificity* (1995); for a practical overview of contemporary serum therapy: Parke, Davis and Co., *Biological Therapy* (1926).

14. Gradmann and Simon, eds., *Evaluating and Standardizing Therapeutic Agents* (2010).

15. For general remarks on the organizational functions of inscriptions, signs, and diagrams in biology and the life sciences: Latour, "Drawing Things Together" (1990); Rheinberger, "Scrips and Scribbles" (2003); Delbourgo and Müller-Wille, "Listmania" (2012).

16. For analyses of paper, writing practices, and their relationships in science, medicine, and their social practices: Hess and Mendelsohn, "Case and Series" (2010); Bittel et al., *Working with Paper* (2019); Jardine, "State of the Field: Paper Tools" (2017).

17. Blood groups became material objects through their inscription as symbols; Ursula Klein analyzes the analogous case of chemical molecules: Klein, *Experiments, Models and Paper Tools* (2003).

18. Whitfield describes the growing distance between givers and beneficiaries: "In merely a few decades, inches stretched to yards, yards to miles, miles to oceans." Whitfield, "Genealogy of the Gift" (2011), 10. On the prior lack of nomenclatural standardization: Pierce and Reid, *Bloody Brilliant!* (2016), 33–34.

19. For understanding blood group serology and genetics in the 1930s and 1940s I rely on the following excellent accounts: Mazumdar, "Two Models for Human Genetics" (1996); Mazumdar, *Eugenics, Human Genetics and Human Failings* (1992). William Schneider authored several insightful histories

of interwar blood group research: Schneider, "Blood Group Research in Great Britain, France and the United States" (1995); Schneider, "Chance and Social Setting" (1983); Myriam Spörri has written a valuable account of the landscape of German blood group research and its meanings and politics between 1900 and 1933: Spörri, *Reines und Gemischtes Blut* (2013).

20. For the functions of paper and inscription work in genetics specifically: Müller-Wille, "Early Mendelism and the Subversion of Taxonomy" (2005); Rheinberger, "Scrips and Scribbles" (2003).

21. On the use of blood groups in forensics in Germany: Okroi and Voswinckel, "'Obviously Impossible'" (2003); Okroi, "Der Blutgruppenforscher Fritz Schiff" (2004); Spörri, *Reines und Gemischtes Blut* (2013).

22. Sellen and Harper have used the concept of "affordances" to analyze how the physical properties of paper invite and condition certain kinds of human actions and activities: Sellen and Harper, *The Myth of the Paperless Office* (2001).

23. On the materiality and functions of bureaucratic documents: Gitelman, *Paper Knowledge* (2014); Hull, *Government of Paper* (2014).

24. Thanks to Erica Milam for her insightful observations of the "wet" and "dry" themes that run throughout this story.

25. For the ways that the changing organization of interwar laboratory medicine depended on and was shaped by the emerging administrative culture of mass medicine: Sturdy and Cooter, "Science, Scientific Management and the Transformation of Medicine in Britain" (1998).

26. Haraway, *Modest_Witness@Second_Millenium.FemaleMan©_Meets_Onco-Mouse™* (2007), 253.

27. Bynum, *Wonderful Blood* (2007); Hart, *Jewish Blood* (2013); Biale, *Blood and Belief* (2007); Moreau, "The Bilineal Transmission of Blood in Ancient Rome" (2013); Bildhauer, *Medieval Blood* (2010); Spörri, *Reines und Gemischtes Blut (2013)*; Efron, *Defenders of the Race* (1994); Daniel, *More Than Black?* (2002); Davis, *Who Is Black?* (2001).

28. Nye, "Kinship, Male Bonds, and Masculinity" (2000); White, *Speaking with Vampires* (2000); for more on the history of vampire narratives: Copeman, *Veins of Devotion* (2009); Law, *The Social Life of Fluids* (2010).

29. Waldby and Mitchell, *Tissue Economics* (2006), 1–10. Whitfield analyzes how, during the Second World War, "the gift" became the dominant way of characterizing donations of blood, and argues that it was vigorously promoted in direct response to wartime technologies of blood banking: Whitfield, "A Genealogy of the Gift" (2011); Whitfield, "Who Is My Donor?" (2013); Whitfield, "Who Is My Stranger?" (2013). Anthropologist Paul Rabinow briefly describes how the act giving blood became understood as an embodied symbol of resistance in occupied France: Rabinow, *French DNA* (1999), 84–85.

30. Waldby and Mitchell, *Tissue Economies* (2006), 1–3; Hernandez, "Donations: Getting Too Much of a Good Thing" (2001); Schmidt, "Blood and Disaster: Supply and Demand" (2002).

31. Bennett, *Banning Queer Blood* (2009).

32. Healy, *Last Best Gifts* (2006); Swanson, *Banking on the Body* (2014); Lederer, *Flesh and Blood* (2008).

33. Susan Lindee has explored the deeply consequential experiences, labor, and identities of human research subjects in genetic medicine: Lindee, *Suffering Made Real* (1994); Lindee, *Moments of Truth* (2005); Lindee and Santos, "The Biological Anthropology of Living Human Populations" (2012).

34. On the formal developments of the treatment of human subjects in US and UK biomedicine: Epstein, *Inclusion* (2007); Stark, *Behind Closed Doors* (2012); Wilson, *The Making of British Bioethics* (2014).

35. For anthropological work on biomedical encounters around blood: Geissler et al., "'He Is Now Like a Brother, I Can Even Give Him Some Blood'" (2008); Reddy, "Citizens in the Commons" (2013).

36. *Oxford English Dictionary Online*, http://www.oed.com/view/Entry/20391. For foundational anthropological analyses of blood-as-relatedness and the dominance of "biological" relatedness in American models of kinship: Schneider, *American Kinship* (1968); Schneider, "Kinship and Biology" (1965). For recent rich commentaries on Schneider's work and legacy: Franklin and McKinnon, eds., *Relative Values* (2001).

37. For further analysis of the relationships between blood and kinship: Carsten, *Blood Work* (2019).

38. Weston, "Kinship, Controversy, and the Sharing of Substance" (2001); Haraway suggestively uses the figure of the vampire—whose drinking of blood infects lineages—as one that "both promises and threatens racialized and sexual mixing": Haraway, *Modest_Witness@Second_Millenium.Female-Man©_Meets_OncoMouse™* (1997), 214.

39. For the cultural history of blood groups and identity in the United States: Lederer, "Bloodlines" (2013).

40. In the film *Emergency Call*, this phrase was uttered by a doctor trying to persuade an unruly donor to give blood. Its sentiment was echoed in the Civil Rights–era novel: Killens, *Youngblood* (1982 [1954]). For analysis of this aspect of the novel: Weston, "Kinship, Controversy, and the Sharing of Substance" (2001).

41. Sarah Chinn analyzes US wartime transfusion as a signifier of modernity, democracy, and citizenship and its exclusions based on race: Chinn, *Technology and the Logic of American Racism* (2000): 93–140.

42. On US blood segregation: Love, *One Blood* (1996); Lederer, *Flesh and Blood* (2008); Guglielmo, "Red Cross, Double Cross" (2010).

43. Quote from Fisher to Landsborough Thomson, July 16, 1941, FD1/5845, NAL. On how notions of gender have impacted paper practices in knowledge production: Bittel et al., *Working with Paper* (2019); Oertzen, "Gender, Skill, and the Politics of Workforce Management" (2016).

44. On gender in early genetics: Kohler, *Lords of the Fly* (1994); Richmond, "Women in the Early History of Genetics" (2001); Richmond, "Opportunities for Women in Early Genetics" (2007).

45. For another discussion of the ways that immune systems make human bodies porous to their social and political environments: Biss, *On Immunity* (2015).
46. Rose, "Calculable Minds and Manageable Individuals" (1988). Analysis by Amade M'charek on the coproduction of sameness and difference: M'charek et al., "Equal before the Law" (2013); M'charek, "Contrasts and Comparisons" (2008).
47. Rose is drawing on Foucault, *Discipline and Punish* (2012) [1977], 190–212.
48. For analyses of corporeal economies and biovalue: Waldby and Mitchell, *Tissue Economies* (2006); Waldby and Cooper, *Clinical Labor* (2014); Murphy, *The Economization of Life* (2017).
49. There is growing scholarship in the social sciences on the "immunopolitics" of human organ economies, including analysis of how racial difference has produced acute scarcity of organs for certain social groups: e.g., Goodwin, *Black Markets* (2013); Jacob, *Matching Organs with Donors* (2012). For histories of immune specificity and its relation to blood economies: Waldby and Mitchell, *Tissue Economies* (2006); Lederer, *Flesh and Blood* (2008); Swanson, *Banking on the Body* (2014).
50. On state-managed bureaucracies and statistics gathering in the public sphere: Porter, *Trust in Numbers* (1995); Crook and O'Hara, *Statistics and the Public Sphere* (2011).
51. These were themes addressed by, for example, the International Union for the Scientific Investigation of Population Problems: Bashford, "Population, Geopolitics, and International Organizations" (2008); on how "population" was made into an object that linked sociology, biology, anthropology, economics, and psychology: Ramsden, "Carving up Population Science" (2002). Murphy studies how Cold War–era data practices made both "population" and "economy" into objects of intervention and governance: Murphy, *The Economization of Life* (2017).
52. Solomon et al., eds., *Shifting Boundaries of Public Health* (2008); Packard, *A History of Global Health* (2016).
53. Bashford, *Global Population* (2014).
54. One of the actors who contributed to this shift was evolutionary biologist Julian Huxley, first president of UNESCO, who moved population planning to center stage in his vision for "material and spiritual betterment." Bashford, "Population, Geopolitics, and International Organizations" (2008), 341; Huxley, *UNESCO: Its Purpose and Its Philosophy* (1947), 12.
55. Paul, *Whitewashing Britain* (1997).
56. For two recent collections connecting human populations to international networks to biological research: Lindee and Santos, eds., "The Biological Anthropology of Living Human Populations" (2012); Bangham and de Chadarevian, "Heredity and the Study of Human Populations after 1945" (2014).

57. This argument was developed in Bangham and de Chadarevian, "Human Heredity after 1945" (2014).

58. Sommer, *History Within* (2016).

59. Haraway, *Primate Visions* (1989); Stocking, *Bones, Bodies, Behavior* (1998); Zimmerman, *Anthropology and Antihumanism in Imperial Germany* (2001); Roque, *Headhunting and Colonialism* (2010).

60. Sommer, "History in the Gene" (2008); Sommer, "DNA and Cultures of Remembrance" (2010); Sommer, *History Within*, 2016, 249–368.

61. Exceptions to this include Landsteiner and Wiener's 1930s attempts to use blood group serology to construct human-primate phylogenies.

62. Paul, *The Politics of Heredity* (1998); Kevles, *In the Name of Eugenics (1995)*; Sommer, *History Within* (2016), 135–248.

63. Stepan, *The Idea of Race in Science* (1992); Weindling, *Health, Race and German Politics* (1989); Barkan, *The Retreat of Scientific Racism* (1992); Wailoo, *Dying in the City of the Blues* (2001); Bland, "British Eugenics and 'Race Crossing'" (2007).

64. Gannett and Griesemer, "The ABO Blood Groups" (2004); Lipphardt, "Isolates and Crosses in Human Population Genetics" (2012).

65. Projit Mukharji draws attention to Ludwig Hirszfeld's narration of blood collection in First World War Salonika, which vividly illustrates his varied encounters with donors, and observes Hirszfeld's prior judgments of the "races" in question: "We had to speak a different way to each nation. It was enough to tell the English that the objectives were scientific. We permitted ourselves to kid our French friends that we would find out with whom they could sin with impunity. We told the Negroes that the blood tests would show who deserved a leave; immediately, they willingly stretched out their black hands to us." Mukharji, "From Serosocial to Sanguinary Identities" (2014), 150. For more on the construction of racial identities of research subjects in blood group mapping: Gannett and Griesemer, "The ABO Blood Groups" (2004).

66. For detailed discussion of precisely what "colored" blood meant in the midcentury United States: Guglielmo, "'Red Cross, Double Cross'" (2010).

67. Wailoo et al., *Genetics and the Unsettled Past* (2010).

68. Turda, "From Craniology to Serology" (2007); Turda, "The Nation as Object" (2007); Mazumdar, "Blood and Soil" (1990); Robertson, "Eugenics in Japan" (2010); Bucur, "Eugenics in Eastern Europe" (2010).

69. Hazard, *Postwar Anti-racism* (2012); Bangham, "What Is Race?" (2015).

70. On the treasured concept of "unity in diversity" and its constructions within the UN project: Selcer, "Patterns of Science" (2012); Selcer, "The View from Everywhere" (2009).

71. E.g., from the *American Journal of Eugenics* to the *American Journal of Human Genetics*, the *Annals of Eugenics* to the *Annals of Human Genetics*.

72. Lindee, *Suffering Made Real* (1994); Creager, *Life Atomic* (2013); Lindee, *Moments of Truth* (2005); Comfort, *The Science of Human Perfection* (2012).

73. Webster, *Problems of Health Care* (1988).
74. The Wellcome named its digitization project "Codebreakers," and its valuable collection of online archives can now be accessed at http://wellcome library.org/collections/digital-collections/makers-of-modern-genetics/. The Wellcome archivists also turned this digitization project into an opportunity to convene a workshop to reflect on the challenges of collecting genomics-related information, which in turn resulted in a series of scholarly papers, including Shaw, "Documenting Genomics" (2016); de Chadarevian, "The Future Historian" (2016); Lindee, "Human Genetics after the Bomb" (2016).
75. For a comprehensive account of the exchange regimes of genomic sequence data in the 1990s: Maxson Jones et al., "The Bermuda Triangle" (2018).
76. This policy was updated in August 2014, and, under the title "Access to Personal Data," it can be found at http://wellcomelibrary.org/content /documents/policy-documents/access-to-personal-data.pdf; accessed February 20, 2020.
77. Other files were available through "restricted access," which means they can be viewed at the Wellcome Library but not photographed, and that personal identifiers must be erased, standards to which this book conforms.
78. E.g., Bangham and Kaplan, eds., *Invisibility and Labour in the Human Sciences* (2016).
79. For rich discussions of these issues: e.g., McKay, *Patient Zero* (2017); Stark, "The Bureaucratic Ethic" (2016); Keuck, "Thinking with Gatekeepers" (2016).
80. Radin, "Collecting Human Subjects" (2014).

CHAPTER ONE

1. On surgeons' experimentation on the Western Front: Pelis, "Taking Credit" (2001); on the changing material characteristics of blood and the effects of this on the uses of transfusion and blood groups: Schneider, "Chance and Social Setting" (1983); Schneider, "Blood Transfusion between the Wars" (2003).
2. Keynes, "Blood Transfusion: Its Theory and Practice" (1920), 1217.
3. Turda and Weindling, eds., *"Blood and Homeland"* (2007); Burton, *Genetic Crossroads* (forthcoming).
4. For in-depth accounts of eugenics and its relationships to nation states and internationalism: Bashford and Levine, eds., *Oxford Handbook of the History of Eugenics* (2010).
5. Further essays on the varied practices of human heredity during this period: Gausemeier et al., eds., *Human Heredity in the Twentieth Century* (2013).
6. For selected histories of genetics during this period: Müller-Wille and Rheinberger, *A Cultural History of Heredity* (2012); Rheinberger and Gaudillière, eds., *Classical Genetic Research and Its Legacy* (2004).

7. Mazumdar, "Two Models for Human Genetics" (1996); Mazumdar, *Eugenics, Human Genetics and Human Failings* (1992).
8. For the long history of the dispute about immunological specificity: Mazumdar, *Species and Specificity* (1995).
9. A "biochemical" science because, as the "golden era" of bacteriology faded around the turn of the century, the study of immunity became aligned with chemistry. Immunologist Paul Ehrlich, for example, worked with chemist Emil Fischer to investigate the relation between chemical antibody structure and biological function: Silverstein, *A History of Immunology* (2009).
10. For a sustained history of serology in relation to the practical field of clinical pathology: Foster, *A Short History of Clinical Pathology* (1961).
11. The major European institutions for serum standards were the State Institute for Serum Research in Berlin (founded 1896) and the Statens Serum Institute in Copenhagen (founded 1902). In Britain, serum standards were set by the Division of Biological Standards at the National Institute for Medical Research (established 1923): Gradmann and Simon, eds., *Evaluating and Standardizing Therapeutic Agents* (2010); Bristow et al., "Standardization of Biological Medicines" (2006).
12. Ludwik Fleck chose a serological test—the Wasserman test for syphilis—as his case study for the collective, social construction of scientific "facts": Fleck, *Genesis and Development of a Scientific Fact* (1979 [1935]). For more on the serological context of Fleck's analysis: Löwy, "'A River Cutting Its Own Bed'" (2004).
13. Lattes, *Individuality of the Blood* (1932), 10.
14. For Nuttall's program, and its development in the United States: Strasser, "Laboratories, Museums, and the Comparative Perspective" (2010).
15. For more on Landsteiner's career and immunological research: Mazumdar, "The Purpose of Immunity" (1975); Mazumdar, *Species and Specificity (1995)*; Silverstein, *A History of Immunology* (2009).
16. Mazumdar observes that technically, the methods of blood group and bacterial serology were identical: Mazumdar, "Blood and Soil" (1990), 188.
17. What became, in Landsteiner's hands, a remarkably neat pattern of reactions had previously been confused by the assumption that agglutinins (antibodies) were always caused by past or present infection. Landsteiner proposed instead that antibodies could be a normal characteristic of immunological individuality: Mazumdar, *Species and Specificity* (1995), 143.
18. Mazumdar, *Species and Specificity* (1995), 337–42.
19. *Individuality of the Blood* (1932), the widely translated book by Italian serologist and forensic specialist Leone Lattes, summarized the wealth of studies on these topics.
20. On the connection between the techniques of transfusion and the significance of blood groups: Schneider, "Chance and Social Setting" (1983); Schneider, "Blood Transfusion between the Wars" (2003).

21. On the variety of early transfusion practices: Pelis, "Blood Standards and Failed Fluids" (2001); Starr, *Blood: An Epic History* (2009); Krementsov, *A Martian Stranded on Earth* (2011); Sunseri, "Blood Trials" (2016).
22. Krementsov, *A Martian Stranded on Earth*, 15–32.
23. Some of the earliest reports of transfusion following tests of blood compatibility were made by American physician and hematologist Reuben Ottenberg, e.g., "Studies in Isoagglutination: I" (1911).
24. Pelis, "Blood Standards and Failed Fluids" (2001); Whitfield, "A Genealogy of the Gift" (2011), 8–31; Schlich, *The Origins of Organ Transplantation* (2010).
25. Pelis, "Taking Credit" (2001).
26. Pelis, "Taking Credit" (2001); Schneider, "Chance and Social Setting" (1983).
27. Whitfield, "A Genealogy of the Gift" (2011), 38.
28. Bernheim, *Blood Transfusion* (1917), 78.
29. Medical staff and students were also often seen as reliable, local sources of blood.
30. Starr, *Blood: An Epic History* (2009), 220–44.
31. Swanson, *Banking on the Body* (2014); Lederer, *Flesh and Blood* (2008).
32. Schneider, "Blood Transfusion Between the Wars" (2003).
33. Krementsov, *A Martian Stranded on Earth* (2011), 104–12.
34. Drawing on this work, doctors further afield experimented with blood from cadavers: Saxton, "Towards Cadaver Blood Transfusions in War" (1938). For more on the story of blood from dead donors: Starr, *Blood: An Epic History* (2009), 78–82.
35. On the London Red Cross Blood Transfusion Service: Pelis, "'A Band of Lunatics down Camberwell Way'" (2007); Gunson and Dodsworth, "Towards a National Blood Transfusion Service" (1996); Whitfield, "A Genealogy of the Gift" (2011). On the export of this model to Red Cross services in other countries: Klugman, *Blood Matters* (2004); Schneider, *The History of Blood Transfusion in Sub-Saharan Africa* (2013).
36. Whitfield convincingly argues that Oliver carefully obscured the relationship between donor and recipient in order to help cultivate donation as an altruistic service to humanity: Whitfield, "A Genealogy of the Gift" (2011), 32–89.
37. On the Rover Scout movement and its involvement in blood transfusion: Whitfield, "A Genealogy of the Gift" (2011). Oliver's service has been highly influential in later British narratives about transfusion in the country as a whole, although it was only one of a number of local services across Britain, some of which relied on paid donors.
38. Krementsov, *A Martian Stranded on Earth (2011)*, 104–12.
39. Palfreeman, *Spain Bleeds* (2015).
40. Lederer, *Flesh and Blood* (2008), 68–106.

41. The Second International Congress of Blood Transfusion devoted a quarter of its presentations to blood preservation: *Bulletin de la société française de la transfusion sanguine* (1939).

42. In the 1930s, the "ABO" blood groups were denoted by at least three distinct nomenclatures. 1935 was the inaugural year of the International Society of Blood Transfusion. Italian expert Lattes presided over the first meeting, which drew delegates from twenty countries and eight Red Cross societies, and which invited as the guest of honor Ludwik Hirszfeld. On this meeting and the society's history: Pierce and Reid, *Bloody Brilliant!* (2016).

43. *Bulletin de la société française de la transfusion sanguine* (1939).

44. Bashford and Levine, *Oxford Handbook of the History of Eugenics* (2010).

45. Müller-Wille and Rheinberger, *A Cultural History of Heredity* (2012); Gausemeier, Müller-Wille, and Ramsden, *Human Heredity in the Twentieth Century* (2013).

46. For diverse examples: Rheinberger and Gaudillière, *Classical Genetic Research and its Legacy* (2004).

47. Von Dungern and Hirszfeld, "Über Vererbung gruppenspezifischer Strukturen des Blutes" (1910); Mazumdar, "Two Models for Human Genetics" (1996).

48. For technical details of Weinberg's work and Bernstein's revisions: Mazumdar, "Two Models for Human Genetics" (1996); for more on Weinberg: Müller-Wille and Rheinberger, *A Cultural History of Heredity* (2012).

49. Mazumdar goes on, alluding to the remarkable genetic research that was being carried out using the *Drosophila* fruit fly: "The blood-grouping laboratory was the fly-room of the human species." Mazumdar, "Two Models for Human Genetics" (1996), 620.

50. Bernstein, "Ergebnisse einer biostatischen zusammenfassenden" (1924).

51. Quotation from Mazumdar's remarkable account of Bernstein's research, and his interactions with and reception among other researchers in Europe, the United States, and Japan: Mazumdar, "Two Models for Human Genetics" (1996), 623.

52. Spörri, *Reines und Gemischtes Blut* (2013); Rudavsky, *Blood Will Tell* (1996); Starr, *Blood: An Epic History* (2009).

53. Adam, *A History of Forensic Science* (2015); Okroi, "Der Blutgruppenforscher Fritz Schiff (1889–1940)" (2004); Spörri, *Reines und Gemischtes Blut* (2013).

54. For accounts of blood grouping in the German courtroom: Okroi and Voswinckel, "'Obviously Impossible'" (2003); Okroi, "Der Blutgruppenforscher Fritz Schiff (1889–1940)"; Spörri, *Reines und Gemischtes Blut* (2013), 41–71.

55. Schiff was celebrated in Germany, but as a German Jew he was forced to emigrate to the United States in 1936. He continued his serological work at Beth Israel Hospital in New York City. In 1942, US immunochemist William Boyd posthumously published Schiff's manual in English: Boyd and Schiff, *Blood Grouping Technic* (1942).

56. Schiff, "The Medico-Legal Significance of Blood Groups" (1929).

57. Lattes noted in the 1930s that legal structures and forms of wording made some countries particularly amenable for the admission of blood group

evidence. German law lent itself particularly well to such tests, while in the United States, lawyers were uncomfortable with the way that the tests were asymmetrical regarding its effects on the genders of the two parties: Lattes, *Individuality of the Blood* (1932).

58. For a captivating account of how this developed in the United States: Rudavsky, *Blood Will Tell* (1996).

59. Lipphardt, "Isolates and Crosses in Human Population Genetics" (2012); Marks, "The Origins of Anthropological Genetics" (2012); Spörri, *Reines und Gemischtes Blut* (2013); Mazumdar, "Blood and Soil" (1990); Turda, "From Craniology to Serology" (2007); Schneider, "The History of Research on Blood Group Genetics" (1996); Schneider, "Introduction to 'The First Genetic Marker' Special Issue" (1996).

60. For more on the vivid recollections of the Hirszfelds during the First World War: Hirszfeld *The Story of One Life* (2010 [1946]).

61. Hirschfeld [Hirszfeld] and Hirschfeld [Hirszfeld], "Serological Differences Between the Blood of Different Races" (1919); Hirszfeld and Hirszfeld, "Essai d'application des méthodes serologiques au problème des races" (1919).

62. Several scholars have analyzed how the Hirszfelds constructed their arresting "biochemical" observations, especially: Schneider, "The History of Research on Blood Group Genetics" (1996); Spörri, *Reines und Gemischtes Blut* (2013); Mukharji, "From Serosocial to Sanguinary Identities" (2014); Gannett and Griesemer, "The ABO Blood Groups" (2004).

63. On such contestations: Zimmerman, *Anthropology and Antihumanism in Imperial Germany* (2001); Roque, *Headhunting and Colonialism* (2010); Burton, *Genetic Crossroads* (forthcoming).

64. On the varied interwar representations of blood group diversity: Gannett and Griesemer "The ABO Blood Groups" (2004); Spörri, *Reines und Gemischtes Blut* (2013).

65. On blood and blood groups in interwar Germany: Mazumdar, "Blood and Soil" (1990); Spörri, *Reines und Gemischtes Blut* (2013); Boaz, *In Search of "Aryan Blood"* (2012).

66. For analysis of the graphical representations in *Zeitschrift für Rassenphysiologie*: Mazumdar, "Blood and Soil" (1990).

67. Reche and Steffen did not represent the tenor of all blood group work in Germany; their claims were resisted by, for example, Jewish serologist Fritz Schiff and physician Lucie Adelsberger (later sacked from her job at the Robert Koch Institute): Spörri, *Reines und Gemischtes Blut* (2013), 106.

68. Turda and Weindling. "Eugenics, Race and Nation" (2006); Turda, "From Craniology to Serology" (2007); Turda, "The Nation as Object" (2007).

69. For a detailed history of craniometry and blood group serology in the Middle East: Burton, *Genetic Crossroads* (forthcoming).

70. Boaz, *In Search of "Aryan Blood"* (2012), 146.

71. Schneider, "Chance and Social Setting" (1983).

72. For example, Armstrong and Matheson, "Blood Groups among Samoans" (1924).
73. On transfusion, "pure" blood, and "bad" blood in Japan: Kim, "The Specter of 'Bad Blood' in Japanese Blood Banks" (2018); on blood and its meanings in Japan: Robertson, "Blood Talks" (2002); Robertson, "Biopower: Blood, Kinship, and Eugenic Marriage" (2005); Robertson, "Eugenics in Japan" (2010). On the incorporation of racial blood grouping into eugenics: Robertson, "Hemato-Nationalism" (2012).
74. For example: Heinbecker and Pauli, "Blood Grouping of the Polar Eskimo," (1927); Coca and Deibert, "A Study of the Occurrence of the Blood Groups among the American Indians" (1923).
75. Krementsov, "Eugenics in Russia and the Soviet Union" (2010); Krementsov, *A Martian Stranded on Earth* (2011).
76. Silverman, "The Blood Group 'Fad'" (2000); Marks, "The Legacy of Serological Studies" (1996); Burton, *Genetic Crossroads* (forthcoming).
77. Prominent eugenicist and anthropologist Eugen Fischer wrote that he had an "instinctive feeling" that blood groups had nothing to do with race, while Hans Günther, author of the popular *Rassenkunde des deutschen Volks* (*Racial Science of the German People*) (1922) and later a leading Nazi authority on race, wrote that it was "a vulgar error to think that one could tell a man's race from his blood." Fischer quoted in Hans-Walter Schmuhl, *The Kaiser Wilhelm Institute for Anthropology, Human Heredity, and Eugenics* (2008), 60; Günther quoted in Mazumdar, "Blood and Soil" (1990), 217.
78. Mazumdar notes that some official race examination forms used in East German border transit camps had a space for blood group, but this was almost always left blank. Mazumdar, "Blood and Soil" (1990), 216.
79. Snyder, "The 'Laws' of Serologic Race-Classification Studies" (1930).
80. Wyman and Boyd, "Human Blood Groups and Anthropology" (1935).
81. Schneider, "The History of Research on Blood Group Genetics" (1996), 277.
82. Palfreeman, *Spain Bleeds* (2015).
83. Rucart, "Séance solenelle d'ouvertue," (1939), 19.

CHAPTER TWO

1. Twenty researchers signed Edwin Smith's appeal in the *Times*: "Racial History of Great Britain (1935).
2. Huxley to Edwin Smith, December 15, 1934, 107/1/6, RAI; Smith requested Huxley's support in a standard letter sent out to many researchers around the country. See also Smith to Lord Raglan, December 11, 1934, 107/1/1, RAI.
3. Both Schneider and Mazumdar agree that there are no clear reasons why British researchers did not become interested in blood groups sooner. The Eugenics Society's *Eugenics Review* published no original articles on blood groups, its card index contained nothing on the topic, and the *Annals*

of Eugenics never mentioned them: Schneider, "Blood Group Research" (1995), 98; Mazumdar, "Blood and Soil" (1990), 187.

4. "Medical Research Council Committee on Human Genetics," minutes of meeting held on February 5, 1932, FD1/3267, NAL.

5. Hogben, Lancelot, "Medical Research Council Human Genetics Committee, Comments on Issues Raised at the First Meeting," March 2, 1932, FD1/3267, NAL.

6. Fisher to O'Brien, July 18, 1934, Rockefeller Foundation Archives, record group 1.1, series 401, box 16, folder 219, RAC.

7. On divergent commitments to eugenics in British academia at this time: Paul, *The Politics of Heredity* (1998), 11–36; Bland and Hall, "Eugenics in Britain" (2010), 213–27; MacKenzie, *Statistics in Britain* (1981).

8. Kevles, *In the Name of Eugenics* (1995), 57–69; Mazumdar, *Eugenics, Human Genetics and Human Failings* (1992), 7–57.

9. Mazumdar, Eugenics, Human Genetics and Human Failings (1992), 40–68.

10. Mazumdar, "Two Models for Human Genetics" (1996).

11. On Fisher's growing frustration on research sponsored by the Eugenics Society: Mazumdar, *Eugenics, Human Genetics and Human Failings* (1992), 94–105.

12. "Editorial: The Scope of *Biometrika*," (1901), 1–2; on the history of the journal: Cox, "*Biometrika*: The First 100 Years" (2001).

13. On the scope of the *Annals of Eugenics*: "Foreword" (1925).

14. On Pearson: MacKenzie, *Statistics in Britain* (1981); Porter, *Karl Pearson* (2004); for more on the relationship between the biometric and Mendelian approaches to heredity: Müller-Wille and Rheinberger, *A Cultural History of Heredity* (2012).

15. Mazumdar, *Eugenics, Human Genetics and Human Failings* (1992), 146–95.

16. The MRC had been established in 1920 with the Ministry of Health Act (1919); it funded medical research using funds generated from the National Insurance Act (1911); for more on its origins: Landsborough Thomson, *Half a Century of Medical Research* (1987).

17. Hurst was friend and ally of Charles Davenport of the Cold Spring Harbor Institute in New York and aligned himself with Davenport's right-leaning eugenics. Hurst to Fletcher, July 14, 1931, FD1/3266, NAL; Hurst to Fletcher, June 15, 1931, FD1/3266, NAL. For more on Davenport and his institution: Kevles, *In the Name of Eugenics* (1995), 41–56.

18. Beveridge was attempting a reform of his own, implementing a new program of "social biology" at LSE in an attempt to make the social sciences more quantitative and "scientific." As part of this program, Beveridge appointed Lancelot Hogben as chair of social biology: Renwick, *British Sociology's Lost Biological Roots* (2012); Renwick, "Completing the Circle of the Social Sciences?" (2014).

19. The LSE meeting initially resulted in an application to the Rockefeller Foundation entitled "The Needs of Research in Human Genetics in Great

Britain—an Appeal to the Rockefeller Foundation of New York." The proposal was apparently drafted by Hogben and contains a large section of Hogben's chapter on blood groups; it is not clear whether it was dropped before or after it was sent to the Rockefeller: Edwards, "Mendelism and Man 1918–1939" (2004); Beveridge to Fletcher, July 8, 1931, FD1/3266, NAL.

20. "Medical Research Council Committee on Human Genetics," minutes of meeting held on February 5, 1932, FD1/3267, NAL.

21. The MN and P blood groups were defined in the late 1920s. The number of human chromosomes was not revised down from twenty-four to twenty-three until the end of the 1950s.

22. On the reception of Bernstein's work among the British geneticists: Mazumdar, *Eugenics, Human Genetics and Human Failings* (1992), 120.

23. Lancelot Hogben, "Confidential: Medical Research Council, Human Genetics Committee, Comments on Issues Raised at the First Meeting," February 1932, FD1/3267, NAL.

24. MRC minutes, "Minutes of the First Meeting of the Committee of Human Genetics," 1932, FD1/3267, NAL.

25. Fisher to Todd, February 5, 1932, FD1/3267, NAL.

26. Lancelot Hogben, "Medical Research Council Human Genetics Committee: Comments on Issues Raised at the First Meeting," March 2, 1932, FD1/3267, NAL.

27. Pearson's preface to the first of these volumes claimed that its pedigrees represented data with no controversial theoretical baggage, which could be used as a resource for researchers whatever their views on Mendelism or biometry: *Treasury of Human Inheritance* (1912), v.

28. Jones, "Bell, Julia (1879–1979)" (2004); Harper, "Julia Bell and the Treasury of Human Inheritance" (2005).

29. Kevles, *In the Name of Eugenics* (1995), 148–163; Mazumdar, *Eugenics, Human Genetics and Human Failings* (1992), 196–255.

30. Penrose, *Mental Defect* (1933).

31. The results of these tests, which Penrose carried out with his wife Margaret, were published not in the context of linkage studies but in a report on the distribution of blood groups in "England," which explained that the "subjects were all drawn from homes in Essex, Suffolk, Norfolk or Cambridgeshire." It noted that "people of Jewish or foreign extraction were excluded." Penrose and Penrose, "The Blood Group Distribution in the Eastern Counties of England" (1933), 160.

32. Werskey, *The Visible College* (1978).

33. Chris Renwick, *British Sociology's Lost Biological Roots* (2012); Renwick, "Completing the Circle of the Social Sciences?" (2014).

34. Mazumdar, *Eugenics, Human Genetics and Human Failings* (1992), 146–95.

35. Hogben, *Genetic Principles* (1931), 69.

36. For biographical accounts of Haldane: Werskey, *The Visible College* (1978); Clark, *J.B.S.* (1984).

37. Fisher's principal biographies: Box, *R. A. Fisher: The Life of a Scientist* (1978); Yates and Mather, "Ronald Aylmer Fisher: 1890–1962" (1963).
38. On Fisher's eugenics: MacKenzie, *Statistics in Britain* (1981).
39. In constructing a genetics of populations, Fisher, Haldane, and US geneticist Sewall Wright built on techniques developed by Pearson, British mathematician H. T. J. Norton, geneticist Reginald Punnett, German gynecologist and mathematician Wilhelm Weinberg, British geneticist G. Harold Hardy, and US geneticist William Castle, among many others. Haldane's early contributions included an extensive series of papers titled "A Mathematical Theory of Natural and Artificial Selection," between 1924 and 1934 in the *Proceedings of the Cambridge Philosophical Society* and *Genetics*. Fisher authored *The Genetical Theory of Natural Selection* (1930), and Wright's early contribution was "Evolution in Mendelian Populations" (1931) in *Genetics*. For a detailed account of the early development of population genetics: Provine, *The Origins of Theoretical Population Genetics* (1971); for Huxley's role in this endeavor, and broader contexts: Sommer, *History Within* (2016); Cain, "Julian Huxley, General Biology and the London Zoo," (2010); Smocovitis, "Unifying Biology" (1992); Cain and Ruse, *Descended from Darwin* (2009).
40. Kevles, *In the Name of Eugenics* (1995), 182–83.
41. Paul, *The Politics of Heredity* (1998), 11–36; Sommer, "Biology as a Technology of Social Justice" (2014).
42. Haldane, "Prehistory in the Light of Genetics" (1931), 361.
43. For more on the shift from race classification to human history within blood group research: Gannett and Griesemer, "The ABO Blood Groups" (2004). With respect to a shift from racial classification to human prehistory, Samuel Redman examines this for collections of skeletal, mummy and fossil human remains: Redman, *Bone Rooms* (2016).
44. Vavilov, "The Problem of the Origin of the World's Agriculture" (1931).
45. For more Vavilov's use of scientific research on biological diversity to recover human history: Aronova, *The Missing Link* (2019); on the broader Soviet field of "genogeography": Bauer, "Virtual Geographies of Belonging" (2014).
46. Haldane, "Prehistory in the Light of Genetics" (1931), 370. Haldane republished his argument in several journals, including the *BMJ* and the journal *Man*: Haldane, "The Blood Groups in Genetics and Anthropology" (1932), 163; Haldane, "Anthropology and Human Biology" (1934). As well as making general claims for the use of blood groups in understanding history, Haldane would soon deploy them in more specific mathematical studies of human population genetics: Haldane and Boyd, "The Blood Group Frequencies of European Peoples, and Racial Origins" (1940); Haldane, "Selection against Heterozygosis in Man" (1941).
47. Haldane and Boyd, "The Blood Group Frequencies of European Peoples, and Racial Origins" (1940), 477. Haldane developed this work with William Boyd, who was dedicated to collecting together worldwide data on the human blood groups. Based on Boyd's data, Haldane developed a

genetic hypothesis for the settlement of Europe: Burton, *Genetic Crossroads* (forthcoming).

48. Royal Anthropological Institute and Institute of Sociology, *Race and Culture* (1936), 1.

49. This question also represented a step in the growing demarcation of "social" from "physical" anthropology.

50. Thackeray, "Leakey, Louis Seymour Bazett (1903–1972)" (2004); Barkan, *The Retreat of Scientific Racism* (1992), 279–340. So discordant were the committee's views that when the report was published two years later it was regarded by many as a "failure" that "satisfied virtually nobody": Crook, *Grafton Elliot Smith* (2012).

51. On the disputes between the personalities involved: Kushner, *We Europeans?* (2004).

52. Royal Anthropological Institute and Institute of Sociology, *Race and Culture* (1936), 2.

53. In the 1920s, universities had begun to endow university positions dedicated to the study of "physical anthropology," which tended to be located in anatomy departments. By then, in Oxford, Leonard Dudley Buxton was demonstrator in physical anthropology in the department of anatomy; in Cambridge, Haddon was lecturer in physical anthropology and William Duckworth reader in human anatomy; and at the University of Aberdeen, the professor of anatomy presided over the Anthropometric Laboratory and Anthropology Museum.

54. Like colleagues elsewhere in Britain and Europe, Pearson commissioned colonial officials and missionaries to collect skulls from distant regions of the world. For more on the collection and circulation of skulls: Roque, *Headhunting and Colonialism* (2010); for more on the anthropometry of Pearson and his associates: Clever, "The Lives and Afterlives of Skulls" (2020).

55. Renwick, *British Sociology's Lost Biological Roots* (2012); Porter, *Karl Pearson* (2004); MacKenzie, *Statistics in Britain* (1981).

56. Stone, "Race in British Eugenics" (2001); Schaffer, "'Like a Baby with a Box of Matches'" (2005); Schaffer, *Racial Science and British Society* (2008).

57. Some of the richest accounts of the politics and practices of race science in interwar Britain are by geographers and historians of geography: Linehan, "Regional Survey and the Economic Geographies of Britain" (2003); Matless, "Regional Surveys and Local Knowledges" (1992); Gruffudd, "Back to the Land" (1994); Evans, "Le Play House and the Regional Survey Movement" (1986).

58. Exceptions to this generalization about Pearson include his development of the "type silhouette" to establish methods for studying the head shape of living people: McLearn et al., "On the Importance of the Type Silhouette" (1928).

59. Fleure was also strongly committed to the public teaching of geography, both to trainee teachers and to the rural people of Wales. And as well as many academic articles on the archaeology of primitive societies, Fleure and his colleague Harold Peake coauthored a nine-volume series for broader audiences of archaeology: *The Corridors of Time*, published between 1927 and 1936.

60. Fleure and James, "Geographical Distribution of Anthropological Types in Wales" (1916); Davies and Fleure, "A Report on an Anthropometric Survey of the Isle of Man" (1936).

61. For recent accounts of Fleure's life and work: Gruffudd, "Back to the Land" (1994); Winlow, *Cartographic Representations of Race* (1999); Winlow, "Anthropometric Cartography" (2001); Brittain, "World War I and the Contribution of Herbert Fleure and Harold Peake" (2008); Brittain, "Herbert Fleure and the League of Nations' (1919) Minorities Treaties" (2009); Rees, "Doing 'Deep Big History'" (2019).

62. On the use of Regional Survey to position the discipline of geography as central to a broader discourse about modernity, urbanization, civilization, and degeneration: Evans, "Le Play House and the Regional Survey Movement" (1986); Matless, "Regional Surveys and Local Knowledges" (1992); Roxby, "The Conference on Regional Survey at Newbury" (1917).

63. Geographer Marion Newbiggin at the British Association 1922, quoted in Bell, "Reshaping Boundaries" (1998), 160.

64. Winlow, *Cartographic Representations of Race* (1999); Brittain, "Herbert Fleure and the League of Nations' (1919) Minorities Treaties" (2009).

65. Fleure, "Some Aspects of Race Study" (1922); Gruffudd, "Back to the Land" (1994).

66. Fleure remarked that it was a commonplace "among Welshmen that one can 'tell' a man from such and such a district anywhere, and such remarks are true in a greater degree in Wales than in most parts of England." Fleure and James, "Geographical Distribution of Anthropological Types in Wales" (1916), 41.

67. Sommer, "Biology as a Technology of Social Justice" (2014).

68. Huxley et al., *We Europeans* (1935), 3. Tony Kushner writes extensively about the objectives and contributions of each of the book's authors. The authors also included Charles Seligman and Charles Singer, who were not named apparently owing to their "obviously" Jewish provenance, which Huxley feared would give the appearance of being biased. Kushner, *We Europeans?* (2004), 49.

69. Huxley et al., *We Europeans* (1935), 7 and 125.

70. On the construction of British and European identities in the book: Kushner, *We Europeans?* (2004). On colonial construction of "the European," see the special issue by Middell, "The Invention of the European" (2015), especially: Lipphardt, "'Europeans' and 'Whites'" (2015); Lipphardt, "Knowing Europe, Europeanizing Knowledge" (2015). For a special issue on the

shifting and varied contexts in which race and identity have been made in Europe: M'charek et al., eds., "Technologies of Belonging" (2014).

71. *We Europeans* was one of several expressions of a 1930s reformist argument that biology—and specifically genetics and evolution—had lessons for human social development. Huxley, Haldane, and Hogben used radio talks, documentary films, popular magazines, pamphlets, and books to push their argument that genetic diversity demanded democratic reforms and greater planning. For detailed accounts of Huxley's arguments in *We Europeans* and their relationship to his scientific humanism: Sommer, "Biology as a Technology of Social Justice" (2014); Sommer, *History Within* (2016).

72. As the *Manchester Guardian* reported, others at the meeting presented opposing views: Pearson's colleague Geoffrey Morant "questioned" Huxley's Mendelian genetics, while Ruggles Gates argued that humankind should be classified into several distinct species. "Unscientific Race Theory: Nazi Conception under Fire," *Manchester Guardian*, September 12, 1936, 19.

73. For the longer trajectory of Fisher's career, especially his statistical work: Parolini, *"Making Sense of Figures"* (2013); Parolini, "The Emergence of Modern Statistics in Agricultural Science" (2014).

74. Box, *R. A. Fisher: The Life of a Scientist* (1978), 344–45.

75. Fisher's pitch to O'Brien was summarized by the latter in a memorandum to a colleague: O'Brien to Gregg, March 1, 1935, Rockefeller Foundation Archives, record group 1.1, series 401, box 16, folder 220, RAC.

76. Testifying to the shared enthusiasm of the two institutions, the Rockefeller Foundation had recently awarded a grant to the MRC to study the inheritance of "mental defects and disorders" among "offspring of consanguineous marriages." Enclosed with: O'Brien to Gregg, March 1, 1935, 5, Rockefeller Foundation Archives, record group 1.1, series 401A, box 16, RAC.

77. Tisedale to Weaver, March 5, 1935, Rockefeller Foundation Archives, record group 1.1, series 401, box 16, folder 220, RAC.

78. For MRC papers on the Galton Serological Laboratory: FD1/3290, NAL.

79. Taylor had taken up this university position after giving up general practice. In that role, Taylor carried out immunological research on the "precipitin reaction," which formed the basis of quantitative assays for determining the titer (concentration) of antibody, and which would be crucially useful for his work with the Galton Serological Laboratory. Wiener, "George Lees Taylor" (1945).

80. Box, *R. A. Fisher: The Life of a Scientist* (1978), 345; Clarke, "Robert Russell Race" (1985).

81. Yates and Mather, "Ronald Aylmer Fisher: 1890–1962" (1963).

82. Alongside blood groups, Fisher investigated other human traits that he thought would have a simple genetic basis and that might therefore be used for mapping, such as earlobe attachment, hair on the second joint of fingers, and the ability to taste phenylthiocarbamide: Box, *R. A. Fisher: The Life of a Scientist* (1978), 261. Fisher was carrying out serological work

alongside numerous other projects, including a substantial program of research on the mathematics of inheritance in relation to agricultural yields, serological studies on fowl, and selection experiments on mice. For a contemporary account of Fisher's activities by a Rockefeller observer: "Official Diary, London," May 16, 1935, Rockefeller Foundation Archives, record group 1.1, series 401A, box 16, RAC.

83. Fisher to the MRC, 1935, quoted in Box, *R. A. Fisher: The Life of a Scientist* (1978), 347.

84. Box, *R. A. Fisher: The Life of a Scientist* (1978), 347.

85. Hogben and Pollack, "A Contribution to the Relation of the Gene Loci" (1935); Mazumdar, *Eugenics, Human Genetics and Human Failings* (1992), 175.

86. For examples of such collaborations: Riddell, "A Pedigree of Blue Sclerotics, Brittle Bones, and Deafness, with Colour Blindness" (1940); Mutch, "Hereditary Corneal Dystrophy" (1940); for a brief summary of these activities: Taylor to MRC, July 28, 1939, FD1/3290, NAL.

87. Taylor and Prior, "Blood Groups in England, I" (1938); Taylor and Prior, "Blood Groups in England, II" (1938); Taylor and Prior, "Blood Groups in England, III" (1939).

88. Fisher to Pearson, May 7, 1935, "Correspondence with Karl Pearson," R. A. Fisher Papers, Barr Smith Library, University of Adelaide, MSS 0013/Series 1, http://hdl.handle.net/2440/67912; also quoted in Box, *R. A. Fisher: The Life of a Scientist* (1978), 346.

89. Craniometry did not disappear completely: Iris Clever notes that the material durability of the skulls persisted, with researchers such as Harvard anthropologist William Howells later using computers to analyze prewar cranial data for patterns of human evolution: Clever, "The Lives and Afterlives of Skulls" (unpublished manuscript).

90. Pearson declared this in response to Edwin Smith's request that he support for the proposed racial survey of Britain. Geoffrey Morant replied on Pearson's behalf: Morant to Smith, January 2, 1935, 107/1/10, RAI.

91. Fisher made these statements in the context of a more general attack on biometry: Fisher, "'The Coefficient of Racial Likeness'" (1936), 63.

CHAPTER THREE

1. Thomson to Fisher, September 26, 1939, FD1/3290, NAL; Fisher to Thomson, September 26, 1939, FD1/3290, NAL.

2. Fisher and Taylor, "Blood Groups in Great Britain" (1939).

3. Star and Ruhleder, "Steps toward an Ecology of Infrastructure" (1996); Creager and Landecker, "Technical Matters" (2009). Rosanna Dent has elaborated the concept of "social infrastructures" that support the collection of scientific data: Dent, "Kinship and Care" (2017).

4. For more on medicine and modernity during this period: Cooter et al., *War, Medicine and Modernity* (1998); more specifically, on the quantitative

and numerical language imported (with blood-storage technologies) from Spain to the EBTS: Whitfield, "A Genealogy of the Gift" (2011), 112–13. Britain was not the only transfusion system reconfigured by the Second World War. Blood banking systems (many based on freely donated blood, some procuring blood in exchange for money) could now be found in the United States and Canada (where the Red Cross ran national programs), as well as Japan, India, and Finland, among others. In Germany, transfusion had been relatively little used among civilians, in part because of an emphasis on whole, fresh blood (and because of restrictions on non-Aryan donors). Spörri, *Reines und Gemischtes Blut* (2013); Pierce and Reid, *Bloody Brilliant!* (2016), 201–20; Kendrick, *Blood Program in World War II* (1964).

5. For Whitfield's argument on how these values were articulated to redress the intrinsically alienating nature of these wartime transfusion technologies: Whitfield, "A Genealogy of the Gift" (2011).

6. On histories of state documentation practices: Caplan and Torpey, *Documenting Individual Identity* (2001); for a longer and global history of registering and documenting people: Breckenridge and Szreter, *Registration and Recognition* (2012). On the color coding of donor cards by the EBTS and its prior use by London's Red Cross Blood Transfusion Service: Giuditta Parolini, *"Making Sense of Figures"* (2013), 150–202.

7. Throughout this chapter I use Whitfield's accounts of the remarkable change in technologies, organization, rhetoric, and propaganda of transfusion as the London Red Cross Blood Transfusion Service was replaced by the EBTS: Whitfield, "A Genealogy of the Gift" (2011).

8. Vaughan was also assistant clinical pathologist at the British Postgraduate Medical School in London: Owen, "Dame Janet Maria Vaughan" (1995).

9. Quotations from her vivid and compelling, unpublished autobiography: Vaughan, "Jogging Along; Or, A Doctor Looks Back" (unpublished manuscript), SC/PO/PP/JV/17, Somerville College Library Archives; for more on Vaughan, her social and family position, her involvement in Spanish Republican blood transfusion, and the Spanish Civil War as a general exemplar for the British Emergency Medical Service: Owen, "Dame Janet Maria Vaughan" (1995); Whitfield, "A Genealogy of the Gift" (2011), 90–132 and references therein; Buchanan, *Britain and the Spanish Civil War* (1997).

10. Vaughan, "Janet Vaughan Draft Manuscript," undated, GC/186/1, WCL.

11. Palfreeman, *Spain Bleeds* (2015).

12. Vaughan, "Jogging Along; Or, A Doctor Looks Back" (unpublished manuscript), SC/PO/PP/JV/17, Somerville College Library Archives.

13. Vaughan, "Janet Vaughan Draft Manuscript," undated, 82, GC/186/1, WCL.

14. On Vaughan's meetings to plan the transfusion service: Whitfield, "A Genealogy of the Gift" (2011), 90–132.

15. "Doctors in Time of War" (1939); Dunn, *The Emergency Medical Services* (1952).

16. Many universities already maintained laboratories for routine public health analysis, and some private pathological laboratories carried out epidemic monitoring alongside routine testing. Now, owing to the special risk of aerial bombardment in London, the MRC assigned the Emergency Public Health Laboratory Service headquarters to Oxford, with regional centers in Cambridge and Cardiff, and multiple, smaller, constituent laboratories in other parts of the country: Wilson, "The Public Health Laboratory Services" (1948).
17. An early announcement of the new service: "Blood Transfusion Service for War" (1939).
18. "Meeting of Depot Officers," April 11, 1939, GC/186/1, WCL.
19. "Sub-Committee Blood Transfusion Emergency Service," May 4, 1939. GC/107, WCL; "Sub-Committee Blood Transfusion Emergency Service," May 12, 1939, GC/107, WCL; "Emergency Blood Transfusion Service," June 15 1939, GC/107, WCL; Whitfield, "A Genealogy of the Gift" (2011); Vaughan, "Janet Vaughan Draft Manuscript," undated, 86, GC/186/1, WCL.
20. Vaughan, "Janet Vaughan Draft Manuscript," undated, 86, GC/186/1, WCL.
21. Vaughan and Panton, "The Civilian Blood Transfusion Service" (1952).
22. On the organization of the NBTS: Proger, "Development of the Emergency Blood Transfusion Scheme" (1942); on the geographical locations of depots and their services: Vaughan and Panton, "The Civilian Blood Transfusion Service" (1952), 334–55; On the Blood Transfusion Research Committee: Kekwick, "Alan Nigel Drury" (1981).
23. For EBTS planning, see papers in GC/186/1, WCL.
24. Vaughan, "Janet Vaughan Draft Manuscript," undated, 82, GC/186/1, WCL; on the qualities expected in laboratory technicians and the gendering of laboratory roles: Casper and Clarke, "Making the Pap Smear into the 'Right Tool' for the Job" (1998).
25. For more on the construction of wartime blood transfusion propaganda: Whitfield, "A Genealogy of the Gift" (2011).
26. Whitfield, "A Genealogy of the Gift" (2011). There were no selection criteria, although particular groups were targeted for their accessibility and potential enthusiasm; e.g., when the committee estimated that they needed 9,000 active donors, they reckoned the most reliable source would be "women working in some organization such as a factory." Infection was a concern, though it was fairly limited; the Wassermann test would be administered at first donation, and the pooling of blood was ruled out to mitigate against increasing infection risk: "Confidential: Emergency Blood Transfusion Service Scheme: Minutes of Meeting Held on 5th April 1939," GC/186/1, WCL
27. "Life Donors," *Times*, July 11, 1939, 11.
28. "News in Brief," *Times*, August 12, 1939, 7.

29. During the Munich crisis, Taylor had written to secretary of the MRC, Sir Edward Mellanby, "If we can be of any use in collecting, preparing and assaying sera or in any other way we shall only be too pleased to help." Taylor to Mellanby, September 28, 1938, FD1/5845, NAL.

30. Thomson to Taylor, December 13, 1938, FD1/5845, NAL; for much more on the London Red Cross Transfusion Service and its director's interest in war work: Whitfield, "A Genealogy of the Gift" (2011).

31. Thomson, "Internal Memo," February 7, 1939, FD1/5845, NAL.

32. Thomson to Brewer, December 5, 1938, FD 1/5845, NAL; Thomson, "Confidential: Emergency Medical Services," July 25, 1939, Rockefeller Foundation Archives, record group 1.1, series 401, box 16, folder 221, RAC.

33. Thomson to Brock, April 4, 1939, FD1/5845, NAL.

34. For the question of consent in this case, see Brock to Thomson, April 11, 1939, FD1/5845, NAL.

35. For discussions about the supply of blood by MRC staff: "Memorandum to All Members of the Staff at the Institute and Farm," June 1939, FD1/5845, NAL; Taylor to Thomson, July 11, 1939, FD1/5845, NAL.

36. Taylor to Thomson, September 26, 1939, FD1/3290, NAL; Topley to Thomson, November 8, 1938, FD1/5845, NAL.

37. Alluding, probably, to the importance of blood transfusion on the battlefield, Taylor added: "We think besides that our testing the groups of what at the present time is certainly the most important group of young men in the country is a thing worth doing for its own sake." Box, *R. A. Fisher: The Life of a Scientist* (1978), 353.

38. Box, *R. A. Fisher: The Life of a Scientist* (1978), 352–53.

39. Taylor, "Non-specific Agglutination Reactions [Instructions Sent with Grouping Serum]," FD1/5845, NAL.

40. Taylor considered all of these part of the unit's "official" duties: Taylor to Thomson, November 15, 1943, FD1/5845, NAL.

41. Taylor to Thomson, November 15, 1943, FD1/5845, NAL.

42. Vaughan, "Janet Vaughan Draft Manuscript," undated, 84–85, GC/186/1, WCL.

43. Vaughan, "Medical Research Council Emergency Blood Transfusion Service, for London and the Home Counties," 1940, folder 13-1-C: Blood Donation, topic collection 13: Health 1939–47, Mass Observation Archive, University of Sussex Special Collections.

44. Vaughan, "Medical Research Council Emergency Blood Transfusion Service, for London and the Home Counties," 1940, 2, folder 13-1-C: Blood Donation, topic collection 13: Health 1939–47, Mass Observation Archive, University of Sussex Special Collections.

45. Vaughan, "Janet Vaughan Draft Manuscript," undated, 84–85, GC/186/1, WCL.

46. Fisher to O'Brien, December 28, 1939, Rockefeller Foundation Archives, record group 6.1, series 1.1, box 2, folder 17, RAC; Fisher, "To the Editor of The Times: London University; Plight of the Galton Laboratory" (1939).

47. Box, *R. A. Fisher: Life of a Scientist* (1978), 273.
48. For more on Fisher's time at Rothamsted: Parolini, *"Making Sense of Figures"* (2013).
49. Box, *R. A. Fisher: The Life of a Scientist* (1978), 350 and 375.
50. Fisher to O'Brien, December 28, 1939, Rockefeller Foundation Archives, record group 6.1, series 1.1, box 2, folder 17, RAC.
51. E.g., Fisher to Copland, September 13, 1939, "Blood Group Survey," R. A. Fisher Papers, Barr Smith Library, University of Adelaide, MSS 0013/Series 1, http://hdl.handle.net/2440/67586.
52. To save time and resources, Vaughan's planning committee, and later the MRC, had decided that it would only test for syphilis those people selected to give blood for transfusion: Whitfield, "A Genealogy of the Gift" (2011), 122.
53. "Message from an Anonymous Blood Donor," July 24, 1940, folder 13-1-C: Blood Donation, topic collection 13: Health 1939–47, Mass Observation Archive, University of Sussex Special Collections, http://www.massobserva tion.amdigital.co.uk.ezp.lib.cam.ac.uk/Documents/Images/TopicCollection -13/86, accessed February 21, 2020.
54. Vaughan, "Medical Research Council Emergency Blood Transfusion Service," 1940, images 90–91, folder 13-1-C: Blood Donation, topic collection 13: Health 1939–47, Mass Observation Archive, University of Sussex Special Collections. http://www.massobservation.amdigital.co.uk.ezp.lib .cam.ac.uk/Documents/Images/TopicCollection-13/86, accessed 21 February 2020.
55. "A Blood Transfusion Depot at Work" (1939).
56. Box, *R. A. Fisher: The Life of a Scientist* (1978), 351.
57. "From many places the material sent to Rothamsted consists of enrolment forms which are there sorted according to the four groups and sex, and counted." Taylor to Thomson, July 3, 1941, FD1/5845, NAL.
58. Fisher to Thomson, July 11, 1941, FD1/5845, NAL. On the range of other work that Fisher and his colleagues continued in their squeezed Rothamsted quarters: Box, *R. A. Fisher: The Life of a Scientist* (1978), 374–77.
59. Memo from Fisher enclosed with letter: O'Brien to Gregg, January 27, 1942, record group 1.1, series 401, box 16, folder 222, RAC.
60. In a large population carrying the alternative alleles A and a, and where the frequency of allele A is p and the frequency of allele a is q, one generation of random mating would put the genotypes AA, Aa, and aa at frequencies p^2, $2pq$, and q^2, respectively. An expanded version of this equation accounted for the three alleles of the ABO blood groups.
61. Fisher to Thomson, July 16, 1941, FD1/5845, NAL.
62. Haldane, *New Paths in Genetics* (1941), 194.
63. These surveys, the authors wrote, were "representative of the English population": "Persons with Scottish, Irish and Welsh names and, in two cases, with names suggestive of a Continental origin were included." Revealing

the prior judgments made of research subjects, they added, "There were two obviously Jewish families." Quotations from Taylor and Prior, "Blood Groups in England, I" (1938); for the remainder of the survey results: Taylor and Prior, "Blood Groups in England, II" (1938); Taylor and Prior, "Blood Groups in England, III" (1939); Ikin et al., "The Distribution of the A_1A_2BO Blood Groups in England" (1939).

64. Fisher to Fraser Roberts, February 17, 1940, "Correspondence with J. A. Fraser Roberts," R. A. Fisher Papers, Barr Smith Library, University of Adelaide, MSS 0013/Series 1, http://hdl.handle.net/2440/67946. For details of the Regional Blood Transfusion Service, established officially in July 1940, see Vaughan and Panton, "The Civilian Blood Transfusion Service" (1952).

65. Quotation from Box, *R. A. Fisher: The Life of a Scientist* (1978), 353.

66. Fisher and Taylor, "Blood Groups in Great Britain" (1939). The idea of using the blood grouping records for this purpose had first been suggested in print by a Dr. Edward Billing, who (for reasons I have been unable to recover) noted in the *BMJ* that "ethnological deductions" might be made from collecting the blood groups of transfusion volunteers: Billing, "Racial Origins from Blood Groupings" (1939).

67. Fisher to Taylor, November 15, 1939, "Correspondence with George Lees Taylor," R. A. Fisher Papers, Barr Smith Library, University of Adelaide, MSS 0013/Series 1, http://hdl.handle.net/2440/68047.

68. Schneider discusses the international status of the journal in "The History of Research on Blood Group Genetics" (1996). For more on the *Zeitschrift für Rassenphysiologie*, the Deutsche Gesellschaft für Blutgruppenforschung, and their relationship to German eugenics: Boaz, *In Search of "Aryan Blood"* (2012); Mazumdar, "Blood and Soil" (1990).

69. Fisher and Vaughan, "Surnames and Blood-Groups" (1939).

70. Vaughan, "Jogging Along; Or, A Doctor Looks Back" (unpublished manuscript), SC/PO/PP/JV/17, Somerville College Library Archives.

71. Fisher and Taylor, "Scandinavian Influence in Scottish Ethnology" (1940). For accounts of the Vikings in Britain, contemporary to Fisher and Taylor: Brøgger, *Ancient Emigrants* (1929); Kendrick, *A History of the Vikings* (1930).

72. Fisher to Vaughan, March 11, 1940, "Correspondence with Blood Group Survey," R. A. Fisher Papers, Barr Smith Library, University of Adelaide, MSS 0013/Series 1, http://hdl.handle.net/2440/67586.

73. The chi-squared test is a statistical method for estimating the chances that observed and expected proportions deviate significantly from one another.

74. Fisher was not the first to suggest this—Leone Lattes had noted in his medico-legal textbook that knowledge about genetic inheritance and data in families and populations could be used to check the efficacy of grouping technique: Lattes, *Individuality of the Blood* (1932), 98–173.

75. Fisher to Taylor, January 17, 1942, "Correspondence with George Lees Taylor," R. A. Fisher Papers, Barr Smith Library, University of Adelaide, MSS 0013/Series 1, http://hdl.handle.net/2440/68047.

76. Fisher to Thomson, July 23, 1941, FD1/5845, NAL.
77. Parolini, *"Making Sense of Figures"* (2013), 150–205.
78. Fisher to Taylor, November 3, 1939, "Correspondence with George Lees Taylor," R. A. Fisher Papers, Barr Smith Library, University of Adelaide, MSS 0013/Series 1, http://hdl.handle.net/2440/68047.
79. Fisher to Taylor, November 11, 1939. SA/BGU/F.1/1/1, WCL.
80. Fisher to Thomson, July 23, 1941, FD1/5845, NAL. Highlighting the dangers of not explicitly maintaining the rhetorical alignment of war work and research, Fisher's clarification came too late, and University College withdrew its support for Simpson, although in the end Fisher pulled together money from other sources to keep her on: Fisher to Taylor, September 25, 1941, "Correspondence with George Lees Taylor," R. A. Fisher Papers, Barr Smith Library, University of Adelaide, MSS 0013/Series 1, http://hdl.handle.net/2440/68047.
81. In the 1930s and 1940s, neither statistics nor genetics were part of medical curricula. Even early-1940s textbooks of medical genetics by Fisher's close friends E. B. Ford and John Fraser Roberts had little to say on how the Hardy-Weinberg protocol might be deployed: Ford, *Genetics for Medical Students* (1942); Roberts, *An Introduction to Medical Genetics* (1940). Not until 1960 was the Hardy-Weinberg equation mentioned in the *BMJ* or *Lancet*. For more on statistics in scientific disciplines, and on "medical opposition to statistics" during the 1940s and 1950s: Porter, *Trust in Numbers* (1995); Magnello and Hardy, *The Road to Medical Statistics* (2002).
82. Fisher to Taylor, November 3, 1939, "Correspondence with George Lees Taylor," R. A. Fisher Papers, Barr Smith Library, University of Adelaide, MSS 0013/Series 1, http://hdl.handle.net/2440/68047; Fisher to Taylor, November 11, 1939, SA/BGU/F.1/1/1, WCL.
83. Fisher to Taylor, November 15, 1939, "Correspondence with George Lees Taylor," R. A. Fisher Papers, Barr Smith Library, University of Adelaide, MSS 0013/Series 1, http://hdl.handle.net/2440/68047.
84. On instructions for how to interpret agglutination reactions in tubes: Boyd and Schiff, *Blood Grouping Technic* (1942).
85. Taylor to Drury, December 29, 1941, FD1/5845, NAL; Taylor et al., "A Reliable Technique for the Diagnosis of the ABO Blood Groups" (1942).
86. Drummond, "Blood Grouping in Tubes" (1943), 118.
87. The memorandum was finally published as Medical Research Council Blood Transfusion Research Committee, "The Determination of Blood Groups" (1943).
88. Drury had trained as a physiologist and qualified as a doctor at St Thomas's Hospital, London: Kekwick, "Alan Nigel Drury" (1981).
89. Drury to Taylor, November 18, 1941, FD1/5901, NAL.
90. Vaughan et al., "Draft Majority and Minority Report of the Sub-Committee of the Blood Transfusion Research Committee," FD1/5901, NAL.
91. Vaughan to Drury, February 11, 1943, FD1/5901, NAL.

92. Whitby, "Criticisms on Blood Group Memorandum," FD1/5901, NAL.
93. Vaughan to Drury, February 11, 1943, FD1/5901, NAL.
94. Hull, *Government of Paper* (2014).
95. Latour, *Science in Action* (1987).

CHAPTER FOUR

1. The researchers used the term "Rhesus factor" to describe the first rhesus antibody and "Rh blood groups" to refer to the antigens, so I use the term "Rh" for the blood groups throughout.
2. On the virtues of studying how knowledge travels: Secord, "Knowledge in Transit," (2004); on the practical uses of blood group nomenclatures: Bangham, "Writing, Printing, Speaking" (2015).
3. On this phase of Landsteiner's career: Mazumdar, *Species and Specificity* (1995), 327–78.
4. Landsteiner and his colleagues thanked for their assistance the superintendents of zoos in New York, Philadelphia, and St Louis: Landsteiner and Miller, "Serological Studies on the Blood of the Primates: I" (1925), 852.
5. On Landsteiner's role as special medical consultant: Starr, *Blood: An Epic History* (2009), 69.
6. Landsteiner and Wiener, "An Agglutinable Factor in Human Blood Recognised by Immune Sera for Rhesus Blood" (1940).
7. Landsteiner and Wiener, "Studies on an Agglutinogen (Rh) in Human Blood" (1941), 309–10.
8. From the very first reports of their discovery, Landsteiner and Wiener disclosed preliminary data on the inheritance of the Rh groups, as well as the groups' varying frequencies in what the researchers called "white" and "negro" populations: Landsteiner and Wiener, "Studies on an Agglutinogen (Rh) in Human Blood" (1941). For Landsteiner, this aspect of the study presumably cohered with a longer-term taxonomic project that included the comparison of "white" and "negro" blood (though he admitted that they had been "unable to demonstrate any characteristic difference"): Landsteiner and Miller, "Serological Studies on the Blood of the Primates: I" (1925), 852.
9. Giblett, "Philip Levine, 1900–1987" (1994).
10. Levine had isolated the antiserum responsible for the condition prior to Landsteiner's and Wiener's Rh discovery but now connected the findings: Levine et al., "The Role of Isoimmunization in the Pathogenesis of Erythroblastosis Fetalis" (1941).
11. Taylor to Fisher, January 3, 1942, R. A. Fisher Papers, Barr Smith Library, University of Adelaide, MSS 0013/Series 1; Taylor to Fisher, January 12, 1942, R. A. Fisher Papers, Barr Smith Library, University of Adelaide, MSS

0013/Series 1; Fisher to Taylor, January 6, 1942, "Correspondence with George Lees Taylor," R. A. Fisher Papers, Barr Smith Library, University of Adelaide, MSS 0013/Series 1, http://hdl.handle.net/2440/68047; Box, *R. A. Fisher: The Life of a Scientist* (1978), 357. Both Race and Prior held licenses to perform experiments on animals: Home Office to Provost of University College, January 2, 1942, FD1/3290, NAL. The monkeys were likely kept in the Cambridge Department of Physiology animal house, a five-minute walk from the Galton Serum Unit.

12. Box, *R. A. Fisher: The Life of a Scientist* (1978), 357.
13. Geneticists at this time used both the terms "gene" and "allelomorph" to refer to variants of a gene, and the term "allelomorph" both for "character" variants and gene variants; so to avoid confusion here I use the admittedly anachronistic term "allele" to mean "gene variant." For contemporary usage: Knight, *Dictionary of Genetics* (1948).
14. For Fisher's own account of this episode: Fisher, "The Rhesus Factor" (1947).
15. Race, "An 'Incomplete' Antibody in Human Serum" (1944); Edwards, "R. A. Fisher's 1943 Unravelling of the Rhesus Blood Group System" (2007).
16. Race, "An 'Incomplete' Antibody in Human Serum" (1944), 772.
17. Mazumdar, *Species and Specificity* (1995), 337–78.
18. Wiener, quoted in Mazumdar, *Species and Specificity* (1995), 366.
19. For Fisher the correspondence between antigen and allele was so direct that to all intents and purposes he saw the antisera as interacting directly with alleles. He explained, for example, that "blood group genotypes" could be "recognised by a test fluid": Fisher, "The Rhesus Factor" (1947), 1; Mazumdar, *Species and Specificity* (1995), 366.
20. Mazumdar, *Species and Specificity* (1995), 305–36.
21. On the algebraic qualities of chemical nomenclatures: Klein, *Experiments, Models, Paper Tools* (2003). Early geneticists themselves noted the analogous functions of chemical symbols and genetic symbols, for example: Carlson, *The Gene* (1966), 29. On the graphic suggestiveness (Klein's term) of Feynman diagrams in physics: Kaiser, "Stick-Figure Realism" (2000).
22. Race et al., "Serological Reactions Caused by the Rare Human Gene Rh_z" (1945).
23. Mollison and Taylor, "Wanted: Anti-Rh Sera" (1942).
24. For Ministry of Health records on the coordinated recruitment of doctors and Rh patients for research: "Blood Transfusion and Rh Factor [Medical Research Council]," 1942–47, FD1/5957, NAL.
25. Mourant, *Blood and Stones* (1995), 57–58.
26. The first example: Haldane, "Two New Allelomorphs for Heterostylism in Primula" (1933). For a captivating exploration of the relationship between research on the complex Rh locus and the genetics of butterfly color patterns: Zallen, "From Butterflies to Blood" (1997).

27. Fisher to Cappell, 29 September 1944, "Correspondence with D. F. Cappell," R. A. Fisher Papers, Barr Smith Library, University of Adelaide, MSS 0013/Series 1, http://hdl.handle.net/2440/67607.
28. Race et al., "The Rh Factor and Erythroblastosis Foetalis" (1943), 4313.
29. Race et al., "The Rh Factor and Erythroblastosis Foetalis" (1943).
30. Hoare, "Occurrence of the Rh Antigen in the Population" (1943).
31. Examples from Wales, Manchester, and Glasgow: Plaut et al., "The Results of Routine Investigation for Rh Factor at the N.W. London Depot" (1945); Stratton et al., "Haemolytic Disease of the Newborn in one of Dizygotic Twins" (1945); Race et al., "The Rh Factor and Erythroblastosis Foetalis" (1943).
32. Whitby, "The Hazards of Transfusion" (1942).
33. Plaut et al., "The Results of Routine Investigation for Rh Factor at the N.W. London Depot" (1945).
34. Vaughan and Panton, "The Civilian Blood Transfusion Service" (1952), 344.
35. The New York Academy of Sciences met in 1946 to discuss the problem; in 1947 the European Committee on Biological Standardization created an "Expert Subcommittee on Rh antigens"; and in 1948 the World Health Organization held a meeting on the topic in Geneva.
36. Mazumdar, *Species and Specificity* (1995); Edwards, "R. A. Fisher's 1943 Unravelling of the Rhesus Blood Group System" (2007).
37. Pierce and Reid, *Bloody Brilliant!* (2016), 345.
38. Cappell, "The Blood Group Rh, Part I" (1946).
39. Cappell, "The Blood Group Rh, Part I" (1946), 604.
40. E.g., Coombs, "Detection of Weak and 'Incomplete' Rh Agglutinins" (1945); Race, "A Summary of Present Knowledge of Human Blood Groups" (1946); Race et al., "Rh Antigens and Antibodies in Man" (1946).
41. Murray, "A Nomenclature of Subgroups of the Rh Factor" (1944).
42. Mollison et al., "The Rh Blood Groups and Their Clinical Effects" (1948).
43. Race and Sanger, *Blood Groups in Man* (1950), 113, 172–73.
44. The "genetic map distance" is an index of the relationships between genes on a chromosome based on recombination frequencies. In fact, Race's calculations are for the "cross-over frequencies" between the genes—the frequency with which new allele combinations are made through recombination.
45. "Blood Hunt Goes On All Night," *Daily Express*, 15662, August 29, 1950, 1.
46. Castle et al., "On the Nomenclature of the Anti-Rh Typing Serums" (1948), 30.
47. Beckman to Mourant, November 5, 1958, PP/AEM/K.4, WCL.
48. However, US geneticist Herluf Strandskov agreed in the *Journal of Heredity* that the Wiener terminology was not easy to present clearly "when written in longhand on paper or on the blackboard." Strandskov, "Blood Group Nomenclature" (1948), 112.
49. For a relevant discussion of what can and cannot be written on blackboards, see Barany and MacKenzie, "Chalk" (2014).

50. Ducey and Modica, "On the Amendment of the Nomenclature of the Rh-CDE System" (1950), 467.
51. Cyril Jenkins Productions Ltd., *Blood Grouping* (1955). View the film at http://catalogue.wellcome.ac.uk/record=b1750596.
52. Haberman and Hill, "Verbal Usage of the CDE Notation for Rh Blood Groups" (1952).
53. DeGowin, *Blood Transfusion* (1949), 83.
54. Castle et al., "On the Nomenclatures of the Anti-Rh Typing Serums" (1948), 30.
55. For example, Schmidt, "Rh-Hr: Alexander Wiener's Last Campaign" (1994).
56. Blood groups Rh+ and Rh– corresponded in the Fisher-Race system to the D and d antigens. For clinicians wanting to identify people at risk from hemolytic disease, they would simply use pure Rh_0/anti-D antisera.
57. Lindee, *Moments of Truth in Genetic Medicine* (2005), 14.
58. Walker, "Refresher Course for General Practitioners" (1951).
59. Mollison, *Blood Transfusion in Clinical Medicine* (1951), 75.
60. Mollison, "Blood Groups" (1951).
61. On Cyril Clarke and Philip Sheppard and their collaboration at Liverpool: Zallen, "From Butterflies to Blood" (1997).
62. E.g., Mollison et al., "The Rh Blood Groups and Their Clinical Effects" (1948).
63. Mazumdar, *Species and Specificity* (1995), 373–74.
64. Rheinberger, *Toward a History of Epistemic Things* (1997).
65. In thinking about the varied experiences and identities of blood groups, I draw on anthropologist Annemarie Mol's analysis of the ontologies of a single disease: Mol, *The Body Multiple* (2002).
66. Blacker, "Medical Genetics" (1950).
67. Ford, "A Uniform Notation for the Human Blood Groups" (1955).
68. Ford to Race, October 16, 1957, SA/BGU/E.11, WCL.
69. Strandskov, "Blood Group Nomenclature" (1948).
70. Race to Diamond, February 11 1955, SA/BGU/E.6, WCL.
71. Kohler, *Lords of the Fly* (1994).

CHAPTER FIVE

1. Thomson to Saunders, June 7, 1945, FD1/3290, NAL.
2. Alan Drury's words were reported by Saunders to Thomson, May 25, 1945, FD1/3290, NAL.
3. The two units were housed at the Lister Institute, but funded and administered by the MRC: "The Lister Institute of Preventive Medicine: Report of the Governing Body," June 14, 1951, SA/BGU/K.5/1, WCL.
4. By the end of 1943, the Regional Transfusion Services and London depots had panels totaling 938,000 donor names, though less than half were bled:

"Annual Report of the Ministry of Health: Summary: War Services," Her Majesty's Stationary Office, 1944, 32 and 45, S4646 1941–1947, WCL.

5. In cross-matching tests, a hospital pathologist would mix a red cell suspension from the donor with serum from the recipient: Cyril Jenkins Productions Ltd., *Blood Grouping* (1955), 00:06:00.

6. As Galton chair, Fisher had run the Galton Laboratory from 1933 to 1943. Penrose succeeded him in both roles: Harris, "Lionel Sharples Penrose, 1898–1972" (1973); Kevles, *In the Name of Eugenics* (1995).

7. For a rich historiographical account of postwar biomedicine, and changing relationships between laboratories, clinical settings, and public health authorities: Quirke and Gaudillière, "The Era of Biomedicine" (2008); on the relationships of biomedical entities (such as oncogenes and DNA markers) to configurations of instruments, people, routine practices, and programs: Keating and Cambrosio, *Biomedical Platforms* (2003); on the interplay between biomedical and clinical research in the culture of clinical experimentation in oncology: Löwy, *Between Bench and Bedside* (1996); on postwar biomedicine as a "new way of knowing" within the broader sweep of the twentieth century: Cooter and Pickstone, eds., *Medicine in the Twentieth Century* (2000).

8. During the last three years of the war, the MoH had convened in London a series of regular planning meetings that (variously) brought together English and Welsh regional transfusion officers, MoH officials, the London transfusion officers, the Belfast transfusion officer, and representatives from the Scottish National Blood Transfusion Service, the Galton Serum Unit, and the Ministry of Information. They agreed that although the EBTS had been designed to respond to air raids, the majority of its work had attended to the civilian sick, and that postwar the MoH should take over the central organization of the service. For rich accounts of these discussions: Gunson and Dodsworth, "Towards a National Blood Transfusion Service" (1996); Whitfield, "A Genealogy of the Gift" (2011), 233–73.

9. The regional hospital boards controlled so-called natural hospital regions, which were defined by the spheres of influence of the teaching hospitals: Webster, *The National Health Service* (2002), 18–19.

10. Some of those fourteen regions shared a center. The London depots soon also became known as "regional transfusion centres"; Gunson and Dodsworth, "The National Blood Transfusion Service (NBTS)" (1996), 17.

11. The NHS in Scotland had its own legislation, defined by the National Health Service (Scotland) Act 1947, and the Parliament of Northern Ireland created its Health and Social Care Service in 1948.

12. For details of materials and routines: Gunson and Dodsworth, "The National Blood Transfusion Service" (1996).

13. Gunson and Dodsworth discuss attempts to maintain uniformity through the publication of booklets on the care of donors and the techniques of transfusion. Repeated attempts to centralize the service continued into the

1980s: Gunson and Dodsworth, "The National Blood Transfusion Service (NBTS)" (1996).

14. Illustrating just how varied practices were in some places, the blood trans-fusion officer for Colchester, Alec Blaxill, had established the service there in 1933, and he continued to supply call-up donors to local hospitals into the 1950s. Blaxill, "Blood for Transfusion" (1948). Lobbying for the contin-ued independence of some panels, the Red Cross argued that many donors were wary of NBTS blood banks because they reduced the personal interest of donors in the transfusion process: Gunson and Dodsworth, "Towards a National Blood Transfusion Service" (1996), 16.

15. "Emergency Blood Transfusion Service, Note of a Meeting," February 14, 1946, 2, BN 13/30, NAL; Gunson and Dodsworth, "Towards a National Blood Transfusion Service" (1996).

16. For the records of the liaison officers: "Emergency Services: Blood Transfu-sion Service Meetings of Donor Panel Liaison Officers [Ministry of Health]," 1946–1948, MH 96/140, NAL.

17. On the ongoing regional specificity of publicity materials: Keren Turton, "Films and Blood Donation Publicity in Mid-twentieth Century Britain" (2019); "Meeting of Publicity Sub-Committee of Regional Donor Organis-ers' Committee," November 23, 1955, MH 55/2180, NAL.

18. Practical service departments at the Lister included "Preparation and Study of Therapeutic Sera," "Preparation and Study of Vaccine Lymph," and "Preparation and Study of Bacterial Vaccines." Kekwick, "Alan Nigel Drury: 3 November 1889–2 August 1980" (1981); Collier, *The Lister Institute of Pre-ventive Medicine* (2000).

19. On Mollison's life and career: Davies, "Patrick Mollison: A Pioneer in Trans-fusion Medicine" (2012).

20. "The Lister Institute of Preventive Medicine: Report of the Governing Body," 1948, 7–8, SA/BGU/K.5/1, WCL.

21. For more on the Morgan lab and on the biochemistry of blood groups: Pierce and Reid, *Bloody Brilliant!*, 137–63.

22. Gunson and Dodsworth, "Towards a National Blood Transfusion Service in England and Wales" (1996), 16.

23. "Their Life a General Mist of Error, or, Hints to Blood Groupers," ca. 1950–1959, SA/BGU/D.1, WCL. The title was from act 3, scene 2 of *The Duchess of Malfi* (1623) by early seventeenth-century dramatist John Webster. The author of the blood grouping guide was keen to impress on readers that lives depended on their work: "Of what is't fools make such vain keeping? / Sin their conception, their birth, weeping: / Their life, a general mist of er-ror, / Their death, a hideous storm of terror."

24. The document is kept in the Research Unit archives of the Wellcome collec-tion and so was most likely authored by Race (it refers informally to "Ruth and I" in the text). But Mourant had a proclivity for literary epigraphs, which head every chapter of his monograph *The Distribution of the Human*

Blood Groups (1954), and it may have been written by him or by multiple authors at the Lister Institute. Certainly, the document was likely used by all blood group serology staff there.

25. Mourant, *Blood and Stones* (1995), 61. More generally, during the 1950s biomedical laboratory technicians were becoming increasingly professionalized. Technicians working in pathology laboratories had their own professional organization, the Institute of Medical Laboratory Sciences, which in 1949 added hematology and blood transfusion to the exams for junior technician: Russell et al., "Missing Links in the History and Practice of Science" (2000). For more on blood group serologist technicians: Pierce and Reid, *Bloody Brilliant!* (2016), 301–326.

26. The British laboratories used wooden blocks with 50 drilled holes, which had been had been made for the Ministry of Agriculture and Fisheries and were introduced into blood laboratories by Taylor in the 1930s: Race and Sanger, *Blood Groups in Man*, 3rd ed. (1958), 243.

27. "Their Life a General Mist of Error, or, Hints to Blood Groupers," ca. 1950–1959, 1, 2 and 7, SA/BGU/D.1, WCL.

28. "Their Life a General Mist of Error, or, Hints to Blood Groupers," ca. 1950–1959, 5, 7, SA/BGU/D.1, WCL.

29. Dunsford and Bowley, *Techniques in Blood Grouping* (1955), 42. Similarly, the *American Association of Blood Banks Bulletin* warned of the "dangerous malady of boredom" during blood grouping: Rymer, "The Editor's Page" (1959), quoted in Pierce and Reid, *Bloody Brilliant!* (2016), 304.

30. "Their Life a General Mist of Error, or, Hints to Blood Groupers," ca. 1950–1959, 9, SA/BGU/D.1, WCL.

31. For example, in the *BMJ* Mourant emphasized that "it practically never happens that a donor has to be grouped or a sample taken for grouping other than at a large transfusion centre of some sort." Mourant, "Clinical Pathology in General Practice: Blood Grouping" (1954), 38; also published as a chapter in British Medical Association, *Clinical Pathology in General Practice* (1955).

32. Mourant, "Clinical Pathology in General Practice: Blood Grouping" (1954), 37.

33. See also Discombe, "Blood Transfusion Accidents" (1953).

34. Mourant, "Clinical Pathology in General Practice: Blood Grouping" (1954), 38.

35. Mourant, "Clinical Pathology in General Practice: Blood Grouping" (1954), 38. The MRC booklets included Medical Research Council Blood Transfusion Research Committee, *The Determination of Blood Groups* (1943), and Mollison et al., *The Rh Blood Groups and their Clinical Effects* (1948).

36. The strategy for exerting control by withholding instructions has been discussed in relation to other medical settings, for example, in analyses of the power dynamics between obstetricians and midwives in the eighteenth century: Wilson, *The Making of Man-Midwifery* (1995), 1–8.

37. The technique's inventor was Danish physician Knud Eldon.
38. Rice-Edwards, "A Simple Blood-Grouping Method" (1955).
39. Rice-Edwards, "A Simple Blood-Grouping Method" (1955).
40. Zeitlin, "A Simple Blood-Grouping Method" (1955), 971; Drummond, "Simple Blood-Grouping Methods" (1956).
41. Pickles, "Simple Blood-Grouping Method" (1955).
42. Pinkerton, "Simple Blood-Grouping Methods" (1956), 289; Drummond, "Simple Blood-Grouping Methods" (1955), 1388.
43. "Meeting of Regional Transfusion Directors," March 14, 1956, 3, BN 13/31, NAL.
44. For another account of the consequences of new techniques that enable "untrained" workers—in this case, the (later) history of pregnancy testing: Olszynko-Gryn, *A Woman's Right to Know* (forthcoming).
45. Drummond, "A Simple Blood-Grouping Method" (1955).
46. One commented sternly: "One would welcome a more positive approach on the part of the National Blood Transfusion Service to the development of simple and rapid techniques more suited to the conditions under which many emergency blood transfusions must be given": Kidd, "Simple Blood-Grouping Methods" (1956).
47. Pickles, "Simple Blood-Grouping Method" (1955).
48. For a biography of Race: Clarke, "Robert Russell Race" (1985)
49. Robson, "Sylvia Lawler" (1996); For more on Lawler's life, including as the first female professor at the Institute of Cancer Research: "Professor Sylvia Lawler," *Times*, January 26, 1996, 19.
50. Hughes-Jones and Tippett, "Ruth Ann Sanger, 6 June 1918–4 June 2001" (2003).
51. Over its first decade, and particularly after *Blood Groups in Man* came out in 1950, the Research Unit slowly transformed from "Rob's lab" to "Rob and Ruth's lab," to quote their correspondence. Letters from friends and colleagues soon treated the two as though they jointly led the laboratory. After Race's wife Monica, with whom he had three children, sadly died after a short illness in 1955, Race and Sanger were married.
52. Race, "MRC Report, October 1945–May 1946," 1946, 2, FD8/18, NAL.
53. I have significantly simplified my description of blood grouping work. The complexities of serology were vast; the actors in this story were grappling with the phenomena of "incomplete" and "blocking" antibodies, the necessity of antigen absorption protocols, the relative virtues of use of saline and albumin suspensions, as well as the multiple subtle complexities of blood grouping. Some antigens differed in the extent to which their genetics related to the doses of the antigen, as determined by titration; while antibodies soon came to be defined not just by their biochemical specificity but also by the plasma fraction in which they were to be found. For a contemporary account of these methods: Dunsford and Bowley, *Techniques in Blood Grouping* (1955); for the history of some of these techniques: Pierce and Reid, *Bloody Brilliant!* (2016), 423–438.

54. The Reference Laboratory was officially the lab to which blood depots would refer their difficult samples, and, indeed, Mourant's official reports to the MRC described the large quantities of serological tests it carried out. Mourant's staff would give to Race and his colleagues particularly interesting samples for further testing, although in a letter to a US colleague, Race complained that the Reference Laboratory did not pass on as many interesting problems as Race would have liked: Race to Diamond, December 15, 1950, SA/BGU/E.6, WCL.

55. Dunsford to Race and Sanger, May 9, 1956, SA/BGU/F.5/3/1, WCL.

56. Robert Race underlined the valuable reach of this network when he wrote about the very rarest Rh genotypes they had picked up, "some of them from remote parts of the country." Race, "Medical Research Council Progress Report 1950–53 of the Blood Group Research Unit," ca. 1953, 4, FD8/67, NAL.

57. The Lister Institute workers announced four novel blood group systems by 1950: Lutheran, Kell, Duffy, and Lewis. For the first announcements of Lutheran: Callender et al., "Hypersensitivity To Transfused Blood" (1945); of Kell: Coombs et al., "In-Vivo Isosensitization of Red Cells in Babies with Haemolytic Disease" (1946); of Duffy: Cutbush et al., "A New Human Blood Group" (1950), 188; of Lewis: Mourant, "A 'New' Human Blood Group Antigen of Frequent Occurrence" (1946).

58. Race to Diamond, December 15, 1950, SA/BGU/E.6, WCL.

59. On the US landscape of blood procurement: Pierce and Reid, *Bloody Brilliant!* (2016), 281–300; Starr, *Blood: An Epic History* (2009); Lederer, *Flesh and Blood* (2008).

60. Sanger and Race to Cahan, October 30, 1956, SA/BGU/E.3/1, WCL; Race to Callender, March 6, 1945, SA/BGU/F.1/27, WCL; Cahan to Race and Sanger, May 15, 1956, SA/BGU/E.3/1, WCL.

61. Cahan's reference to "pointed heads" was a colloquialism derived from the phrenological notion that a higher forehead signified high intellectual capacity: Cahan to Race, July 6, 1954, SA/BGU/E.3/1, WCL.

62. Cahan to Race, March 25 1954, SA/BGU/E.3/1, WCL.

63. Writing about the 1970s, Jenny Stanton describes the social connections forged by sharing blood samples: Stanton, "Blood Brotherhood" (1994).

64. Race and Sanger to Cahan, February 23, 1955, SA/BGU/E.3/1, WCL.

65. Race, "Letter," October 23, 1953, SA/BGU/G.3/2 [restricted access], WCL.

66. Race, "Template Letter to Donors," February 1950, SA/BGU/G.3/2 [restricted access], WCL.

67. Race, "Medical Research Council Progress Report, 1953–55," ca. 1955, 1–2, FD8/18, NAL.

68. For a useful table of when blood groups were first identified: Dunsford and Bowley, *Techniques of Blood Grouping* (1967), 4.

69. Kevles, *In the Name of Eugenics* (1995), 213; Kevles interview with Sylvia Lawler, June 29, 1982, Daniel J. Kevles papers, Oral History Interview Transcripts, 1982–1984, box 1, folder 17, RAC.

70. Kevles, interview with Ruth Sanger, June 23, 1982, 23, Daniel J. Kevles papers, Oral History Interview Transcripts, 1982–1984, box 2, folder 31, RAC.
71. Kevles, interview with Sylvia Lawler, June 29, 1982, 16, Daniel J. Kevles papers, Oral History Interview Transcripts, 1982–1984, box 1, folder 17, RAC; more on Penrose, his lab and the postwar London atmosphere for human genetics: Kevles, *In the Name of Eugenics* (1995), 212–222.
72. In recalling this move, Lawler herself noted, "What I went to do was to use the blood groups as markers for the chromosomes, looking for genetical linkages between diseases and the markers, or the markers themselves"; Kevles, interview with Sylvia Lawler, June 29, 1982, 4, Daniel J. Kevles papers, Oral History Interview Transcripts, 1982–1984, box 1, folder 17, RAC. In 1954 Lawler published, for example, the first linkage between ellipto-cytosis (the condition of oval-shaped blood cells) and the Rh blood group system: Lawler, "Family Studies Showing Linkage between Elliptocytosis and the Rhesus Blood Group System" (1954); Lawler and Sandler, "Data on Linkage in Man" (1954).
73. Race, "Draft: Blood Group Research Unit MRC Report (October 1951–September 1952)," ca. 1952, FD8/67, NAL.

CHAPTER SIX

1. Gilbert, *Emergency Call* (1952), 00:04:23 and 00:06:04 (my transcription).
2. "Reviews," *Observer*, May 18, 1952, 6.
3. All three plays, broadcast on the BBC's Light Programme, were written by BBC scriptwriter Stephen Grenfell: *Life-Blood* on October 28, 1954; *The Miracle Baby* on January 14, 1957; and *Precious Cargo* on August 22, 1960. The Home Service also broadcast a theater adaptation of *Emergency Call* on June 12, 1954, and October 27, 1955. For information on past BBC broadcasts: https://genome.ch.bbc.co.uk, accessed September 22, 2019. In 1962, *Emergency Call* was remade under the title *Emergency*; except this time, in a Cold War twist, one of the would-be donors was an "atomic scientist selling secrets." Rare blood was also the central theme of a famous episode of the BBC television situation comedy "Hancock's Half Hour," broadcast on June 23 1961.
4. On plasma stockpiling, "Meeting of Regional Transfusion Organisers," March 3, 1943, 3, MH123/187, NAL; statistics from "Annual Report of the Ministry of Health: Summary: War Services," Her Majesty's Stationary Office, 1944, 32, S4646 1941–1947, WCL; "Annual Report of the Ministry of Health: Year Ended 31 March 1949," Her Majesty's Stationary Office, 1950, S4646 1941–1947, WCL. Quote from "Emergency Blood Transfusion Service," February 14, 1947, 61, BN 13/30, NAL.
5. To watch this remarkable wartime film produced by Paul Rotha: http://wellcomelibrary.org/item/b16758651. Nieter, *Blood Transfusion Service* (1941).
6. On Paul Rotha and postwar documentaries and the British history of science documentary films: Boon, *Films of Fact* (2008); for an overview of

similar themes in relation to the United States: Lederer and Rogers, "Media" (2013).

7. On the development of these films and their wide range of uses: Turton, "Films and Blood Donation Publicity in Mid-twentieth Century Britain" (2019).

8. "Search for Rare Blood," *Times*, August 29, 1950, 2.

9. "Blood Hunt Goes On All Night," *Daily Express*, 15662, August 29, 1950, 1.

10. "Rare Blood Search Continuing: Supply Flown from Copenhagen," *Times*, August 30, 1950, 4.

11. "Search for Rare Blood," *Times*, August 29, 1950, 2.

12. "Search for Rare Blood," *Times*, August 31, 1950, 4.

13. "Rare Blood Transfusion," *Times*, September 4, 1950, 2; "Rare Blood Transfusion," *Times*, September 5, 1950, 3.

14. "He's One in 20,000," *Courier and Advertiser* (Dundee), September 4, 1950, 3; "Rare Blood Woman Says 'Thank You,'" *Gloucester Citizen*, September 7, 1950, 1.

15. For the NBTS publicity photograph of Freddie Mills: SA/HHC/D/3/6/3, WCL. Thanks to Keren Turton for bringing this to my attention.

16. "Meetings of Regional Transfusion Directors: Minutes," September 27, 1950, BN 13/31, NAL.

17. "Meetings of Regional Transfusion Directors: Minutes," September 27, 1950, BN 13/31, NAL; on the management of the public communication of science, health and medicine: Loughlin, "Networks of Mass Communication" (2005).

18. "Examples of Types Which Might Be Required for Special Cases," ca. 1950, BN 13/65, NAL; Maycock to Drummond, November 22, 1950, BN 13/65, NAL.

19. "Meetings of Regional Transfusion Directors: Minutes," December 6, 1950, 2, BN 13/31, NAL.

20. For plans for the testing of all known blood groups of large panels, from which a "national panel" would be assembled: Mourant, "Medical Research Council," January 10, 1951, BN 13/65, NAL. The South London depot director was also involved in the implementation of techniques for identifying rare blood groups, and establishing the Rare Blood Panel: Contreras, "Thomas Edward Cleghorn" (1992).

21. Maycock to Nicole, "Re: Blood Group Reference Laboratory," April 4, 1952, BN 13/65, NAL.

22. "Meetings of Regional Transfusion Directors: Minutes," December 6, 1950, 2, BN 13/31, NAL.

23. "A New Blood-Donor Service" (1952), 670.

24. "Central Register of Donors with Rare Blood Group (Extract from Minutes of RTDs Meeting)," February 26, 1952, BN 13/65, NAL.

25. "Meetings of Regional Transfusion Directors: Minutes," December 7, 1955, 2, BN 13/31, NAL.

26. E.g., "Meetings of Regional Transfusion Directors: Minutes," April 23, 1955, BN 13/31, NAL.
27. In the United States there was far greater potential for rare donors to reap financial rewards: Lederer, *Flesh and Blood* (2008) 158–61.
28. "National Panel of Donors: Draft Letter to Donors Resigning for Reasons Other Than Age or Ill Health," ca. 1956, BN 13/31, NAL.
29. On the preparation of various antisera: Dunsford and Bowley, *Techniques in Blood Grouping* (1955), 116–39.
30. For the longer history of "normal" antibodies as constitutional: Keating, "Holistic Bacteriology" (1998). More recently, such "naturally occurring" antibodies have been understood as stimulated by exposure to proteins similar to the ABO antigens via ingestion of plants and microorganisms.
31. Unknown to Drummond, January 17, 1951, BN 13/65, NAL.
32. Stratton, "Ministry of Health Template Letter," ca. 1950, BN 13/65, NAL.
33. Designs for donor cards, ca. 1950, BN 13/65, NAL.
34. Ministry of Health Memorandum circulated to RTDs and Mourant: "High Titre Donors in the R.A.F.," February 25, 1950, BN 13/65, NAL; and other papers in the collection.
35. "Shortage of Rh Testing Sera" (1950).
36. Kevles, *In the Name of Eugenics* (1995), 213.
37. *The Rh Factor: A Leaflet for Midwives, Nurses and Health Visitors* (Ministry of Health, 1949), BD 18/206, NAL.
38. By 1950 it was apparently customary for doctors to take between half and a whole pint of blood from each such person for this purpose: "Shortage of Rh Testing Sera" (1950), 108.
39. "Shortage of Rh Testing Sera" (1950), 109.
40. Callender et al., "Hypersensitivity to Transfused Blood" (1945); Callender and Race, "A Serological and Genetical Study of Multiple Antibodies" (1946).
41. Callender and Paykoç, "Irregular Haemagglutinins after Transfusion" (1946).
42. Callender et al., "Hypersensitivity To Transfused Blood" (1945), 83.
43. Callender and Race, "A Serological and Genetical Study of Multiple Antibodies" (1946).
44. The authors noted their gratitude to the "donors for permission to use their names and for providing repeated samples of blood": Callender and Race, "A Serological and Genetical Study of Multiple Antibodies" (1946), 117.
45. Rare instances of diseases being named for patients include Christmas disease, a blood clotting disorder named for patient Stephen Christmas, and Hartnup disease—named for the family in which it was first defined. On controversies over disease eponyms in the late twentieth century: Hogan, "Medical Eponyms" (2016). Nick Hopwood observes that embryologist William His named the embryos for the donors—that is, not the women whose bodies they came from, but the medical men who had supplied them to His: Hopwood, "Producing Development" (2000).

46. Of course "F.M." herself had multiple novel antigens circulating in their blood, which is perhaps why new groups were not named after transfused patients like her. Moreover, "F.M." may have had her anonymity preserved to maintain her medical confidentiality. In another case, the "Kidd" blood group was named after a child suffering from erythoblastosis fetalis with this novel antigen: "Medical Research Council Progress Report, 1950–53, of the Blood Group Research Unit," 2–3, FD8/18, NAL.
47. Mourant, *Blood and Stones* (1995), 54.
48. Sanger to Cahan, April 13, 1954, SA/BGU/E.3/1, WCL.
49. E.g., Sanger to Cahan, April 13, 1954, SA/BGU/E.3/1, WCL; Race to Cahan, July 10, 1954, SA/BGU/E.3/1, WCL.
50. Cahan, "A Fluid for Shipping Whole Blood Specimens," June 9, 1955, SA/BGU/E.3/1, WCL.
51. The innovation of using antibiotics for this purpose was apparently Cahan's: Race, "Medical Research Council Progress Report, 1953–55, of the Blood Group Research Unit," 2, FD8/18, NAL.
52. When Race, Sanger, and their colleagues sent back their agglutination results, Cahan replied, "At long last we have the comforting knowledge that our panel is definitive as a result of your work." Now that the panel of red cells had been characterized by the most advanced blood group serology laboratory in the world, the blood bank began marketing the "Panocell" to hospitals as a testing service. When a hospital "subscribed" to the Panocell, they would receive regular shipments of the red cells, which they could use themselves in-house. Cahan to Race and Sanger, October 23, 1954, SA/BGU/E.3/1, WCL; Cahan to Race and Sanger, July 11, 1956, SA/BGU/E.3/1, WCL.
53. For a rich account of the politics of Red Cross segregation and labeling, and the changing category of "negro" blood during the Second World War: Guglielmo, "Red Cross, Double Cross" (2010); For more on US blood transfusion and race: Love, *One Blood* (1996); Lederer, *Flesh and Blood* (2008).
54. Even as the official records of blood donors ceased to carry these racial designations, blood banks of all kinds may have used them for private use within the institution.
55. On race in Britain in the postwar era: Paul, *Whitewashing Britain* (1997); on race in the NHS: Bivins, "Picturing Race in the British National Health Service" (2017); For a broad and rich account of race in the NHS, including the use of racialized clinical material presented by New Commonwealth migrants in inner-city hospitals and clinics: Bivins, *Contagious Communities* (2015).
56. Race, "Medical Research Council Progress Report, 1953–55, of the Blood Group Research Unit," ca. 1955, 2–3, FD8/18, NAL.
57. Race, "Draft Abstract, Sent to MD Publications," November 1, 1955, SA/BGU/E.3/1, WCL.
58. Race to Cahan, January 16, 1955, SA/BGU/E.3/1, WCL.

59. In their correspondence, Race and Cahan seem to use the terms "colored" and "negro" interchangeably. Race, "Medical Research Council Progress Report, 1953–55, of the Blood Group Research Unit," ca.1955, 2, FD8/18, NAL.

60. In a 1956 lecture to the Genetical Society of Great Britain, authored by both researchers, Sanger reflected on this new interest. She explained that because so many specimens were arriving from the United States, the lab "*had* to become interested in negro blood for so many of the donors there are negroes" (my emphasis). Sanger went on: "Until last year we had not done any original work on negro blood through distrust of the state of antigens in samples of blood sent from far away. This difficulty has now been remove by the use of suspending fluid containing antibiotics." Sanger and Race, "Lecture to the Genetical Society, Edinburgh: Some Recent Blood Group Investigations between Negros and Whites," April 10, 1956, 1, FD8/35, NAL.

61. Gilbert, *Emergency Call* (1952), 00:57:55.

62. Despite the cessation of racial labeling on blood processed and collected by the American Red Cross, Cahan evidently believed that knowledge of the varied blood group frequencies of "white" and "negro" donors was important to blood banks, perhaps as a check on grouping technique. Indeed, an early version of his Panocell recording card prominently summarized Sanger's own statistics on the frequencies of all known blood groups among "whites and negros" from a lecture Sanger had delivered at Theodosius Dobzhansky's Institute for the Study of Human Variation in New York in 1955: Cahan, "Approximate US Blood Groups (Whites and Negroes) [Printed on Reverse of 'Pan "O" Cell Master List.']," May 1956, SA/BGU/E.3/1, WCL.

63. On contemporary claims about hemoglobin: Tapper, *In the Blood* (1999), 13; Wailoo, *Drawing Blood* (1999).

64. Sanger and Race, "Some Recent Blood Group Distinctions between Negroes and Whites II," April 10, 1956, FD 8/35, NAL.

65. Race, "Medical Research Council Progress Report, 1953–55, of the Blood Group Research Unit," 2, FD8/18, NAL.

66. Race, "Medical Research Council Progress Report," 3.

67. On the construction of whiteness in a different time and place: Anderson, *The Cultivation Of Whiteness* (2003).

68. Race and Sanger to Cahan, February 23, 1955, SA/BGU/E.3.1, WCL; Race, "Medical Research Council Progress Report, 1953–55, of the Blood Group Research Unit," 3, FD8/18, NAL.

69. Paul, *Whitewashing Britain* (1997); Bivins, *Contagious Communities* (2015).

70. Sanger and Race, "Lecture to the Genetical Society, Edinburgh: Some Recent Blood Group Investigations between Negros and Whites," April 10, 1956, 4, FD8/35, NA.

71. Race, "Medical Research Council Progress Report, 1953–55, of the Blood Group Research Unit," 3, FD8/18, NAL.

72. Maycock, "National Panel of Donors," September 13, 1956, BN 13/65, NAL.
73. Maycock to Bowley, January 13, 1960, BN 13/65, NAL.
74. Mourant, "The Establishment of an International Panel of Blood Donors of Rare Types" (1965); Mourant, "Template Letter Regarding the International Panel of Donors of Rare Blood Types," March 21, 1968, BN 13/66, NAL. On the international rare blood panels today: International Rare Donor Panel, https://www.nhsbt.nhs.uk/ibgrl/services/international-rare-donor-panel/, accessed September 22, 2019.
75. Scotland Yard's blood grouping laboratory was likely established in the early 1950s. For the development of forensic science laboratories in Britain: Adam, *A History of Forensic Science* (2015); for records of blood groups in police work: "Metropolitan Police: Forensic Science Laboratory [Photographs]," (date unknown), MEPO 13/308, NAL; "Electrophoresis Apparatus to Determine Blood Group [Photographs]," (date unknown), MEPO 13/312, NAL; "Blood Group Sampling, Payment of Fees, Metropolitan Police," 1946–68, MEPO 2/10919, NAL; "Forensic Science Laboratories: Directors' Meetings," 1951–59, HO 287/197, NAL; "Forensic Science Laboratories: Directors' Meetings," 1963–66, HO 287/407, NAL.
76. "Alleged Double Murder: Son-in-Law Charged," *Times*, Friday October 28, 1949, 6.
77. "'Rare' Blood on Accused Man's Clothing," *Gloucester Citizen*, November 22, 1949, 1.
78. "Blood Types Trapping Criminals," *New York Times*, September 6, 1953, 4.
79. Lederer, "Bloodlines" (2013).
80. For more on the Indiana initiative and its context: "Atomic Tattoos." January 19, 2019, episode 337 of *99% Invisible* podcast, https://99percentinvisible.org/episode/atomic-tattoos/, accessed September 22, 2019.
81. Further on the cultural "fetishization of rare blood": Lederer, "Bloodlines" (2013).

CHAPTER SEVEN

1. Mourant, "Nuffield Blood Group Centre, Report to Council," April 22, 1952, 91/2/14, RAI.
2. "Map Man's Blood Groups," *Science News-Letter* 59, no. 15 (April 14, 1951): 237. *Science News-Letter* was a US magazine published from 1922 by the Society for Science and the Public.
3. Mourant, "Blood Groups" (1961).
4. For histories of total archives, and the specificity of this postwar moment: Jardine and Drage, "The Total Archive" (2018).
5. For the principal biographical accounts of Mourant's life: Misson et al., "Arthur Ernest Mourant: 11 April 1904–29 August 1994" (1999); Mourant, *Blood and Stones* (1995).

6. Mourant, *Blood and Stones* (1995), 52–53.
7. For more on the "Coombs test," and Mourant's involvement in the collection and testing of blood from local obstetric hospitals: Lachmann, "Robert Royston Amos (Robin) Coombs" (2009); Mourant, *Blood and Stones* (1995), 55–58.
8. Mourant, *Blood and Stones* (1995), 61–62.
9. Mourant, *Blood and Stones* (1995), 61.
10. Steven Pierce and Marion Reid note that the technical handling of blood for transfusion generally was completely dominated by women: Pierce and Reid, *Bloody Brilliant!* (2016), 301–26; on the professionalization of technical laboratory staff in Britain: Russell et al., "Missing Links in the History and Practice of Science" (2000).
11. Quoted from Mourant, "Progress Report 1949–50 of the Blood Group Reference Laboratory," 1950, 3, PP/AEM/C.1, WCL.
12. Hematology and blood transfusion were made specialty areas for the qualification of senior technicians in 1949, but for several years Mourant's staff continued to teach serological blood grouping to hematologists and doctors across the country.
13. For the UK history of freeze-drying blood: Gunson and Dodsworth, "The Drying and Fractionation of Plasma, 1935–55" (1996). For in-depth US accounts: Creager, "Biotechnology and Blood" (1998); Creager, "Producing Molecular Therapies from Human Blood" (1998); Creager, "'What Blood Told Dr. Cohn'" (1999).
14. Bristow et al., "Standardization of Biological Medicines" (2006).
15. Gradmann and Simon, *Evaluating and Standardizing Therapeutic Agents, 1890–1950* (2010).
16. Miles, "International Standards for Anti-A and Anti-B Blood-Grouping Sera" (1950).
17. Mourant, *Blood and Stones* (1995), 62.
18. Between the wars, the League of Nations Health Organisation had coordinated standards and units for thirty-four substances assayed by biological methods. Fitzgerald, "The Work of the Health Organisation of the League of Nations" (1933); World Health Organization, *The First Ten Years of the World Health Organization* (1958).
19. Miles, "International Standards for Anti-A and Anti-B Blood-Grouping Sera" (1950), 5.
20. In that capacity, Miles was overseeing the standardization of drugs and sera, including penicillin, thyrotropin, insulin, and many vaccines.
21. In the United States it was common practice to immunize donors artificially to increase their A or B antibody concentrations, something that was not done in Britain. On this occasion the researchers found that this practice did not make much difference to the antibody titers of the US pool.

22. All of these methods are described in Miles, "International Standards for Anti-A and Anti-B Blood-Grouping Sera" (1950).

23. The final titer of that standard was agreed to be 1/256, which defined the "international unit" of antibody concentration. Although this was a "reference" for determining antibody concentration, Mourant also asked the labs to measure the "avidity" of the samples—that is, the time taken for agglutination to happen, and their specificity to A_1, A_2, A_2B, and B cells. This indicated how effective the antisera were at producing the expected reactions. Miles, "The International Standards for Anti-A and Anti-B Blood-Grouping Sera" (1950).

24. Fraser Roberts, "Surnames and Blood Groups" (1942); Fraser Roberts, "Blood Group Frequencies in North Wales" (1942); Fisher and Fraser Roberts, "A Sex Difference in Blood Group Frequencies" (1943); Fraser Roberts, "The Frequencies of the ABO Blood Groups in South-Western England" (1948); Polani, "John Alexander Fraser Roberts" (1992).

25. The *Times* was reporting a joint meeting of the zoology, anthropology, and physiology sections of the British Association: "Relations of Race and Language: Genetic Foundations," *Times*, September 14, 1948, 2.

26. Lewis, "Cyril Dean Darlington" (1983).

27. Founded in 1943, the Nuffield Foundation was a charitable trust founded by William Morris (Lord Nuffield, founder of Morris Motors) to support projects in education and social policy. For the application: Darlington, "Letter to General Bullen Smith (of the Nuffield Foundation)," March 31, 1948, A91/2/5, RAI.

28. "Proposed Survey of the British Isles [Enclosed in Letter for the Nuffield Foundation]," May 30, 1951, 91/2/2, RAI.

29. Paul, *Whitewashing Britain* (1997).

30. Darlington, "Letter to General Bullen Smith (of the Nuffield Foundation)," March 31, 1948, A91/2/5, RAI.

31. Fraser Roberts, *History in Your Blood* (1952), 28.

32. Fraser Roberts, *History in Your Blood* (1952), 29.

33. Fraser Roberts, "An Analysis of the ABO Blood Group Records of the North of England" (1953), 386.

34. Later published as Watkin, "122. Blood Groups in Wales and the Marches" (1952).

35. Hartmann et al., "The Rh Genotypes of a Series of Oslo Blood Donors" (1947); Race and Mourant, "The Rh Chromosome Frequencies in England" (1948); Mourant, "The Blood Groups of the Basques" (1947).

36. Mourant, "World Health Organization, Expert Committee on Biological Standardization: International Blood Group Reference Laboratory Proposal," November 10, 1950, 3, WHO/BS/111, WHO Archives, Geneva.

37. Darlington, "The Genetic Component of Language" (1947).

38. "Letter for the Nuffield Foundation, Draft Only," May 30, 1951, A91/2/1, RAI.

39. "Blood Groups and Anthropology: Special Meeting, March 17, 1951," 1951, 1, 91/1/7, RAI.

40. Felicia Stallman, "[Template Letter Inviting Members of the New Blood Group Committee]," April 11, 1951, 91/1/8, RAI.

41. Fleure's lecture dealt with the state and future of the Institute, and with "the scope of anthropology," especially its relation to archaeology, prehistory, and the improvement of human welfare: Fleure, "The Institute and Its Development" (1946).

42. Fleure, "The Institute and Its Development" (1946), 2.

43. The Nuffield Foundation's decision to fund the work was probably helped by trustee Janet Vaughan, an orchestrator of the wartime blood transfusion service and a close colleague to several blood group researchers: Zallen et al., "The Rhesus Factor and Disease Prevention" (2004).

44. On that "anthropological" testing: Mourant, "World Health Organization, Expert Committee on Biological Standardization: International Blood Group Reference Laboratory Proposal," November 10, 1950, 3, WHO/BS/111, WHO Archives, Geneva.

45. Mourant, *Blood and Stones* (1995), 77.

46. Kopeć carried out her calculations on a Monroe calculating machine. In addition to the clerical and statistical work, the center was responsible for workers encouraged and advised on blood group research in both field and laboratory. Mourant, "Nuffield Blood Group Centre, Report to Council," April 23, 1952, A91/2/14, RAI.

47. "Meetings of Regional Transfusion Directors: Minutes," October 3, 1951, BN 13/31, NAL; Mourant, "Proposal to Nuffield Foundation," June 6, 1951, PP/AEM/C.9, WCL.

48. Kopeć, *The Distribution of the Blood Groups in the United Kingdom* (1970), vii.

49. Even by 1953, only six of the twelve regional centers regularly complied, perhaps testifying to the perceived risk to the cards. Mourant reprimanded uncooperative transfusion directors and complained that some cards were sent in incomplete. "Minutes, Regional Transfusion Officers Meeting, Ministry of Health," September 30, 1953, BN 13/31, NAL.

50. Maycock to Malone, "Sheffield Regional Transfusion Laboratory," April 5, 1951, BN 13/31, NAL.

51. Mourant, A. E. "The Nuffield Blood Group Centre of the Royal Anthropological Institute [Memorandum on Progress]," ca. 1956, A91/2/24, RAI.

52. Mourant, "Memorandum on the History, the Work and the Future Plans of the Centre," ca. 1955, PP/AEM/C.11, WCL.

53. The Red Cross had a powerful hold on transfusion in many countries, building on its experiences of recruiting donors and collecting blood during the war. The United States had a dual system: the American Red Cross

relaunched its nationwide program in 1947, the same year that the new American Association of Blood Banks forged a coalition among commercial blood enterprises, allowing the tracking of blood between organizations. US occupation forces imported this dual system to Japan, which had Red Cross and commercial provision. On some of these national and Red Cross programs: Kim, "The Specter of 'Bad Blood' in Japanese Blood Banks" (2018); Starr, *Blood: An Epic History* (2009); Pierce and Reid, *Bloody Brilliant!* (2016); Chauveau, "Du don à l'industrie: La transfusion sanguine en France depuis les années 1940" (2011).

54. Schneider, *The History of Blood Transfusion in Sub-Saharan Africa* (2013), 42–43.

55. See papers of William Maycock, Consultant Advisor to the Department on Blood Transfusion: "Correspondence: India [Ministry of Health and Department of Health and Social Security]," 1949–1978, BN 13/58, NAL; "Correspondence: New Zealand [Ministry of Health and Department of Health and Social Security]," 1948–1977, BN 13/57, NAL; "Correspondence: Canada [Ministry of Health and Department of Health and Social Security]," 1946–1955, BN 13/50, NAL.

56. All major British colonies in Africa had organized blood transfusion by 1953. For more on this, the history of transfusion in the Belgian Congo, and data on the rising numbers of blood transfusions reported by British colonial government medical departments and the Red Cross between 1947 and 1962: Schneider, *The History of Blood Transfusion in Sub-Saharan Africa* (2013), 28–64.

57. Mourant had recently collaborated with Hartmann on a study of Rh genotypes among Oslo blood donors, suggesting that Mourant had helped to plan this proposal.

58. Cueto et al., *The World Health Organization: A History* (2019).

59. On the administrative infrastructure of advisory groups and technical reports: Radin, "Unfolding Epidemiological Stories" (2014).

60. Executive Board, "Biological Standardization—International Centres," December 27, 1951, 2, EB9/50, WHO Archives, Geneva.

61. On the negotiation of national laboratories for blood grouping: ARC010-3, centralized files, 3rd generation, sub-fonds 3, 1952–1980, H5-286-3(A), Designation of Blood Grouping Laboratories, WHO Archives, Geneva; ARC010-3, centralized files, 3rd generation, sub-fonds 3, 1952–1980, H5-181-2, CTS, WHO Archives, Geneva.

62. This model of designating international centers in charge of biological standards was applied by the WHO to all sorts of other biological substances: Executive Board, "Biological Standardization—International Centres," December 27, 1951, EB9/50, WHO Archives, Geneva.

63. A "red-cell panel" was an array of samples containing known antigens, which could be used for probing the antibody composition in as yet unknown samples of blood.

64. Mourant, "World Health Organization, Expert Committee on Biological

Standardization: International Blood Group Reference Laboratory Proposal," November 10, 1950, 3, WHO/BS/111, WHO Archives, Geneva.

65. In reports to the MRC, the Reference Laboratory "incorporated" the International Blood Group Reference Laboratory—implying that the work was a distinctive thread nested within the broader activities of the laboratory. The WHO estimated that $2,800 would be awarded to the Reference Laboratory annually: Executive Board, "Biological Standardization—International Centres," 2.

66. E.g., Lourie to Mourant, 15 March 1954, ARC010-3, centralized files, 3rd generation, sub-fonds 3, 1952–1980, H5-286-3(A), Designation of Blood Grouping Laboratories, WHO Archives, Geneva.

67. Mourant cultivated these connections over many years. Some surviving folders of correspondence span two or three decades.

68. Mourant, "World Health Organization, Expert Committee on Biological Standardization: International Blood Group Reference Laboratory: Proposal," November 10, 1950, WHO/BS/111, WHO Archives, Geneva.

69. Mourant, "Progress Report 1949–50 of the Blood Group Reference Laboratory," 1950, PP/AEM/C.1, WCL.

70. Mourant's archive, given to the Wellcome Library in 1994, contains rather few letters from the 1940s and 1950s: the bulk of his correspondence dates from the mid-1960s, when he began work on the second edition of *The Distribution of the Human Blood Groups* and moved from the Reference Laboratory to the new Serological Population Genetics Laboratory.

71. Hatcher to Mourant, October 15, 1968, PP/AEM/K.57, WCL.

72. Ikin regularly tested arrays of anthropological samples, appearing as a co-author (with Mourant) on papers published from collections made in India, Spain, Iceland, Egypt, Sudan, Nigeria, Italy, and many more.

73. On the function of gifts as establishing bonds of reciprocity between donors and recipients: Mauss, *The Gift* (2002), 10.

74. Examples include: Arneaud to Mourant, March 20, 1956, PP/AEM/K.1, WCL; Bird to Mourant, December 19, 1958, PP/AEM/K.6, WCL; Reid to Mourant, March 13, 1950, PP/AEM/K.114, WCL; Pei-en Chen to Mourant, December 9, 1957, PP/AEM/K.15, WCL. The Connaught Hospital was founded by the Duke of Connaught in 1912; the University College in Trinidad was founded in 1948 as an external college of the University of London.

75. Burton, "'Essential Collaborators'" (2018).

76. Julien to Mourant," April 29, 1952, PP/AEM/K.90, WCL.

77. Mourant to Chappel, May 9, 1962, PP/AEM/E.8, WCL. Mourant had a whole folder of correspondents devoted to "student expeditions" (mostly from Oxford and Cambridge) to India, Brazil, Ethiopia, the People's Republic of Mongolia, Bhutan, and Madagascar, among others.

78. Roberts et al., "Blood Groups of the Northern Nilotes" (1955).

79. Before Allison set off on this first expedition, he also met Mourant's colleague Marshall Chambers for training in "field" sickle cell tests, an element of his project that eventually became the focus of his early research.

Later, Allison proposed that the sickle cell allele had been selected to confer an evolutionary advantage in malarial environments.

80. The Oxford Exploration Club turned out to be a remarkably productive institution for Mourant. As well as several club expeditions listed in Mourant's WCL archive, Burton notes at least two other blood collecting expeditions that it supported in the early 1950s—one to Iran, another to Socotra in the Arabian Sea: Burton, *Genetic Crossroads* (forthcoming). For biographical notes on Allison: Carroll, *Into the Jungle* (2008), 149–65.

81. The Colonial Office Research Fund provided £650; the rest came from the University of Oxford, the Oxford colleges, and members themselves, but they also solicited £30 (of a budget of approximately £1,300) from the British Museum, and £100 from the Royal Geographical Society, which routinely contributed money to Oxford University expeditions. OUEC to Oxford University Registry, February 14, 1949, MSS Dep c. 991, University of Oxford, Special Collections. For more on the elevation of scientific research by the Colonial Office after the Second World War: Clarke, "A Technocratic Imperial State?" (2007): 453–80.

82. On the early stages of the Mau Mau resistance moment, and the unsustainable population pressure experienced on Kikuyu "reserves" and other parts of Kenya: Anderson, *Histories of the Hanged* (2005).

83. "Alexander Allan Paton Memorial Fund: Report for 1949," n.d., 3, MSS Dep c. 991, University of Oxford, Special Collections.

84. Allison, personal communication by email to the author, March 5, 2012.

85. Allison et al., "Blood Groups in Some East African Tribes" (1952), 55.

86. Allison et al., "Blood Groups in Some East African Tribes" (1952); Allison, personal communication by email to the author, March 12, 2012.

87. After his apparent successes in Kenya, Allison took a similar approach to an expedition to northern Scandinavia to test the blood groups of the "Lapps"—a term used in the 1950s for Sami people, consisting of communities in far northern Sweden, Norway, and Finland, and the Kola Peninsula of Russia: Bangham, "Blood Groups and Human Groups" (2014).

88. On the "isolates" of human genetics: Lipphardt, "From 'Races' to 'Isolates' to 'Endogamous Communities'" (2013); Lipphardt, "'Geographical Distribution Patterns of Various Genes'" (2014); Widmer, "Making Blood 'Melanesian'" (2014); Germann, "'Nature's Laboratories of Human Genetics'" (2017).

89. The local health inspector was a "Mr. Albertyn": Zoutendyk et al., "The Blood Groups of the Hottentots," (1955).

90. By the 1950s, almost all "reserves" in South Africa were suffering from extreme poverty; for more on this history: Posel, "The Apartheid Project, 1948–1970" (2011).

91. The regional health officer was a Werner Kuschke: Zoutendyk et al., "The Blood Groups of the Bushmen" (1953).

92. Roberts cited help from a "Dr. Clarke" and "Mr. Bailey": Roberts et al., "Blood Groups of the Northern Nilotes" (1955).
93. Tilley, *Africa as a Living Laboratory* (2011); Widmer and Lipphardt, eds., *Health and Difference* (2016).
94. Here, I use the terms "research subject" and "donor" for those who gave blood; but "donors" were also often research subjects, and both kinds of donation almost always entailed a stranger with a needle.
95. On the subjectification of people, bodies, and their tissues, and the "tournaments of value" that constitute negotiations for body parts: Anderson, "Objectivity and Its Discontents" (2013); for an expansive discussion of the exchanges that accompany the scientists' acquisition of cells, blood, and other tissues: Lock, "The Alienation of Body Tissue" (2001).
96. Blood extraction in medical settings has multiple social and technical histories. There is no single account of blood extraction or of the needle, but for partial and brief accounts of medical bloodletting: Rosenfeld, "A Golden Age of Clinical Chemistry" (2000); Parapia, "History of Bloodletting by Phlebotomy" (2008): 490–95; Schmidt and Ness, "Hemotherapy" (2006).
97. Allison recalled his assistants as credible medical practitioners who "spoke the local languages and were respected for the excellent work they had done to promote health": Allison et al., "Further Observations on Blood Groups in East African Tribes" (1954). Another of Mourant's correspondents, William Laughlin, described blood collections among Basque communities in Oregon, and the importance of assistant Anthony Yturri, a local "Basque university graduate": Laughlin to Mourant, April 3, 1951, PP/AEM/K.99, WCL.
98. Obscuring the identities of informants was a familiar practice among Western anthropologists; for a rich account of the work, motivation, and professional identities of such informants: Schumaker, *Africanizing Anthropology* (2001).
99. Royal Anthropological Institute of Great Britain and Ireland, *Notes and Queries on Anthropology*, 6th ed. (1951), 21.
100. A venule was a glass tube, with a rubber stopper punctured by a glass needle. Blood samples were typically then refrigerated using ice, at a temperature just above 0°C (to prevent their freezing), and transported as quickly as possible to a blood grouping laboratory. Labeling was paramount, as nothing useful could be done with the blood without it. Writing later about the problem of labeling, Mourant advised that the ice be kept separated from specimens by plastic bags, and that the "vacutainers be indelibly numbered preferably with a diamond." Mourant to Boyd, December 10, 1962, PP/AEM/K.13, WCL.
101. Royal Anthropological Institute of Great Britain and Ireland, *Notes and Queries on Anthropology* (1951), 21.
102. Royal Anthropological Institute of Great Britain and Ireland, *Notes and Queries on Anthropology* (1951), 19.

103. Burton, "Essential Collaborators'" (2018). For other insightful historical accounts of how research subjects cultivated and used biomedical researchers to leverage goods and power: Anderson, "Objectivity and Its Discontents" (2013); Dent, "Invisible Infrastructures" (2016); Widmer, "Making Blood 'Melanesian'" (2014). For a vivid account of the exchanges around postwar tissue collection: Anderson, *Collectors of Lost Souls* (2008).

104. The paper in which this quote appeared was one of several published alongside Mourant's Basque work in a single issue of *American Journal of Physical Anthropology*: Da Silva, "Blood Groups of Indians, Whites and White-Indian Mixtures" (1949).

105. For an in-depth analysis of the power dynamics between Mourant, his correspondents, and their research subjects: Burton, "Essential Collaborators'" (2018); Burton, *Genetic Crossroads* (forthcoming).

106. The phrase "village bleed" had been used by depot director Janet Vaughan during the Second World War: Vaughan, "Medical Research Council Emergency Blood Transfusion Service, for London and the Home Counties," 1940, folder 13-1-C: Blood Donation, topic collection 13: Health 1939–47, Mass Observation Archive, University of Sussex Special Collections.

107. For a fascinating historical analysis of rumors about blood-sucking firemen in colonial East Africa: White, "Cars out of Place" (1993); White, "Tsetse Visions" (1995); White, *Speaking with Vampires* (2000).

108. Anthropologist Wenzel Geissler has written about the diverse experiences of medical researchers in Kenya and Uganda in the 1950s: Geissler, "'Kachinja Are Coming!'" (2005).

109. For anthropological work on biomedical encounters around blood: Geissler et al., "'He Is Now Like a Brother, I Can Even Give Him Some Blood' (2008); Reddy, "Citizens in the Commons" (2013).

110. Kohler, *Lords of the Fly*, 133–72.

111. Thanks to Elise Burton for helping to develop this characterization of Mourant. On the postwar internationalist construction of this way of seeing: Selcer, "The View from Everywhere" (2009).

CHAPTER EIGHT

1. This trilogy of blood books for Blackwell was instigated by publisher John Grant, whose wife, Jean Malcolm Grant, was the transfusion officer for Oxford: Mourant, *Blood and Stones* (1995), 68.

2. Birdsell, "Review of *The Distribution of the Human Blood Groups*" (1956); Boyd, "*The Distribution of the Human Blood Groups*" (1955); Kherumian, "Review: *The Distribution of the Human Blood Groups*" (1954).

3. On the midcentury construction of genetic population categories, and the relationship of these to political projects in the Middle East, United States, and Europe: Burton, *Genetic Crossroads* (forthcoming); Chresfield, *What Lies Between* (unpublished manuscript); Abu El-Haj, *The Genealogical Sci-*

ence (2012); Kirsh, "Population Genetics in Israel in the 1950s" (2003). For rich accounts of how local administrative practices ordered the biological, health, and social sciences conducted in colonial settings: Widmer and Lipphardt, eds., *Health and Difference* (2016); for analyses of the taxonomic categories used by midcentury geneticists: Lipphardt, "'Geographical Distribution Patterns of Various Genes'" (2014); Lipphardt, "From 'Races' to 'Isolates' to 'Endogamous Communities'" (2013).

4. For a more extensive discussion of the ramifications of "a priori" classifications in early blood group mapping: Gannett and Griesemer, "The ABO Blood Groups" (2004), 128–39.

5. For these quotes on the Basques: Araquistain, "21. Some Survivals of Ancient Iberia in Modern Spain" (1945), 33; Barandiarán, "On the Conservation of the Basque Peoples" (1946), 96; Taylor, "The Evolution and Distribution of Race, Culture, and Language" (1921), 83.

6. Morant, "A Contribution to Basque Craniometry" (1929), 67.

7. Mourant also had a technical pretext for his interest in the Basques. Taking an evolutionary view of Rh incompatibility, geneticist Haldane and serologist Wiener had independently pointed out that the lethal consequences of this condition should have caused the rarer of the Rh alleles (*Rh+* or *rh–*) to be selected out of human populations; yet they coexisted. Haldane posited that the existence of this polymorphism might be explained if modern Europeans were descended from two ancestral groups, one carrying only *Rh+* alleles, and the other only *rh–* alleles. In one of the few instances in which Mourant himself collected data to address a specific hypothesis, he sought to test the idea that the Basques were the source of the unexpected Rh-negative alleles in Europe. Haldane, "Selection against Heterozygotes in Man" (1942); Wiener, "The Rh Factor and Racial Origins" (1942).

8. Barandiarán, "On the Conservation of the Basque Peoples" (1946), 97.

9. Chalmers et al., "The ABO, MN and Rh Blood Groups of the Basque People" (1949), 531.

10. Chalmers et al., "The ABO, MN and Rh Blood Groups of the Basque People" (1949), 531–32.

11. Mourant, *The Distribution of the Human Blood Groups* (1954), 41.

12. Chalmers et al., "The ABO, MN and Rh Blood Groups of the Basque People" (1949), 530.

13. "Rh Factor Clue to Race," *Science News-Letter* 54, no. 10 (1948): 149.

14. Allison, personal communication by email to the author, March 23, 2012.

15. Allison, personal communication by email to the author, March 12, 2012.

16. Laughlin to Mourant, April 3, 1951, PP/AEM/K.99, WCL.

17. Here, anthropologically "relevant particulars" included (1) a serial number (so that the punch card could be related to an individual donor); (2) whether the donor was on the "live" or resigned panel (which might introduce bias); (3) donor surname; (4) sex and marital status; (5) ABO blood group; and (6) whether the donor was Rh-positive or Rh-negative.

Fraser Roberts, "An Analysis of the ABO Blood Group Records of the North of England" (1953), 362–63.

18. Fraser Roberts, "An Analysis of the ABO Blood Group Records of the North of England" (1953), 363.

19. Fraser Roberts, "An Analysis of the ABO Blood Group Records of the North of England" (1953), 387.

20. Before 1959, large towns and cities had been divided into several independent areas to facilitate postal sorting, but these were not fine-grained enough for the mapping project.

21. Fraser Roberts, "The Coding of Postal Addresses," October 15, 1953, A91/2/16, RAI.

22. "A Proposed Survey of the British Isles (Draft)," ca. 1951, A91/2/2, RAI.

23. Fraser Roberts, "An Analysis of the ABO Blood Group Records of the North of England" (1953), 363–65.

24. The chi-squared test was a statistical method for estimating the chances that the observed and expected proportions deviated significantly from one another.

25. Kopeć, *The Distribution of the Blood Groups in the United Kingdom* (1970), 1.

26. The war had been the backdrop of the country's influential Population Health Survey, which later became a foundational model for population health analysis: Szreter, "The Population Health Approach in Historical Perspective" (2003). On statistics gathering in postwar Britain in relation to public health: Alex Mold et al., *Placing the Public in Public Health* (2019); for a brief mention of statistic gathering on hospital management within the NHS: Rivett, *From Cradle to Grave* (1998), 97; on postwar cohort studies and the Population Investigation Committee: Edmund Ramsden, "Surveying the Meritocracy" (2014); on a proposal to bring cytogenetic (chromosome analysis) facilities into the framework of the NHS, with a view to studying its public health dimension: de Chadarevian, "Chromosome Surveys" (2014).

27. Blumberg, *Proceedings of the Conference on Genetic Polymorphisms* (1962).

28. This new role for blood in hospitals had apparently dramatically improved the quality of blood group testing there; the authors of one association study remarked that, for that reason, they dealt only with cases treated since 1948: Aird, "A Relationship between Cancer of Stomach and the ABO Blood Groups" (1953).

29. Aird et al., "The Blood Groups in Relation to Peptic Ulceration" (1954); Grahame, "The ABO Blood Groups and Peptic Ulceration" (1961).

30. Maxwell and Maxwell, "ABO Blood Groups and Hypertension" (1955).

31. Mourant, "Memorandum on the History, the Work and the Future Plans of the Centre," ca. 1955–1956, 3, PP/AEM/C.11, WCL.

32. This collection of cards was recently in the custody of the Division of Biological Anthropology, University of Cambridge, and is now being catalogued by the Wellcome.

33. Mourant, "Memorandum on the History, the Work and the Future Plans of the Centre," ca. 1955–1956, 4, PP/AEM/C.11, WCL.
34. Mourant, "Memorandum on the History, the Work and the Future Plans of the Centre," ca. 1955–1956, 3, PP/AEM/C.11, WCL.
35. Mourant, "Memorandum on the History, the Work and the Future Plans of the Centre," ca. 1955–1956, 4, PP/AEM/C.11, WCL.
36. Mourant, *The Distribution of the Human Blood Groups*, 70. Having grown up in a religious family, Mourant claimed that he felt a special affinity with "the Jews." In his 1978 book, *The Genetics of the Jews*, he ascribed this to an influential school teacher who, as a British Israelite, believed that all British people, if not all Europeans, were descended from the lost tribes of Israel. Mourant wrote: "I believed that I, and all the people among whom I lived, were really Jews, and that the biblical prophecies about the return from captivity, and the eschatological prophecies about the Jews in the book of Revelation applied to us." He went on: "For some of my most formative years, I considered myself a Jew, and this sense of identity has to some extent persisted and, though I can now see no evidence that I have any Jewish ancestors, I have maintained a deep interest in the Jewish peoples." Mourant, *The Genetics of the Jews* (1978), v.
37. The calculations needed to transform a population "blood group frequency" into "allele frequency" were different for each blood group system, and depended on the genetic architecture of those systems. There was also a choice of mathematical techniques for making those calculations; the most sophisticated was Fisher's "maximum likelihood method."
38. Nelly Oudshoorn makes a similar observation about the representation of women and menstrual cycles in pill trials: Oudshoorn, *Beyond the Natural Body* (1994), 132.
39. So, for example, in a section of his book on Tibet, Mourant could write "Tennant's observations on 187 Tibetans must be viewed with considerable caution since he found 24.1 per cent of AB's [sic], a very much higher frequency than would be expected on the basis of genetical equilibrium": Mourant, *The Distribution of the Human Blood Groups* (1954), 117.
40. On Mourant's apparent indifference to sample sizes: Burton, *Genetic Crossroads* (forthcoming).
41. Mourant, *The Distribution of the Human Blood Groups* (1954), vii.
42. On the postwar debates over the uses of blood groups to defend racial taxonomies: Silverman, "The Blood Group 'Fad'" (2000).
43. I take the term "calibration" in this sense from Nick Jardine, who uses it to describe how new methods and standards might successfully substitute for old ones: Jardine, *The Scenes of Inquiry*, 155–67.
44. Mourant offers no explanation of why he chose to include maps of the following alleles: *A*, *B*, *O* and *Rh* (*D*) in Europe; *A*, *B*, *Rh* (*C*), *Rh* (*E*), and *M* across the world.

45. Alongside these archival materials, the cartographer for the second edition, John Hunt, has given a detailed account of his collaboration with Mourant: Bangham, interview with Hunt, Milton Keynes, 2017.
46. Although Mourant was working in a radically more uncertain domain, this recalls the "trained judgment" that characterized scientific atlas making in the 1950s: Daston and Galison, *Objectivity* (2007).
47. Dunn, "The Coming of Age of Blood Group Research" (1955), 37.
48. Mourant, "Memorandum on the History, the Work and the Future Plans of the Centre," ca. 1955, 8, PP/AEM/C.11, WCL.
49. Mourant, *The Distribution of the Human Blood Groups* (1954), 1.
50. Mourant, *The Distribution of the Human Blood Groups* (1954), xv; Birdsell, "Review of *The Distribution of the Human Blood Groups*" (1956), 208.
51. Mourant, *The Distribution of the Human Blood Groups* (1954), 148.
52. Marks, "'We're Going to Tell These People Who They Really Are'" (2001); Marks et al., "Human Biodiversity" (1995); Silverman, "The Blood Group 'Fad'" (2000); Lipphardt, "The Jewish Community of Rome"; Lipphardt, "Isolates and Crosses in Human Population Genetics" (2012).
53. Not everyone took population categories for granted, and in the 1920s Lawrence Snyder had drawn attention to the arbitrariness of the population boundaries that he used, but by and large such categories were simply assumed: Gannett and Griesemer, "ABO Blood Groups" (2004), 130–31.
54. For a detailed account of Dunn's journey to Rome: Lipphardt, "Jewish Community of Rome—an Isolated Population?" (2010).
55. Roberts et al., "Blood Groups of the Northern Nilotes" (1955), 135.
56. Kraus and White, "Microevolution in a Human Population" (1956), 1017, 1019.
57. Chown, "Problems in Blood Group Analysis" (1957), 888.
58. Taking an example from another detailed study of genetic fieldwork, geneticists James Neel and Francisco Salzano took with them social anthropologist David Maybury-Lewis on their first field expedition to study Xavante people in Brazil in 1962: Dent, *Studying Indigenous Brazil* (2017).
59. Birdsell, "The ABO Blood Groups" (1960).
60. Gannett and Griesemer, "ABO Blood Groups" (2004); Reardon, *Race to the Finish* (2004).
61. For detailed analyses of the topographies of genetic mapping in a different national context (the Soviet Union): Bauer, "Population Genetics, Cybernetics of Difference, and Pasts in the Present" (2015); Bauer, "Mutations in Soviet Public Health Science" (2014).

CHAPTER NINE

1. Karen-Sue Taussig remarks that human genetics is "readily decontextualized as a symbol of the universal and a sign of the unmarked cosmopolitan": Taussig, *Ordinary Genomes* (2009), 3.

2. For histories of the UN and its specialized agencies: Amrith and Sluga, "New Histories of the United Nations" (2008).

3. For a review of new scholarship on history of the UN and its contradictions: Amrith and Sluga, "New Histories of the United Nations" (2008).

4. On the very partial vision of what "antiracism" could be: Hazard, *Postwar Anti-racism* (2012); Amrith and Sluga, "New Histories of the United Nations" (2008).

5. Kevles, *In the Name of Eugenics* (1995), 210; on postwar German "whitewashing" of human genetics: Weiss, "After the Fall" (2010).

6. UNESCO, *Creation of UNESCO* (1945).

7. UNESCO, "UNESCO Constitution" (1945).

8. UNESCO, *The Race Concept* (1952), 6.

9. On the ambition of the UN to "speak for all peoples": Amrith, *Decolonizing International Health* (2006); Amrith and Sluga, "New Histories of the United Nations" (2008).

10. Petitjean, "The Joint Establishment of the World Federation of Scientific Workers and of UNESCO" (2008). For scientific humanism in Britain: Harman, "C. D. Darlington and the British and American Reaction to Lysenko" (2003); Mayer, "Reluctant Technocrats" (2005); Smith, "Biology and Values in Interwar Britain" (2003); Sommer, "Biology as a Technology of Social Justice" (2014). For science and social and economic planning: McGucken, "On Freedom and Planning in Science" (1978): 42–72; Werskey, *The Visible College* (1978).

11. Sommer, *History Within* (2016), 135–248.

12. Donna Haraway, *Primate Visions* (1989), 186–230, esp. 198.

13. Hazard, *Postwar Anti-racism* (2012); Miller, "'An Effective Instrument of Peace'" (2006); UNESCO's image of neutrality was made and promoted through its organizational structures: Rangil, "The Politics of Neutrality" (2011); Selcer, "The View from Everywhere" (2009).

14. In Huxley, *UNESCO* (1947). For Huxley's "world community" and influence on UNESCO: Sluga, "UNESCO and the (One) World of Julian Huxley" (2010).

15. This project was represented by books such as Dobzhansky's *Genetics and the Origin of Species* (1937), Mayr's *Systematics and the Origin of Species* (1942), and George Gaylord Simpson's *Tempo and Mode in Evolution* (1944).

16. Mayr and Provine, eds., *The Evolutionary Synthesis* (1980); Smocovitis, "Unifying Biology" (1992).

17. Smocovitis, "Humanizing Evolution" (2012).

18. UNESCO, *Report to the United Nations* (1949), 44–45.

19. UN General Assembly, "Universal Declaration of Human Rights," General Assembly Resolution 217 A (III), December 10, 1948, https://www.un.org/en/ga/search/view_doc.asp?symbol=A/RES/217(III). On the UN's concern with equality and its coherence (or lack thereof) with a US-led modernization agenda: Hazard, *Postwar Anti-racism* (2012) and references therein.

20. "Resolution of Economic and Social Council on 'The Prevention of Discrimination and the Protection of Minorities,'" October 28, 1949, 323.1, UNESCO Archives.

21. Brodersen and Klineberg. "Memo: Project on Dissemination of Scientific Facts Regarding Race," March 25, 1949, 323.1, UNESCO Archives.

22. Brattain, "Race, Racism, and Antiracism" (2007); Gormley, "Scientific Discrimination and the Activist Scientist" (2009); Hazard, "A Racialized Deconstruction?" (2011); Müller-Wille, "Race et appartenance ethnique" (2007); Selcer, "The View from Everywhere" (2009).

23. Participant Ashley Montagu explained that science could oppose the irrationality and ignorance that had led to a "great and terrible war": Reardon, *Race to the Finish* (2004), 26.

24. On official US opposition to the race campaign: Duedahl, "UNESCO Man" (2008); on UNESCO's projection of science to the public, and on opposition to the second statement by Fisher: Brattain, "Race, Racism, and Antiracism" (2007); Gormley, "Scientific Discrimination and the Activist Scientist" (2009); on the engagement with the statements on race by civil rights activists and critics: Hazard, "A Racialized Deconstruction?" (2011). For analysis of the campaign in relation to late colonial, postcolonial, and Cold War theories of socioeconomic development: Gil-Riano, "Historicizing Anti-racism" (2014); Gil-Riano, "Relocating Anti-racist Science" (2018).

25. These quotes are derived from the 1950 statement and were restated in the 1951 statement: UNESCO, "Four Statements on the Race Question" (1969), 30–35 (1950 statement) and 36–43 (1951 statement).

26. UNESCO, *What Is Race?* (1952), 6.

27. During the course of the campaign on race, the department also published a series of pamphlets by distinguished scholars that included *Race and Psychology* (1951) by Otto Klineberg, *Race and Biology* (1951) by Dunn, and *Race and History* (1952) by Claude Levi-Strauss. These were aimed at an "educated public already familiar with the major themes of culture and science," an aspiration reflected in the pamphlets' sober and muted design and their absence of images. Quote from Delaveney, Head, Documents and Publications Service to Department of Mass Communication, from March 15, 1951, 323.1 (094.4), UNESCO Archives.

28. To illustrate the book Tead chose Paris-based US painter and magazine illustrator Jane Eakin.

29. Both quotations from UNESCO, *What Is Race?* (1952), 4.

30. The inside cover of *What Is Race?* explained that the booklet sought to convey "provisional conclusions reached by scientists" in "non-technical language and illustrated—often in an oversimplified form—so as to make them more easily intelligible to the layman."

31. UNESCO, *What Is Race?* (1952), 5, 42.

32. UNESCO, *What Is Race?* (1952), 51.

33. For two analyses of the contradictory cultural meanings of transfusion in

relation to race: Weston, "Kinship, Controversy, and the Sharing of Substance" (2001); Chinn, *Technology and the Logic of American Racism* (2000).

34. Gilbert, *Emergency Call* (1952), 00:57:55.
35. Killens, *Youngblood* (1982), 460. For a close reading of the politics and kinships in this novel: Weston, "Kinship, Controversy, and the Sharing of Substance" (2001).
36. Titmuss, *The Gift Relationship* (1970), 15.
37. On World Blood Donor Day in 2016, the WHO used the slogan "Blood Connects us All." See World Health Organization, "World Blood Donor Day 2016," http://www9.who.int/campaigns/world-blood-donor-day/2016/en/.
38. Mourant's collecting project was devoted to the notion that the different races of the world had different frequencies of blood groups, and Robert Race was deeply interested in the potential for certain new groups to disclose the true racial identities of donors.
39. On the history of the BBC: Briggs, *The BBC* (1985).
40. On the BBC in relation to science: Boon *Films of Fact* (2008); other useful works on the BBC and science in slightly later periods include de Chadarevian, *Designs for Life*, 136–60; Nathoo, *Hearts Exposed* (2009).
41. The memo—"Science and the BBC"—was forwarded by Huxley in 1941. Quotation from Boon, *Films of Fact* (2008), 186.
42. Boon, *Films of Fact* (2008), 195.
43. Cardiff and Scannell, "Broadcasting and National Unity" (1987).
44. Waters, "'Dark Strangers' in Our Midst" (1997); Paul, *Whitewashing Britain* (1997).
45. George Noordhof, "'Race' Programme," May 1, 1952, T32/209/1, BBC Written Archives. For more on BBC television and race during this period: Newton, "Calling the West Indies" (2008); Newton, *Paving the Empire Road* (2012).
46. George Noordhof, "'Race' Programme," May 1, 1952, T32/209/1, BBC Written Archives.
47. James Bredin, "Memo: Africa Series: Scientific Programme, to Wyndham Goldie," October 15, 1952, T32/209/1, BBC Written Archives.
48. Bredin, "Memo: Africa Series."
49. "Race and Colour: Richie Calder and Dr. Trevor," November 9, 1952, T32/209/1, BBC Written Archives.
50. Attenborough had recently started his first job at the BBC, *Race and Colour* producers had approached him in the BBC canteen to step in and act as the "Caucasian" representative: Attenborough, February 15, 2012, personal communication to the author.
51. "Race and Colour: Draft Script (for Cameras)," n.d., T32/209/1, BBC Written Archives.
52. "Unesco Radio: Science and Racial Barriers," 1951, T32/209/1, BBC Written Archives.
53. "A Viewer Research Report: 'Race and Colour,'" November 26, 1952, T32/209/1, BBC Written Archives.

54. "A Viewer Research Report," 1.
55. "A Viewer Research Report," 1.
56. Mourant, *The Distribution of the Human Blood Groups* (1954), 1.
57. Comfort, *The Science of Human Perfection* (2012); Lindee, *Suffering Made Real* (1994).
58. "Chromosomes in Medicine" (1962).

CHAPTER TEN

1. Mourant, "Blood Groups" (1961), 5.
2. Lemov, *The Database of Dreams* (2015); Erickson et al., *How Reason Almost Lost Its Mind* (2015).
3. Aronova et al., "Big Science and Big Data in Biology" (2010).
4. Radin, *Life on Ice* (2017).
5. Comfort, *The Science of Human Perfection* (2012); Lindee, *Suffering Made Real* (1994).
6. Lindee, *Moments of Truth* (2005); Hogan, *Life Histories of Genetic Disease* (2016); Comfort, *The Science of Human Perfection* (2012); Kevles, *In the Name of Eugenics* (1995).
7. Radin, "Unfolding Epidemiological Stories" (2014).
8. For a riveting account of the midcentury labor of storing and managing data: Lemov, *Database of Dreams* (2015).
9. For analysis of more recent efforts to maintain the vitality of blood samples: Kowal, "Orphan DNA" (2013). For discussion of the complex and hard-to-manage fates of scientific collections: Jardine et al., "How Collections End" (2019); in relation to political and disciplinary change: Roque, "The Blood That Remains (2019); in relation to shifting disciplines and paradigms: Kakaliouras, "The Repatriation of the Palaeoamericans (2019); in relation to retirement: Hopwood, "The Tragedy of the Emeritus and the Fates of Anatomical Collections" (2019).
10. Pauling et al., "Sickle-Cell Anemia: A Molecular Disease" (1949); Neel, "The Inheritance of Sickle Cell Anemia" (1949).
11. Kay, "Laboratory Technology and Biological Knowledge" (1988); Kay, *Molecular Vision of Life* (1993).
12. Allison, "The Distribution of the Sickle-Cell Trait" (1954). During the same year Allison published work in the *BMJ* in which he demonstrated an apparent resistance to malaria among sickle cell carriers by deliberately infecting people with and without the trait: Allison, "Protection Afforded by Sickle Cell Trait against Subtertian Malarial Infection" (1954).
13. De Chadarevian, "Following Molecules" (1998).
14. Lehmann and Smith, "Separation of Different Haemoglobins by Paper Electrophoresis" (1954), 12; see also Motulsky et al., "Paper Electrophoresis of Abnormal Hemoglobins and Its Clinical Applications" (1954).

15. For an contemporary review: Allison, "Abnormal Haemoglobins and Erythrocyte Enzyme-Deficiency Traits" (1961); on Lehmann and research on sickle cell anemia in Britain: Redhead, "Histories of Sickle Cell Anaemia"; on sickle cell studies in the Middle East: Burton, "Red Crescents" (2019).

16. Radin, *Life on Ice* (2017).

17. Smithies, "Grouped Variations in the Occurrence of New Protein Components" (1955); Barnicot, "Haptoglobins and Transferrins" (1961), 41; Giblett, "Haptoglobins and Transferrins" (1961).

18. Hockwald et al., "Toxicity of Primaquine in Negroes" (1952); Beutler, "Glucose-6-Phosphate Dehydrogenase Deficiency" (2008); Motulsky and Campbell-Kraut, "Population Genetics of Glucose-6-Phosphate Dehyrogenase Deficiency of the Red Cell" (1961), 159–91. For historical analysis of racial genetic studies in US prisons: Comfort, "The Prisoner as Model Organism" (2009).

19. In a paper that later became known as the first clear hypothesis that varied responses to drugs might be related to subtle genetic differences: Motulsky, "Drug Reactions, Enzymes, and Biochemical Genetics" (1957).

20. Motulsky et al., "Population Genetics in the Congo. I" (1966).

21. De Chadarevian, "Chromosome Surveys" (2014); de Chadarevian, "Putting Human Genetics on a Solid Basis" (2013).

22. De Chadarevian, "Chromosome Surveys" (2014); de Chadarevian, *Heredity under the Microscope* (2020).

23. Jean Dausset of the Centre National de Transfusion Sanguine is credited with the discovery of the first HLA antigen: Dausset, "Iso-Leuco-Anticorps" (1958); Van Rood et al., "Leucocyte Antibodies in Sera from Pregnant Women (1958); Payne and Rolfs, "Fetomaternal Leukocyte Incompatibility" (1958).

24. Thorsby, "A Short History of HLA" (2009).

25. Race and Sanger, *Blood Groups in Man* (1962), preface.

26. Ceppellini et al., "Genetics of Leukocyte Antigens" (1967).

27. "The more extensive our knowledge of the differences in biochemical make-up between different normal individuals . . . the more likely we are to begin to understand why . . . some people are more susceptible than others to particular types of clinical disorder." "Biochemical Individuality" (1961): 650–51; Blumberg, ed., *Proceedings of the Conference on Genetic Polymorphisms* (1961).

28. Fibrin is a fibrous protein involved in the clotting of blood. The "foam" and "film" fractions are so called according to their distinct textures. The products made from the preparation of blood plasma would, by the 1970s, constitute a multimillion-dollar industry worldwide. For a riveting account of the long history of blood fractionation, see "Blood Cracks Like Oil," in Starr, *Blood: An Epic History*, 118–42.

29. E.g., Mollison and Sloviter, "Successful Transfusion of Previously Frozen Human Red Cells" (1951). On the uses of glycerol to successively freeze and thaw semen, ovarian tissue, cell cultures, and blood: Landecker, "Living

Differently in Biological Time" (2005); Parry, "Technologies of Immortality" (2007).

30. Radin, *Life on Ice* (2017); specifically on serum preservation: Radin, "Latent Life" (2013).

31. In the fourth edition of *Blood Groups in Man* (1962), Race and Sanger noted in their preface with some bemusement: "Though, in 1950, 'Blood Groups' was appropriate in the title of a book dealing almost entirely with antigens of the red cells, we now feel that we owe an apology to those who have discovered other inherited groups in man, scarcely mentioned in this book, such as the various haemoglobins, the haptoglobins, the Gm serum groups and the splendidly complicated antigens of the white cells."

32. Race and Sanger, "Recent Applications of Blood Groups to Human Genetics" (1969). Sylvia Lawler, in a later interview, remarked that all of the researchers had been "absolutely delighted" with the progress they made with blood groups and linkage: "We were quite happy that we found as much as we [did]." Kevles, Daniel J., Sylvia Lawler, June 29, 1982, 7, Daniel J. Kevles papers, Oral History Interview Transcripts, 1982–1984, box 1, folder 17, RAC.

33. Washburn, "The New Physical Anthropology" (1951); Strandskov and Washburn, "Genetics and Physical Anthropology" (1951).

34. On Washburn's field-building work from 1950, and on this larger shift: Haraway, "Remodelling the Human Way of Life" (1988), 206–60; Smocovitis, "Humanizing Evolution" (2012); Lindee and Radin, "Patrons of the Human Experience" (2016); Lindee and Santos, "The Biological Anthropology of Living Human Populations" (2012); Little, "Human Population Biology in the Second Half of the Twentieth Century" (2012).

35. Planning for this society was shaped in 1957 by a Ciba Foundation meeting convened by Weiner and Roberts in London to discuss "the scope of physical anthropology," and its relationship with a broad range of other disciplines. Meeting participants discussed the need to develop the field of physical anthropology in relation to "practical" problems, such as clothing, industrial design, and school furniture. Weiner, "International Biological Programme (IBP) 'Biology of Human Adaptability' [Letter of Invitation to Mourant]," ca. 1962, PP/AEM/K.394, WCL; Weiner, "Physical Anthropology . . . an Appraisal" (1957).

36. On the Human Adaptability section, Radin, *Life on Ice* (2017), 86–117; for an early outline: Weiner, "International Biological Programme (IBP) 'Biology of Human Adaptability' [Letter of Invitation to Mourant]," ca. 1962, PP/AEM/K.394, WCL.

37. For more on the links between these data-driven initiatives: Aronova et al., "Big Science and Big Data in Biology" (2010).

38. The Royal Society was the British national body for the International Council of Scientific Unions—an organization devoted to promoting international cooperation for the advancement of science; on its postwar status

from a British perspective: Stratton, "International Council of Scientific Unions" (1946).

39. Clapham, "A Review of the United Kingdom Contribution to the International Biological Programme: Introductory Remarks" (1976); Mourant, "Some Aspects of the International Biological Programme," ca. 1968, [uncatalogued, box "Subject R-W + Misc"], Division of Biological Anthropology, [now accessioned by Wellcome Collections, London].

40. Mourant, "Some Aspects of the International Biological Programme," ca. 1968, [uncatalogued, box "Subject R-W + Misc"], Division of Biological Anthropology, [now accessioned by Wellcome Collections, London].

41. Other IBP program areas: Conservation of Terrestrial Communities, Productivity of Freshwater Communities, Productivity of Marine Communities, Production Processes, Productivity of Terrestrial Communities, Use and Management of Biological Resources. For the first outline of the HA section ambitions: Weiner, "International Biological Programme (IBP) 'Biology of Human Adaptability' [Letter of Invitation to Mourant]," ca. 1962, PP/AEM/K.394, WCL.

42. For an in-depth account of the difficulties faced by the IBP in standardizing data collection: Aronova et al., "Big Science and Big Data in Biology" (2010).

43. The IBP's scientific director circulated a memo in 1966 that defined the criteria of IBP "stations" ("an existing or proposed research institution . . . where an IBP research team of major importance is based") and IBP "centres" ("an existing research institution of high standing, which undertakes to collect, process, store and retrieve material or data for IBP purposes"): Worthington, "IBP Centres and Stations [Definitions and Criteria]," August 10, 1966, PP/AEMK.396, WCL.

44. "ICSU—International Biological Programme, Report of Conference Project D—Human Adaptability, Section on Population Genetics," ca. 1963, PP/AEM/K.394, WCL.

45. Weiner, "ICSU—International Biological Programme Conference on Project D—'Human Adaptability,' London, December 4–6, 1962," 1962, PP/AEM/K.394, WCL. Weiner's conference began just a day after a meeting at WHO headquarters in Geneva that would be deeply consequential for the HA section of the IBP: Mourant to Weiner, October 26, 1962, PP/AEM/K.394, WCL. Building on discussions a few years before, the WHO meeting outlined the need for an extensive program of genetic and physiological research on humans, which would be based on large-scale human blood serum banking. Its focus on "primitive groups" was imported into the IBP/HA program and became the basis of collaboration between the two. Quotes from *Research in Population Genetics of Primitive Groups* (1964). For more on the WHO program of serum banking and its relationship to salvage anthropology and the IBP: Radin, *Life on Ice* (2017), 55–117; on the WHO technical reports: Radin, "Latent Life" (2013); Radin, "Unfolding Epidemiological Stories" (2014).

46. Mourant, "The International Biological Programme" (1964).
47. Wiener also served as president of the RAI from 1963 to 1965.
48. Mourant, "No Title [Mourant's Nomination of Weiner to Fellow of the Royal Society]," 1970, PP/AEM/K.398, WCL.
49. Mourant to Himsworth, June 24, 1964, PP/AEM/D.2, WCL.
50. Mourant to Weiner, January 22, 1959, PP/AEM/K.392, WCL; Mourant to Weiner, August 3, 1961, PP/AEM/K.393, WCL.
51. Mourant, "Proposal for the Foundation of a New Unit for Research in Serological Population Genetics," ca. 1965, PP/AEM/D.2, WCL.
52. When the Reference Laboratory had been founded, "tests of population samples were usually performed with antisera of nine or ten specificities"; now "about 25 [were] commonly used and more would be included if supplies of reagents were sufficient." Mourant, "A Unit for Research in Serological Population Genetics," ca. 1962, 7, PP/AEM/D.1, WCL.
53. Mourant, "A Unit for Research in Serological Population Genetics," ca. 1962, 7, PP/AEM/D.1, WCL.
54. For correspondence about the new unit: PP/AEM/D.2, WCL.
55. Mourant chose "serological" because there were already too many labs beginning with "blood group," and "there is already far too much misdirecting of specimens among the existing ones." Mourant to Bunjé, January 27, 1965, PP/AEM/D.2, WCL. On the Oxford unit: Harper et al., *Clinical Genetics in Britain* (2010). On that lab's involvement in a WHO population project to karyotype newborns: de Chadarevian, "Putting Human Genetics on a Solid Basis" (2013).
56. Maycock, "Kenneth Leslie Grant Goldsmith" (1976); Drury to Mourant, March 17, 1964, PP/AEM/D.2, WCL. For more on the later history of the Reference Laboratory: Gunson and Dodsworth, "The International Blood Group Reference Laboratory" (1996).
57. Their most recent volume had been *The ABO Blood Groups: Comprehensive Tables and Maps of World Distribution* (1958), which brought together six million pieces of data on the frequencies of the ABO blood groups up to the end of 1957—something that Mourant had been unable to do in *The Distribution of the Human Blood Groups* four years previously. The volume was printed in A4 landscape format—to more easily accommodate its maps—and contained very little text. This supplement also included data given to Mourant by William Boyd, who had published his last compilation of blood group data in 1939. Mourant et al., *The ABO Blood Groups: Comprehensive Tables and Maps of World Distribution* (1958); Lehmann, "Blood Groups [Review of *The ABO Blood Groups*, Mourant et al.]" (1959).
58. Mourant, "'The Nuffield Blood Group Centre,'" 1968, PP/AEM/C.13, WCL; The clerical offices were in the corporate heart of London; the rooms were sublet to the MRC by the Corporation of the City of London, who, in turn, leased them from the Abbey Life Assurance Company Limited. Mourant to

Lush, April 20, 1970, PP/AEM/D.15, WCL. Mourant also had the resources, by this time, to hire another female staff member, Sylvia Heath.

59. For this copious correspondence: PP/AEM/K.395, K.396, and K.397, WCL. Mourant also advised on the political dynamics of blood group laboratories, advising Weiner when his contacts were in danger of slipping up, for example sending samples of blood collected in China to Korea for testing. Mourant to Weiner, January 20, 1961, PP/AEM/K.392, WCL.

60. Although Mourant remained centrally interested in blood groups, he also became a local source of advice for some other tests, including the use of G6PD field testing. He encouraged the IBP to keep abreast of new gene polymorphisms, himself advising Weiner in 1968 that the IBP extend its gene-related purview to the exciting work being carried out on platelet and leucocyte antigens, "which are now being used as indices of compatibility in tissue grafts." Mourant to Weiner, November 14, 1968, PP/AEM/K.397, WCL; Mourant, "Unit for Research in Serological Population Genetics," ca. 1964, PP/AEM/D.1, WCL.

61. Mourant and Tills, "The International Biological Programme with Particular Reference to the Human Adaptability Section" (1967).

62. Weiner to Mourant, July 31, 1967, PP/AEM/K.397, WCL.

63. The section "Blood Collection and Subdivision" explained how samples should be obtained so that they could be subdivided for the purposes of different tests: Weiner and Lourie, eds., *Human Biology* (1969), 80–81.

64. Mourant to Bunjé, February 4, 1971, PP/AEM/D.18, WCL; Radin, "Unfolding Epidemiological Stories" (2014).

65. For a succinct primary summary of this work: Edwards, "Studying Human Evolution by Computer" (1966); for a rich historical account of the research carried out by Cavalli-Sforza and Edwards: Sommer, *History Within* (2016), 257–84.

66. Mourant to Himsworth, June 10, 1964, PP/AEM/D.2, WCL.

67. Edwards, "Blood Group Tabulation System [unpublished manuscript]," November 24, 1966, A. W. F. Edwards, private collection.

68. Edwards's proposal offered a detailed account of efficiently organizing how punch cards (for "library reference," "population and sera," and "phenotype"). His report was prepared with PhD students Christopher Cannings and Ann Eyland and programmer Agnes Corfield. US geneticist James Neel was also deeply interested in Edwards's system, hoping that he would broaden its scope to include other markers. Neel to Edwards, August 11, 1965, James V. Neel papers, MS coll. 96, series I, Edwards A. W. F. 1995–1993, folder 10, box 20, APS.

69. Edwards to Neel, January 11, 1967, James V. Neel papers, MS coll. 96, series I, Edwards A. W. F. 1995–1993, folder 10, box 20, APS.

70. For these transformations, Mourant and his colleagues contacted Edwards's medical geneticist brother, John, who, at the University of Birmingham, had developed some software appropriate for calculating gene frequencies

for the more complicated blood group data: Mourant et al., *Distribution of the Human Blood Groups and Other Polymorphisms* (1976), xiii. Kopeć tabulated and coded this blood group data and sent it to be transcribed onto punch cards by the MRC's Computer Services Centre, then sent it on to Birmingham. John Edwards sent successive printouts of the results back to Mourant in London, which Kopeć then tabulated by hand: Mourant to Osmundsen, November 13, 1970, PP/AEM/D.43, WCL.

71. In an awkward clarification of his cutoff date, Mourant wrote: "In extracting data from the literature we originally planned to end with material published in 1968, but this was subsequently extended to 1969. From the beginning of 1970 we have included only data which were accessible with ease, or which made a substantial contribution to known world distribution patterns, especially those of newly discovered factors." Mourant et al., *Distribution of the Human Blood Groups and Other Polymorphisms* (1976), xiii.

72. On the changing technologies and infrastructures for collecting and circulating biological data more generally, especially from the 1970s: Strasser, "Genbank: Natural History in the 21st Century?" (2008); Strasser, "Collecting, Comparing, and Computing Sequences" (2010).

73. Edwards, "[Untitled]," July 16, 1963, A. W. F. Edwards, private collection.

74. Mourant to Bunjé, February 4, 1971, PP/AEM/D.18, WCL.

75. By then, the budget of the statistical laboratory was about £13,500 per year, supporting five full-time workers. Initially Mourant asked for a single year of extra funding, up to mid-1972. In making the case for the completion of the book, Mourant calculated that the MRC had, by 1971, "spent at least £50,000 on the collection and computation of data." Mourant, "Draft Grant Proposal: Preparation of a Second Edition of 'The Distribution of the Human Blood Groups,'" ca. 1971, PP/AEM/D.15, WCL.

76. For papers on the future of the SPGL: PP/AEM/D.15, WCL. On the US Army European Research Office: May to Hoyt, October 23, 1970, PP/AEM/D.26, WCL; Mourant to Learmonth, October 6, 1970, PP/AEM/D.26, WCL.

77. Lush to Mourant, July 21, 1970, PP/AEM/D.16, WCL.

78. Reported in Kopeć, *The Distribution of the Blood Groups in the United Kingdom* (1970).

79. Tills to Beardmore, March 29, 1984, DF140/5/2, NHM; Tills to Beardmore, May 14, 1984. DF140/5/2. NHM.

80. Ball to Mourant, February 5, 1977, DF/140/5, NHM.

81. Mourant to Bunjé, February 4, 1971, PP/AEM/D.18, WCL.

82. Tills to Race and Sanger, October 4, 1973, DF/140/5, NHM.

83. Ball to Mourant, January 31, 1977, DF/140/5, NHM.

84. Perhaps because of its willingness to take Mourant's materials, the Natural History Museum also offered a solution to the problem of what to do with IBP/HA data that had originated in other British labs. The Natural History Museum had initially been reluctant to take the HA data, but the British Royal Society eventually brokered an agreement whereby all "basic" HA

data originating in the United Kingdom would be accommodated at the museum. IBP/HA leaders around the country were instructed by the Royal Society to send to the subdepartment of anthropology copies of their data—"in the form of computer cards or magnetic tape . . . with the appropriate coding." Martin to "All UK IBP/HA Project Leaders," February 12, 1975, DF140/5, NHM. On the difficulties of the IBP in establishing enduring data centers: Aronova, Baker, and Oreskes, "Big Science and Big Data in Biology" (2010).

85. Tills to Beardmore, May 14, 1984, DF140/5/2, NHM; Tills to Beardmore, March 29, 1984, DF140/5/2, NHM.

86. Mourant to Bunjé, February 4, 1971, PP/AEM/D.18, WCL.

87. For extensive analysis of such frozen collections: Radin, "Latent Life" (2013); Radin, *Life on Ice* (2017); Radin and Kowal, "A Comparative Study" (2015); Radin and Kowal, eds., *Cryopolitics* (2017).

88. Livingstone, *Abnormal Hemoglobins in Human Populations* (1967); Steinberg and Cook, *The Distribution of the Human Immunoglobulin Allotypes* (1981).

89. Cavalli-Sforza, "The DNA Revolution in Population Genetics" (1998).

90. On the Human Genome Diversity Project: Reardon, *Race to the Finish* (2004); M'charek, *The Human Genome Diversity Project* (2005).

CONCLUSION

1. For another story of how human genetics was been shaped by people in different social and professional locations: Lindee, *Moments of Truth* (2005).

2. Lindee, "Scaling up" (2014); Bangham and de Chadarevian, "Human Heredity After 1945" (2014); Lindee and Santos, "The Biological Anthropology of Living Human Populations" (2012).

3. Fortun, *Promising Genomics* (2008).

4. Quoted in Reardon, *Race to the Finish* (2004), 1.

5. E.g., Macintyre, "Opening the Book of Life" (2000).

6. Abu-El Haj, *The Genealogical Science* (2012); Reardon, *The Postgenomic Condition* (2017); Burton, *Genetic Crossroads* (forthcoming); Schwartz-Marín and Silva-Zolezzi, "'The Map of the Mexican's Genome'" (2010).

7. Reardon and TallBear, "'Your DNA Is Our History'" (2012).

8. 23andMe website, https://www.23andme.com/dna-ancestry/; accessed November 15, 2019.

9. The YouTube video "Momondo—the DNA Journey" is a marketing collaboration between AncestryDNA and the travel company Momondo. The campaign asked participants to record a video of themselves reacting to their DNA results, and to send it in to be judged in a competition—the winner would be granted a free trip to the countries corresponding to their DNA ancestry. It has been watched tens of millions of times: https://www.youtube.com/watch?v=tyaEQEmt5ls/; accessed February 22, 2020.

10. The Human Genome Diversity Project came to a premature end after groups advocating for indigenous rights condemned the project. For a fascinating

analysis of why this happened, and why the scientists involved were so surprised by the controversy: Reardon, *Race to the Finish* (2004).

11. For contemporary critiques of the use of race in genetics, many of which delve deep into the technical details of how genetic indices are constructed: Bolnick et al., "The Science and Business of Genetic Ancestry Testing" (2007); Duster, "Medicalisation of Race" (2007); Kahn, "How Not to Talk about Race and Genetics" (2018). For highly informative contextual analyses: Duster, *Backdoor to Eugenics* (2003); Wailoo and Pemberton, *The Troubled Dream of Genetic Medicine* (2006); Fujimura et al., "Introduction: Race, Genetics, and Disease" (2008); El-Haj, "The Genetic Reinscription of Race" (2007).

12. For analyses of "biogeographic ancestry," or BGA: Gannett, "Biogeographical Ancestry and Race" (2014); for "ancestry-informative markers," or AIMs: Fullwiley, "The Biologistical Construction of Race" (2008); for "single nucleotide polymorphisms," or SNPs: Rajagopalan and Fujimura, "Variations on a Chip" (2018).

13. Fullwiley, "The Biologistical Construction of Race" (2008); Gannett, "Biogeographical Ancestry and Race" (2014); Sommer, "History in the Gene" (2008).

14. Kakaliouras and Radin, "Archiving Anthropos" (2014).

15. Daston, "The Sciences of the Archive" (2012).

16. A Natural History Museum report explained that the collection sample was made up of eighty separate collections in the form of plasma fractions and as hemolysates (whole blood with lysed red cells).

17. For example: Monsalve and Hagelberg, "Mitochondrial DNA Polymorphisms in Carib People of Belize" (1997); Ricaut et al., "Mitochondrial DNA Variation in Karkar Islanders" (2008).

18. Kivisild, "Mourant Collection: Draft of the Report, September 2016," Leverhulme Centre for Human Evolutionary Studies, private collection.

19. For a rich history of these collections, and including why and how they were kept for future uses: Radin, *Life on Ice* (2017).

20. Radin and Kowal, "A Comparative Study" (2015).

21. World Health Organization Department of Vaccines and Biologicals, *WHO Global Action Plan for Laboratory Containment of Wild Polioviruses* (2004).

22. Under the Human Tissue Act, tissue "consists of or includes human cells." Many of the Cambridge samples are of serum, or they contain frozen whole blood, which would have hemolyzed the cells. The "Duckworth Collection" is an anthropological collection kept, as the website describes it, "solely as a source for scientific research." It comprises human remains acquired by the University of Cambridge from an estimated 18,000 people. Leverhulme Centre for Human Evolutionary Studies, "The Duckworth Laboratory," http://www.human-evol.cam.ac.uk/duckworth.html.

23. On the consent practices of some IBP researchers: Radin, *Life on Ice* (2017), 121–52; for a historical overview of consent protocols: Radin and Kowal, "A Comparative Study" (2015).

24. Radin and Kowal, "A Comparative Study" (2015).
25. The researcher I spoke to was concerned that the identification of risky genetic markers through population genomics research had the potential to compromise a community's access to medical care.
26. Emma Kowal describes frozen blood that has become divorced from documented relationships with donors as "orphaned." Efforts to keep old, frozen blood scientifically useful require ongoing affective and bureaucratic ties between blood, its "guardians," and indigenous donors: Kowal, "Orphan DNA" (2013).
27. On ethical regimes: Radin, *Life on Ice* (2017); Kowal, "Orphan DNA" (2013); on changing systems of governance, Reardon, *Race to the Finish* (2004); Reardon, *The Postgenomic Condition* (2017); on changing customs of exchange and openness: Reardon et al., "Bermuda 2.0" (2016).
28. For the endings and afterlife of another anthropological blood and paper collection, this time of blood soaked into paper cards: Roque, "The Blood That Remains" (2019).
29. On the affective relationships of trust that shape the transformation of scientific records into historical material: Radin, "Collecting Human Subjects" (2014).
30. On the ongoing coproduction of social and technical orders that shape the meanings and uses of frozen blood and other materials: Kowal et al., "Indigenous Body Parts, Mutating Temporalities, and the Half-Lives of Postcolonial Technoscience" (2013).

Bibliography

Abu El-Haj, Nadia. *The Genealogical Science: Genetics, the Origins of the Jews, and the Politics of Epistemology.* Chicago: University of Chicago Press, 2012.

Abu El-Haj, Nadia. "The Genetic Reinscription of Race." *Annual Review of Anthropology* 36 (2007): 283–300.

Adam, Alison. *A History of Forensic Science: British Beginnings in the Twentieth Century.* New York: Routledge, 2015.

Aird, Ian, H. H. Bentall, and John A. Fraser Roberts. "A Relationship between Cancer of Stomach and the ABO Blood Groups." *British Medical Journal* 4814 (1953): 799–801.

Aird, Ian, H. H. Bentall, J. A. Mehigan, and John A. Fraser Roberts. "The Blood Groups in Relation to Peptic Ulceration and Carcinoma of Colon, Rectum, Breast, and Bronchus." *British Medical Journal* 4883 (1954): 315–21.

Allison, Anthony C. "Abnormal Haemoglobins and Erythrocyte Enzyme-Deficiency Traits." In *Genetical Variation in Human Populations,* edited by G. A. Harrison, 16–40. Oxford: Pergamon Press, 1961.

Allison, Anthony C. "The Distribution of the Sickle-Cell Trait in East Africa and Elsewhere, and Its Apparent Relationship to the Incidence of Subtertian Malaria." *Transactions of the Royal Society of Tropical Medicine and Hygiene* 48, no. 4 (1954): 312–18.

Allison, Anthony C. "Protection Afforded by Sickle-Cell Trait against Subtertian Malarial Infection." *British Medical Journal* 4857 (1954): 290–94.

Allison, Anthony C., Elizabeth W. Ikin, and Arthur E. Mourant. "Further Observations on Blood Groups in East African Tribes." *Journal of the Royal Anthropological Institute of Great Britain and Ireland* 84 (1954): 158–62.

Allison, Anthony C., Elizabeth W. Ikin, Arthur E. Mourant, and A. B. Raper. "Blood Groups in Some East African Tribes." *Journal of the Royal Anthropological Institute of Great Britain and Ireland* 82 (1952): 55–61.

Amrith, Sunil S. *Decolonizing International Health: India and Southeast Asia, 1930–65.* London: Palgrave Macmillan, 2006.

Amrith, Sunil S., and Glenda Sluga. "New Histories of the United Nations." *Journal of World History* 19, no. 3 (2008): 251–74.

Anderson, David. *Histories of the Hanged: Britain's Dirty War in Kenya and the End of Empire.* London: Weidenfeld & Nicolson, 2005.

Anderson, Warwick. *Collectors of Lost Souls: Turning Kuru Scientists into Whitemen.* Baltimore: Johns Hopkins University Press, 2008.

Anderson, Warwick. *The Cultivation of Whiteness: Science, Health, and Racial Destiny in Australia.* New York: Basic Books, 2003.

Anderson, Warwick. "Objectivity and Its Discontents." *Social Studies of Science* 43, no. 4 (2013): 557–76.

Araquistain, Luis. "Some Survivals of Ancient Iberia in Modern Spain." *Man* 45 (1945): 30–38.

Armstrong, J. S., and Farquhar Matheson. "Blood Groups among Samoans." *British Medical Journal* 3326 (1924): 575.

Aronova, Elena. "The Missing Link: Nikolai Vavilov, Genogeography, and History's Past Future." Paper presented at History of Science Society annual meeting, Utrecht, July 2019.

Aronova, Elena, Karen S. Baker, and Naomi Oreskes. "Big Science and Big Data in Biology: From the International Geophysical Year through the International Biological Program to the Long Term Ecological Research (LTER) Network, 1957–Present." *Historical Studies in the Natural Sciences* 40, no. 2 (2010): 183–224.

Bangham, Jenny. "Blood Groups and Human Groups: Collecting and Calibrating Genetic Data after World War Two." *Studies in History and Philosophy of Biological and Biomedical Sciences* 47, part A (2014): 74–86.

Bangham, Jenny. "What Is Race? UNESCO, Mass Communication and Human Genetics in the Early 1950s." *History of the Human Sciences* 28, no. 5 (2015): 80–107.

Bangham, Jenny. "Writing, Printing, Speaking: Rhesus Blood-Group Genetics and Nomenclatures in the Mid-Twentieth Century." *British Journal for the History of Science* 47, no. 2 (2014), 335–61.

Bangham, Jenny, and Soraya de Chadarevian, "Human Heredity after 1945: Moving Populations Centre Stage." *Studies in History and Philosophy of Biological and Biomedical Sciences*, 47, part A (2014): 45–49.

Bangham, Jenny, and Judith Kaplan, eds., *Invisibility and Labour in the Human Sciences*, preprint 484. Berlin: Max Planck Institute for the History of Science, 2016.

Barandiarán, J. M. de. "On the Conservation of the Basque Peoples." *Man* 46 (1946): 96–97.

Barany, Michael, and Donald MacKenzie. "Chalk: Materials and Concepts in Mathematics Research." In *Representation in Scientific Practice Revisited*, edited by Catelijne Coopmans, Michael Lynch, Janet Vertesi, and Steve Woolgar, 107–29. Boston: MIT Press, 2014.

Barkan, Elazar. *The Retreat of Scientific Racism: Changing Concepts of Race in Britain and the United States Between the World Wars.* Cambridge: Cambridge University Press, 1992.

Barnicot, Nigel A. "Haptoglobins and Transferrins." In *Genetical Variation in Human Populations,* edited by G. A. Harrison, 41–61. Oxford: Pergamon Press, 1961.

Bashford, Alison. *Global Population: History, Geopolitics and Life on Earth.* New York: Columbia University Press, 2014.

Bashford, Alison. "Population, Geopolitics, and International Organizations in the Mid Twentieth Century." *Journal of World History* 19, no. 3 (2008): 327–47.

Bashford, Alison, and Philippa Levine. "Introduction: Eugenics and the Modern World." In *Oxford Handbook of the History of Eugenics,* edited by Alison Bashford and Philippa Levine. 3–26. Oxford: Oxford University Press, 2010.

Bashford, Alison, and Philippa Levine, eds. *Oxford Handbook of the History of Eugenics.* Oxford: Oxford University Press, 2010.

Bauer, Susanne. "Mutations in Soviet Public Health Science: Post-Lysenko Medical Genetics, 1969–1991." *Studies in History and Philosophy of Biological and Biomedical Sciences* 47, part A (2014): 163–72.

Bauer, Susanne. "Population Genetics, Cybernetics of Difference, and Pasts in the Present: Soviet and Post-Soviet Maps on Human Variation." *History of the Human Sciences* 28, no. 5 (2015): 146–67.

Bauer, Susanne. "Virtual Geographies of Belonging: The Case of Soviet and Post-Soviet Human Genetic Diversity Research." *Science, Technology, & Human Values* 39, no. 4 (2014): 511–37.

Beattie, John, L. H. Dudley Buxton, R. A. Fisher, Herbert J. Fleure, Cyril Fox, R. Ruggles Gates, R. A. Gregory, Julian S. Huxley, Arthur Keith, Alex Low, Meston, Geoffrey M. Morant, John L. Myers, Onslow, Raglan, R. U. Sayce, Henry Wellcome, Matthew Young, and Edwin W. Smith. "Racial History of Great Britain: An Anthropometric Survey." *Times,* March 13, 1935, 10.

Bell, Morag. "Reshaping Boundaries: International Ethics and Environmental Consciousness in the Early Twentieth Century." *Transactions of the Institute of British Geographers* 23, no. 2 (1998): 151–75.

Bennett, Brett M., and Joseph M. Hodge, eds. *Science and Empire: Knowledge and Networks of Science across the British Empire, 1800–1970.* Basingstoke: Palgrave Macmillan, 2011.

Bennett, J. H., and C. B. V. Walker. "Fertility and Blood Groups of Some East Anglian Blood Donors." *Annals of Human Genetics* 20, no. 4 (1956): 299–308.

Bennett, Jeffrey A. *Banning Queer Blood: Rhetorics of Citizenship, Contagion, and Resistance.* Tuscaloosa: University of Alabama Press, 2009.

Bernheim, Bertram M. *Blood Transfusion, Hemorrhage and the Anaemias.* Philadelphia: Lippincott, 1917.

Bernstein, Felix. "Ergebnisse einer biostatischen zusammenfassenden Betrachtung über die erblichen Blutstrukturen des Menschen." *Klinische Wochenschrift* 3 (1924): 1495–97.

Beutler, Ernest. "Glucose-6-Phosphate Dehydrogenase Deficiency: A Historical Perspective." *Blood* 111, no. 1 (2008): 16–24.

Bhende, Y. M., C. K. Deshpande, H. M. Bhatia, Ruth Sanger, Robert R. Race, W. T. J. Morgan, and W. M. Watkins. "A 'New' Blood-Group Character Related to the ABO System." *Lancet* 259, no. 6714 (1952): 903–4.

Biale, David. *Blood and Belief: The Circulation of a Symbol between Jews and Christians*. Berkeley: University of California Press, 2007.

Bildhauer, Bettina. *Medieval Blood*. Cardiff: University of Wales Press, 2010.

Billing, Edward. "Racial Origins from Blood Groupings." *British Medical Journal* 4108 (1939): 712.

"Biochemical Individuality." *Lancet* 277, no. 7178 (1961): 650–51.

Birdsell, Joseph B. "The ABO Blood Groups [Review]." *American Journal of Physical Anthropology* 18, no. 1 (1960), 75.

Birdsell, Joseph B. "Review of *The Distribution of the Human Blood Groups*." *American Anthropologist* 58, no. 1 (1956): 206–8.

Biss, Eula. *On Immunity: An Innoculation*. London: Fitzcarraldo, 2015.

Bittel, Carla, Elaine Leong, and Christine von Oertzen. *Working with Paper: Gendered Practices in the History of Knowledge*. Pittsburgh: University of Pittsburgh Press, 2019.

Bivins, Roberta. *Contagious Communities: Medicine, Migration, and the NHS in Post War Britain*. Oxford: Oxford University Press, 2015.

Bivins, Roberta. "Picturing Race in the British National Health Service, 1948–1988." *Twentieth Century British History* 28, no. 1 (2017): 83–109.

Blacker, Carlos P. "Medical Genetics." *Lancet* 256, no. 6623 (1950), 221–22.

Bland, Lucy. "British Eugenics and "Race Crossing": An Interwar Investigation." *New Formations* 60 (2007): 66–78.

Bland, Lucy, and Lesley Hall. "Eugenics in Britain: The View from the Metropole." In *The Oxford Handbook of the History of Eugenics*, edited by Alison Bashford and Philippa Levine, 213–27. Oxford: Oxford University Press, 2010.

Blaxill, Alec E. "Blood for Transfusion." *Lancet* 252, no. 6534 (1948): 828.

"A Blood Transfusion Depot at Work." *British Medical Journal* 4109 (1939): 730.

"Blood Transfusion Service for War." *British Medical Journal* 4095 (1939), 35.

Blumberg, Baruch S., ed. *Proceedings of the Conference on Genetic Polymorphisms and Geographic Variations in Disease*. New York: Grune and Stratton, 1962.

Boaz, Rachel E. *In Search of "Aryan Blood": Serology in Interwar and National Socialist Germany*. New York: Central European University Press, 2012.

Bolnick, Deborah A., et al. "The Science and Business of Genetic Ancestry Testing," *Science* 318, no. 5849 (2007): 399–400.

Boon, Tim. *Films of Fact*. London: Wallflower Press, 2008.

Boyd, William C. *Blood Groups*. The Hague: W. Junk, 1939.

Boyd, William C. "*The Distribution of the Human Blood Groups* by A. E. Mourant." *American Journal of Physical Anthropology* 13, no. 1 (1955): 153–58.

Boyd, William C., and F. Schiff. *Blood Grouping Technic: A Manual for Clinicians, Serologists, Anthropologists and Students of Legal and Military Medicine*. New York: Interscience, 1942.

Box, Joan Fisher. *R. A. Fisher: The Life of a Scientist*. New York: Wiley, 1978.

Brattain, Michelle. "Race, Racism, and Antiracism: UNESCO and the Politics of Presenting Science to the Postwar Public." *American Historical Review* 112, no. 5 (2007), 1386–1413.

Breckenridge, Keith, and Simon Szreter. *Registration and Recognition: Documenting the Person in World History*. Oxford: Oxford University Press, 2012.

Bristow, Adrian F., Trevor Barrowcliffe, and Derek R. Bangham. "Standardization of Biological Medicines: The First Hundred Years, 1900–2000." *Notes and Records of the Royal Society* 60, no. 3 (2006): 271–89.

British Medical Association. *Clinical Pathology in General Practice*. London: British Medical Association, 1955.

Brittain, Marcus. "Herbert Fleure and the League of Nations' (1919) Minorities Treaties: A Study in Archaeology and Post-Conflict Reconstruction after WWI." Paper presented at Histories of Archaeology Research Network conference, Cambridge, UK, March 14, 2009.

Brittain, Marcus. "World War I and the Contribution of Herbert Fleure & Harold Peake to Post-War Reconstruction & Urban Planning." Paper presented at Theoretical Anthropology Group conference, Southampton, UK, December 15–17, 2008.

Brøgger, A. W. *Ancient Emigrants: A History of the Norse Settlements of Scotland*. Oxford: Clarendon, 1929.

Buchanan, Tom. *Britain and the Spanish Civil War*. Cambridge: Cambridge University Press, 1997.

Bucur, Maria. "Eugenics in Eastern Europe, 1870s–1945." In *The Oxford Handbook of the History of Eugenics*, edited by Alison Bashford and Philippa Levine, 258–73. Oxford: Oxford University Press, 2010.

Bulletin de la société française de la transfusion sanguine: IIe congrès international de la transfusion sanguine. Paris: Baillère, 1939.

Burton, Elise, K. "'Essential Collaborators': Locating Middle Eastern Geneticists in the Global Scientific Infrastructure, 1950s–1970s." *Comparative Studies in Society and History* 60, no. 1 (2018): 119–49.

Burton, Elise, K. *Genetic Crossroads: The Middle East and the Science of Human Heredity*. Stanford, CA: Stanford University Press, forthcoming.

Burton, Elise K. "Red Crescents: Race, Genetics, and Sickle Cell Disease in the Middle East." *Isis* 110, no. 2 (2019): 250–69.

Bynum, Caroline Walker. *Wonderful Blood: Theology and Practice in Late Medieval Northern Germany and Beyond*. Philadelphia: University of Pennsylvania Press, 2007.

Cain, Joe. "Julian Huxley, General Biology and the London Zoo, 1935–42." *Notes and Records of the Royal Society of London* 64 (2010): 359–78.

Cain, Joe, and Michael Ruse, eds. *Descended from Darwin: Insights into the History of Evolutionary Studies, 1900–1970*. Philadelphia: American Philosophical Society, 2009.

Callender, Sheila, and Z. V. Paykoç. "Irregular Haemagglutinins after Transfusion." *British Medical Journal* 4438 (1946): 119–21.

Callender, Sheila, and Robert R. Race. "A Serological and Genetical Study of Multiple Antibodies Formed in Response to Blood Transfusion by a Patient with Lupus Erythematosus Diffusus." *Annals of Eugenics* 13 (1946): 102–17.

Callender, Sheila, Robert R. Race, and Zafer V. Paykoç. "Hypersensitivity To Transfused Blood." *British Medical Journal* 4411 (1945): 83–84.

Caplan, Jane, and John Torpey, eds. *Documenting Individual Identity: The Development of State Practices in the Modern World*. Princeton, NJ: Princeton University Press, 2001.

Cappell, D. F. "The Blood Group Rh. Part I. A Review of The Antigenic Structure and Serological Reactions of the Rh Subtypes." *British Medical Journal* 4477 (1946): 601–5.

Cardiff, David, and Paddy Scannell. "Broadcasting and National Unity." In *Impacts and Influences: Essays on Media Power in the Twentieth Century*, edited by James Curran, Anthony Smith, and Pauline Wingate, 157–73. London: Routledge, 1987.

Carlson, Elof Axel. *The Gene: A Critical History*. Philadelphia: W. B. Saunders, 1966.

Carroll, Sean B. *Into the Jungle: Great Adventures in the Search for Evolution*. San Francisco: Pearson Education, 2008.

Carsten, Janet. *Blood Work: Life and Laboratories in Penang*. Durham, NC: Duke University Press, 2019.

Casper, Monica, and Adele Clarke. "Making the Pap Smear into the 'Right Tool' for the Job: Cervical Cancer Screening in the USA, circa 1940–95." *Social Studies of Science* 28, no. 2 (1998): 255–90.

Castle, William B., Maxwell M. Wintrobe, and Laurence H. Snyder. "On the Nomenclature of the Anti-Rh Typing Serums: Report of the Advisory Review Board." *Science* 107, no. 2767 (1948): 27–31.

Cavalli-Sforza, Luca L. "The DNA Revolution in Population Genetics." *Trends in Genetics* 14, no. 2 (1998): 60–65.

Ceppellini, R., E. S. Curtoni, P. L. Mattiuz, V. Miggiano, G. Scudeller, and A. Serra. "Genetics of Leukocyte Antigens: A Family Study of Segregation and Linkage." In *Histocompatibility Testing 1967*, 149–87. Copenhagen: Munksgaard, 1967.

Chauveau, Sophie. "Du don à l'industrie: La transfusion sanguine en France depuis les années 1940." *Terrain: Anthropologie & sciences humaines* 56 (2011): 74–89.

Chinn, Sarah E. *Technology and the Logic of American Racism*. London: Continuum, 2000.

Chown, Bruce. "Problems in Blood Group Analysis." *American Anthropologist* 59, no. 5 (1957): 885–88.

Chresfield, Michell. *What Lies Between: Race, Science, and the Prehistory of Multiracial America*. Unpublished manuscript.

"Chromosomes in Medicine." *British Medical Journal* 5317 (1962): 1453–54.

Clapham, A. R. "A Review of the United Kingdom Contribution to the International Biological Programme: Introductory Remarks." *Philosophical Transactions of the Royal Society of London, Series B* 274 (1976): 277–81.

Clark, Ronald W. *J.B.S.: The Life and Work of J.B.S. Haldane*. Oxford: Oxford University Press, 1984.

Clarke, Adele E., and Joan H. Fujimura, eds. *The Right Tools for the Job: At Work in Twentieth-Century Life Sciences*. Princeton, NJ: Princeton University Press, 1992.

Clarke, Cyril. "Robert Russell Race: 28 November 1907–15 April 1984." *Biographical Memoirs of Fellows of the Royal Society* 31 (1985): 455–92.

Clarke, Sabine. "A Technocratic Imperial State? The Colonial Office and Scientific Research, 1940–1960." *Twentieth Century British History* 18, no. 4 (2007): 453–80.

Clever, Iris. "The Lives and Afterlives of Skulls: The Development of Biometric Methods of Measuring Race (1880–1950)." PhD diss., University of California, Los Angeles, 2020.

Coca, Arthur F., and Olin Deibert. "A Study of the Occurrence of the Blood Groups among the American Indians." *Journal of Immunology* 8, no. 6 (1923): 487–91.

Collier, L. H. *The Lister Institute of Preventive Medicine: A Concise History*. Bushey Heath, UK: Lister Institute of Preventive Medicine, 2000.

Comfort, Nathaniel. "The Prisoner as Model Organism: Malaria Research at Stateville Penitentiary." *Studies in History and Philosophy of Biological and Biomedical Sciences* 40, no. 3 (2009): 190–203.

Comfort, Nathaniel. *The Science of Human Perfection: How Genes Became the Heart of American Medicine*. New Haven, CT: Yale University Press, 2012.

Contreras, M. "Thomas Edward Cleghorn." *British Medical Journal* 305 (1992): 580.

Coombs, Robert R. A. "Detection of Weak and 'Incomplete' Rh Agglutinins: A New Test." *Lancet* 246, no. 6358 (1945): 15–16.

Coombs, Robert R. A., Arthur E. Mourant, and Robert R. Race. "In-Vivo Isosensitization of Red Cells in Babies with Haemolytic Disease." *Lancet* 247, no. 6391 (1946): 264–66.

Cooter, Roger, and John V. Pickstone, eds. *Medicine in the Twentieth Century*. Amsterdam: Harwood Academic, 2000.

Cooter, Roger, Steve Sturdy, and Mark Harrison, eds. *War, Medicine and Modernity*. Stroud: Sutton, 1998.

Copeman, Jacob, *Veins of Devotion: Blood Donation and Religious Experience in North India*. New Brunswick, NJ: Rutgers University Press, 2009.

Cox, D. R. "*Biometrika*: The First 100 Years." *Biometrika* 88, no. 1 (2001): 3–11.

Creager, Angela N. H. "Biotechnology and Blood: Edwin Cohn's Plasma Fractionation Project, 1940–1953." In *Private Science: Biotechnology and the Rise of the Molecular Sciences*, edited by Arnold Thackray, 39–62. Philadelphia: University of Pennsylvania Press, 1998.

Creager, Angela N. H. *Life Atomic: A History of Radioisotopes in Science and Medicine*. Chicago: University of Chicago Press, 2013.

Creager, Angela N. H. "Producing Molecular Therapies from Human Blood: Edwin Cohn's Wartime Enterprise." In *Molecularizing Biology and Medicine: New Practices and Alliances, 1910s–1970s,* edited by Soraya de Chadarevian and Harmke Kamminga, 107–38. Amsterdam: Harwood, 1998.

Creager, Angela N. H. "'What Blood Told Dr. Cohn': World War II, Plasma Fractionation, and the Growth of Human Blood Research." *Studies in History and Philosophy of Biological and Biomedical Sciences* 30, no. 3 (1999): 377–405.

Creager, Angela N. H., and Hannah Landecker. "Technical Matters: Method, Knowledge and Infrastructure in Twentieth-Century Life Science." *Nature Methods* 6, no. 10 (2009): 701–5.

Crook, David P. *Grafton Elliot Smith, Egyptology & the Diffusion of Culture: A Biographical Perspective.* Brighton: Sussex Academic Press, 2012.

Crook, Tom, and Glen O'Hara, eds. *Statistics and the Public Sphere in Modern Britain, c. 1800–2000.* New York: Routledge, 2011.

Cueto, Marcos, Theodore M. Brown, and Elizabeth Fee. *The World Health Organization: A History.* Cambridge, UK: Cambridge University Press, 2019.

Cutbush, Marie, Patrick L. Mollison, and Dorothy M. Parkin. "A New Human Blood Group." *Nature* 165 (February 4, 1950): 188–89.

Cyril Jenkins Productions Ltd. *Blood Grouping.* 20:33 min, sound, color. Imperial Chemical Industries Limited, 1955.

Daniel, Reginald G. *More Than Black? Multiracial Identity and the New Racial Order.* Philadelphia: Temple University Press, 2002.

Darlington, Cyril. "The Genetic Component of Language." *Heredity* 1 (1947): 269–86.

Da Silva, E. M. "Blood Groups of Indians, Whites and White-Indian Mixtures in Southern Mato Grosso, Brazil." *American Journal of Physical Anthropology* 7 no. 4 (1949): 575–86.

Daston, Lorraine. "The Sciences of the Archive." *Osiris* 27 (2012): 156–87.

Daston, Lorraine, ed. *Science in the Archives: Pasts, Presents, Futures.* Chicago: University of Chicago Press, 2017.

Daston, Lorraine, and Peter Galison. *Objectivity.* New York: Zone Books, 2007.

Dausset, J. "Iso-Leuco-Anticorps." *Acta Haematology* 20 (1958): 156–66.

Davies, Elwyn, and Herbert J. Fleure. "A Report on an Anthropometric Survey of the Isle of Man." *Journal of the Royal Anthropological Institute of Great Britain and Ireland* 66 (1936): 129–87.

Davies, Peter. "Patrick Mollison: A Pioneer in Transfusion Medicine." *British Medical Journal* 344, no. 7845 (2012): e1233.

Davis, F. James. *Who Is Black? One Nation's Definition.* University Park: Pennsylvania State University Press, 2001 [1991].

de Chadarevian, Soraya. "Chromosome Surveys of Human Populations: Between Epidemiology and Anthropology." *Studies in History and Philosophy of Biological and Biomedical Sciences* 47, part A (2014): 87–96.

de Chadarevian, Soraya. *Designs for Life: Molecular Biology after World War II.* Cambridge: Cambridge University Press, 2002.

de Chadarevian, Soraya. "Following Molecules: Hemoglobin between the Clinic and the Laboratory." In *Molecularizing Biology and Medicine: New Practices and Alliances, 1910s–1970s*, edited by Soraya de Chadarevian and Harmke Kamminga, 171–201. Amsterdam: Harwood, 1998.

de Chadarevian, Soraya. "The Future Historian: Reflections on the Archives of Contemporary Sciences." *Studies in History and Philosophy of Biological and Biomedical Sciences* 55 (2016): 54–60.

de Chadarevian, Soraya. *Heredity under the Microscope: Chromosomes and the Study of the Human Genome*. Chicago: University of Chicago Press, 2020.

de Chadarevian, Soraya. "Putting Human Genetics on a Solid Basis: Human Chromosome Research, 1950s–1970s." In *Human Heredity in the Twentieth Century*, edited by Bernd Gausemeier, 141–52. London: Pickering & Chatto, 2013.

DeGowin, Elmer Louis. *Blood Transfusion*. Philadelphia: W. B. Saunders, 1949.

Delbourgo, James, and Staffan Müller-Wille. "Listmania: How Lists Can Open up Fresh Possibilities for Research in the History of Science." *Isis* 103, no. 4 (2012): 710–15.

Dent, Rosanna. "Invisible Infrastructures: Xavante Strategies to Enrol and Manage Warazú Researchers." In *Invisibility and Labour in the Human Sciences*, edited by Jenny Bangham and Judith Kaplan, 65–74. Berlin: Max Planck Institute for the History of Science, 2016.

Dent, Rosanna. "Kinship and Care: Social Infrastructures for Maintaining Research in Terra Indígena Xavante." Paper presented at International Congress of History of Science and Technology, Rio de Janeiro, 2017.

Dent, Rosanna. "Studying Indigenous Brazil: The Xavante and the Human Sciences, 1958–2015." PhD diss., University of Pennsylvania, 2017.

Discombe, George. "Blood Transfusion Accidents." *British Medical Journal* 4835 (1953): 569.

Dobzhansky, Theodosius Grigorievich. *Genetics and the Origin of Species*. New York: Columbia University Press, 1941.

"Doctors in Time of War: Scales of Pay for a National Emergency Medical Service." *British Medical Journal* 4099 (1939): 238–39.

Drummond, R. "Blood Grouping in Tubes." *British Medical Journal* 4307 (1943): 118.

Drummond, R. "A Simple Blood-Grouping Method." *British Medical Journal* 4952 (1955): 1388–89.

Drummond, R. "Simple Blood-Grouping Methods." *British Medical Journal* 4965 (1956): 514–15.

Ducey, Edward F., and Robert I. Modica. "On the Amendment of the Nomenclature of the Rh-CDE System." *Science* 111, no. 2887 (1950): 466–67.

Duedahl, Poul. "UNESCO Man: Changing the Concept of Race, 1945–65." Paper presented at American Anthropological Association 107th annual meeting, San Francisco, 2008.

Dukepoo, Frank C. "It's More Than the Human Genome Diversity Project." *Politics and the Life Sciences* 18, no. 2 (1999): 293–97.

Dunn, C. L. *The Emergency Medical Services*, vol. 1, *England and Wales*. London: Her Majesty's Stationary Office, 1952.

Dunn, Leslie C. "The Coming of Age of Blood Group Research." *American Naturalist* 89, no. 884 (1955): 55–60.

Dunn, Leslie C. *Race and Biology*. Paris: UNESCO, 1951.

Dunsford, Ivor, and C. Christopher Bowley. *Techniques in Blood Grouping*. Edinburgh: Oliver and Boyd, 1955.

Duster, Troy. *Backdoor to Eugenics*, 2nd ed. New York: Routledge, 2003.

Duster, Troy. "Medicalisation of Race." *Lancet* 369, no. 9562 (2007): 702–4.

"Editorial: The Scope of *Biometrika*." *Biometrika* 1, no. 1 (1901): 1–2.

Edwards, Anthony W. F. "Mendelism and Man 1918–1939." In *A Century of Mendelism in Human Genetics*, edited by W. Milo Keynes, Anthony W. F. Edwards, and Robert Peel, 33–46. Boca Raton, FL: CRC Press, 2004.

Edwards, Anthony W. F. "R. A. Fisher's 1943 Unravelling of the Rhesus Blood-Group System." *Genetics* 175, no. 2 (2007): 471–76.

Edwards, Anthony W. F. "Studying Human Evolution by Computer." *New Scientist* 30 (May 19, 1966): 438–40.

Efron, John M. *Defenders of the Race: Jewish Doctors and Race Science in Fin-de-Siècle Europe*. New Haven, CT: Yale University Press, 1994.

Epstein, Steven. *Inclusion: The Politics of Difference in Medical Research*. Chicago: University of Chicago Press, 2007.

Erickson, Paul, Judy L. Klein, Lorraine Daston, Thomas Sturm, and Michael Gordin. *How Reason Almost Lost Its Mind: The Strange Career of Cold War Rationality*. Chicago: University of Chicago Press, 2015.

Evans, D. F. T. "Le Play House and the Regional Survey Movement in British Sociology 1920–1955." MPhil thesis, City of Birmingham Polytechnic, 1986, http://www.dfte.co.uk/ios.

Fabian, Ann. *The Skull Collectors: Race, Science, and America's Unburied Dead*. Chicago: University of Chicago Press, 2010.

Fisher, R. A. "'The Coefficient of Racial Likeness' and the Future of Craniometry." *Journal of the Royal Anthropological Institute of Great Britain and Ireland* 66 (1936): 57–63.

Fisher, R. A. *The Genetical Theory of Natural Selection*. Oxford: Clarendon Press, 1930.

Fisher, R. A. "To the Editor of The Times: London University; Plight of the Galton Laboratory." *Times*, October 3, 1939, 6.

Fisher, R. A. "The Rhesus Factor: A Study in Scientific Method." *American Scientist* 35, no. 1 (1947): 95–103.

Fisher, R. A., and John A. Fraser Roberts. "A Sex Difference in Blood-Group Frequencies." *Nature* 151 (June 5, 1943): 640–41.

Fisher, R. A., and George L. Taylor. "Blood Groups in Great Britain." *British Medical Journal* 4111 (1939): 826.

Fisher, R. A., and George L. Taylor. "Scandinavian Influence in Scottish Ethnology." *Nature* 145 (April 13, 1940): 590–92.

Fisher, R. A., and Janet M. Vaughan. "Surnames and Blood-Groups." *Nature* 144, no. 3660 (December 23, 1939): 1047–1048.

Fitzgerald, J. G. "The Work of the Health Organisation of the League of Nations." *Canadian Public Health Journal* 24, no. 8 (1933): 368–72.

Fleck, Ludwik. *Genesis and Development of a Scientific Fact*, translated by Frederick Bradley and Thaddeus J. Trenn. Chicago: University of Chicago Press, 1979. First published by Schwabe & Co. as *Entstehung and Entwicklung einer wissenschaftlichen Tatsache: Einführung in die Lehre vom Denkstil und Denkkollektiv*, 1935.

Fleure, Herbert J. "The Institute and Its Development." *Journal of the Royal Anthropological Institute of Great Britain and Ireland* 76 (1946): 1–4.

Fleure, Herbert J. "Some Aspects of Race Study." *Eugenics Review* 14, no. 2 (1922): 93–102.

Fleure, Herbert J., and T. C. James. "Geographical Distribution of Anthropological Types in Wales." *Journal of the Royal Anthropological Institute of Great Britain and Ireland* 46 (1916): 35–153.

Ford, E. B. *Genetics for Medical Students*. London: Methuen, 1942.

Ford, E. B. "A Uniform Notation for the Human Blood Groups." *Heredity* 9 (1955): 135–42.

"Foreword." *Annals of Eugenics*, 1 (1925), 1–4.

Fortun, Michael. *Promising Genomics: Iceland and DeCODE Genetics in a World of Speculation*. Berkeley: University of California Press, 2008.

Foster, William Derek. *A Short History of Clinical Pathology*. Edinburgh: E&S Livingstone, 1961.

Foucault, Michel. *Discipline and Punish: The Birth of the Prison*, translated by Alan Sheridan. London: Allen Lane, 2012 [1977].

Franklin, Sarah, and Susan McKinnon, eds. *Relative Values: Refiguring Kinship Studies*. Durham, NC: Duke University Press, 2001.

Fraser Roberts, John A. "An Analysis of the ABO Blood-Group Records of the North of England." *Heredity* 7 (1953): 361–88.

Fraser Roberts, John A. "Blood Group Frequencies in North Wales." *Annals of Eugenics* 11 (1942): 260–71.

Fraser Roberts, John A. "The Frequencies of the ABO Blood Groups in South-Western England." *Nature* 14, no. 2 (1948): 109–16.

Fraser Roberts, John A. "History in Your Blood." *Eugenics Review* 44 (1952): 28–30.

Fraser Roberts, John A. *An Introduction to Medical Genetics*. London: Oxford University Press, 1940.

Fraser Roberts, John A. "Surnames and Blood Groups, with a Note on a Probable Remarkable Difference between North and South Wales." *Nature* 149 (1942): 138.

Fujimura, Joan H., Troy Duster, and Ramya Rajagopalan, "Introduction: Race, Genetics, and Disease: Questions of Evidence, Matters of Consequence." *Social Studies of Science* 38 (2008): 643–56

Fullwiley, Duana. "The Biologistical Construction of Race: 'Admixture' Technology and the New Genetic Medicine." *Social Studies of Science* 38, no. 5 (2008): 695–735.

Gannett, Lisa, "Biogeographical Ancestry and Race." *Studies in History and Philosophy of Biological and Biomedical Sciences* 47, part A (2014): 173–84.

Gannett, Lisa, and James R. Griesemer. "The ABO Blood Groups: Mapping the History and Geography of Genes in Homo Sapiens." In *Classical Genetic Research and Its Legacy: The Mapping Cultures of Twentieth-Century Genetics*, edited by Hans-Jörg Rheinberger and Jean-Paul Gaudillière, 119–72. London: Routledge, 2004.

Gannett, Lisa, and James R. Greisemer. "Classical Genetics and the Geography of Genes." In *Classical Genetic Research and Its Legacy: The Mapping Cultures of Twentieth-Century Genetics*, edited by Hans-Jörg Rheinberger and Jean-Paul Gaudillière, 57–88. London: Routledge, 2004.

Gaudillière, Jean-Paul, and Hans-Jörg Rheinberger, eds. *Classical Genetic Research and Its Legacy: The Mapping Cultures of Twentieth Century Genetics*. London: Routledge, 2004.

Gausemeier, Bernd, Staffan Müller-Wille, and Edmund Ramsden, eds. *Human Heredity in the Twentieth Century*. London: Pickering & Chatto, 2013.

Geissler, P. Wenzel. "'Kachinja Are Coming!' Encounters around Medical Research Work in a Kenyan Village." *Africa* 75, no. 2 (2005): 173–202.

Geissler, P. Wenzel, Ann Kelly, Babatunde Imoukhuede, and Robert Pool. "'He Is Now Like a Brother, I Can Even Give Him Some Blood': Relational Ethics and Material Exchanges in a Malaria Vaccine 'Trial Community' in the Gambia." *Social Science & Medicine* 67 (2008): 696–707.

Germann, Pascal. "'Nature's Laboratories of Human Genetics': Alpine Isolates, Hereditary Diseases and Medical Genetic Fieldwork, 1920–1970." In *History of Human Genetics: Important Discoveries and Global Perspectives*, edited by Heike I. Petermann, Peter S. Harper, and Susanne Doetz, 145–66. Cham, Switzerland: Springer, 2017.

Giblett, Eloise, R., "Haptoglobins and Transferrins." In *Proceedings of the Conference on Genetic Polymorphisms and Geographic Variations in Disease*, edited by Baruch S. Blumberg, 132–58. New York: Grune and Stratton, 1961.

Giblett, Eloise, R., "Philip Levine, 1900–1987." National Academy of Sciences, 1994, http://www.nasonline.org/publications/biographical-memoirs/memoir -pdfs/levine-philip.pdf.

Gilbert, Lewis, dir. *Emergency Call*. Nettlefold Films, 1952.

Gil-Riano, Sebastián. "Historicizing Anti-racism: UNESCO's Campaigns against Race Prejudice in the 1950s." PhD diss., University of Toronto, 2014.

Gil-Riano, Sebastián. "Relocating Anti-racist Science: The 1950 UNESCO Statement on Race and Economic Development in the Global South." *British Journal for the History of Science* 51, no. 2 (2018): 281–303.

Gitelman, Lisa. *Paper Knowledge: Toward a Media History of Documents*. Durham: Duke University Press, 2014.

Goodwin, Michele. *Black Markets: The Supply and Demand of Body Parts*. Cambridge: Cambridge University Press, 2013.

Gormley, Melinda. "Scientific Discrimination and the Activist Scientist: L. C. Dunn and the Professionalization of Genetics and Human Genetics in the United States." *Journal of the History of Biology* 42, no. 1 (2009): 33–72.

Gradmann, Christoph, and Jonathan Simon, eds. *Evaluating and Standardizing Therapeutic Agents, 1890–1950*. London: Palgrave Macmillan, 2010.

Grahame, Ernest W. "The ABO Blood Groups and Peptic Ulceration: A Survey of 1,080 Cases on South Tees-Side." *British Medical Journal* 5219 (1961): 95–96.

Gruffudd, Pyrs. "Back to the Land: Historiography, Rurality and the Nation in Interwar Wales." *Transactions of the Institute of British Geographers* 19 (1994): 61–77.

Guglielmo, Thomas A. "'Red Cross, Double Cross': Race and America's World War II-Era Blood Donor Service." *Journal of American History* 97, no. 1 (2010): 63–90.

Gunson, Harold H., and Helen Dodsworth. "The Drying and Fractionation of Plasma, 1935–55." *Transfusion Medicine* 6, suppl. 1 (1996): 37–41.

Gunson, Harold H., and Helen Dodsworth. "The National Blood Transfusion Service (NBTS), 1946–1988." *Transfusion Medicine* 6, suppl. 1 (1996): 17–24.

Gunson, Harold H., and Helen Dodsworth. "Towards a National Blood Transfusion Service in England and Wales, 1900–1946." *Transfusion Medicine* 6, suppl. 1 (1996): 4–16.

Haberman, Sol, and Joseph M. Hill. "Verbal Usage of the CDE Notation for Rh Blood Groups." *British Medical Journal* 4736 (1952): 851.

Haldane, J. B. S. "Anthropology and Human Biology." *Man* 34 (1934): 142–43.

Haldane, J. B. S. "The Blood Groups in Genetics and Anthropology." *British Medical Journal* 3730 (1932): 26–27.

Haldane, J. B. S. *New Paths in Genetics*. London: Allen & Unwin, 1941.

Haldane, J. B. S. "Prehistory in the Light of Genetics." *Proceedings of the Royal Institution of Great Britain* 26 (1931): S355–70.

Haldane, J. B. S. "Selection against Heterozygosis in Man." *Annals of Eugenics* 11 (1941): 333–40.

Haldane, J. B. S. "Two New Allelomorphs for Heterostylism in Primula." *American Naturalist* 67 (1933): S59–60.

Haldane, J. B. S., and William C. Boyd. "The Blood-Group Frequencies of European Peoples, and Racial Origins." *Human Biology* 12 (1940): 457–80.

Haraway, Donna. *Modest_Witness@Second_Millenium.FemaleMan©_Meets_Onco-Mouse™*. New York: Routledge, 1997.

Haraway, Donna. *Primate Visions: Gender, Race and Nature in the World of Modern Science*. New York: Routledge, 1989.

Haraway, Donna. "Remodelling the Human Way of Life: Sherwood Washburn and the New Physical Anthropology, 1950–1980." In *Bones, Bodies and Behavior: Essays in Behavioral Anthropology*, edited by George W. Stocking Jr., 206–60. Madison: University of Wisconsin Press, 1988.

Haraway, Donna. "Universal Donors in a Vampire Culture: It's All in the Family: Biological Kinship Categories in the Twentieth-Century United States." In *Uncommon Ground: Toward Reinventing Nature*, edited by William Cronon, 321–66. New York: W. W. Norton, 1995.

Harman, Oren Solomon. "C. D. Darlington and the British and American Reaction to Lysenko and the Soviet Conception of Science." *Journal of the History of Biology* 36, no. 2 (2003): 309–52.

Harper, Peter S. "Julia Bell and the Treasury of Human Inheritance." *Human Genetics* 116, no. 5 (2005): 422–32.

Harper, Peter S. *A Short History of Medical Genetics*. Oxford: Oxford University Press, 2008.

Harper, Peter S., L. A. Reynolds, and E. M. Tansey. *Clinical Genetics in Britain: Origins and Development*. Witness Seminar Held by the Wellcome Trust Centre for the History of Medicine at University College London, 2010.

Harris, Harry. "Lionel Sharples Penrose. 1898–1972." *Biographical Memoirs of Fellows of the Royal Society* 19 (1973): 521–61.

Hart, Mitchell. *Jewish Blood: Reality and Metaphor in History, Religion and Culture*. London: Routledge, 2013.

Hartmann, Otto, Arthur E. Mourant, and Robert R. Race. "The Rh Genotypes of a Series of Oslo Blood Donors." *Acta Pathologica et Microbiologica Scandinavica* 24 (1947): 330–33.

Hazard, Anthony Q., Jr. *Postwar Anti-racism: The U.S., UNESCO, and "Race," 1945–1968*. London: Palgrave Macmillan, 2012.

Hazard, Anthony Q., Jr. "A Racialized Deconstruction? Ashley Montagu and the 1950 UNESCO Statement on Race." *Transforming Anthropology* 19, no. 2 (2011): 174–86.

Healy, K. *Last Best Gifts: Altruism and the Market for Human Blood and Organs*. Chicago: University of Chicago Press, 2006.

Heinbecker, Peter, and Ruth H. Pauli. "Blood Grouping of the Polar Eskimo." *Journal of Immunology* 13, no. 4 (1927): 279–83.

Hernandez, Raymond. "Donations: Getting Too Much of a Good Thing." *New York Times*. Monday, November 12, 2001, sec. G, 3.

Hirschfeld [Hirszfeld], Ludwik, and Hanka Hirschfeld [Hirszfeld]. "Serological Differences between the Blood of Different Races: The Result of Researches on the Macedonian Front." *Lancet* 194, no. 5016 (1919): 675–79.

Hirszfeld, Hanna, and Ludwik Hirszfeld. "Essai d'application des méthodes sérologiques au problème des races." *L'Anthropologie* 29 (1919): 505–37.

Hirszfeld, Ludwik. *The Story of One Life*, edited by William H. Schneider and translated by Marta A. Balińska. Rochester, NY: University of Rochester Press, 2010.

Hoare, Edward D. "Occurrence of the Rh Antigen in the Population: Notes on 5 Cases of Erythroblastosis Foetalis." *British Medical Journal* 4313 (1943): 297–98.

Hockwald, Robert S., John Arnold, Charles B. Clayman, and Alf S. Alving. "Toxicity of Primaquine in Negroes." *Journal of the American Medical Association* 149, no. 17 (1952): 1568–70.

Hogan, Andrew J. *Life Histories of Genetic Disease: Patterns and Prevention in Postwar Medical Genetics*. Baltimore: Johns Hopkins University Press, 2016.

Hogan, Andrew J. "Medical Eponyms: Patient Advocates, Professional Interests and the Persistence of Honorary Naming." *Social History of Medicine* 29 (2016): 534–56.

Hogben, Lancelot. *Genetic Principles in Medicine and Social Science*. London: Williams and Norgate, 1931.

Hogben, Lancelot, and Ray Pollack. "A Contribution to the Relation of the Gene

Loci Involved in the Isoagglutinin Reaction, Taste Blindness, Friedreich's Ataxia and Major Brachydactyly of Man." *Journal of Genetics* 31, no. 3 (1935): 353–62.

Hopwood, Nick. "Producing Development: The Anatomy of Human Embryos and the Norms of William His." *Bulletin of the History of Medicine* 74 (2000): 29–79.

Hopwood, Nick. "The Tragedy of the Emeritus and the Fates of Anatomical Collections: Alfred Benninghoff's Memoir of Ferdinand Count Spee." *BJHS Themes* 4 (2019): 169–94.

Hughes-Jones, Nevin, and Patricia Tippett. "Ruth Ann Sanger: 6 June 1918–4 June 2001." *Biographical Memoirs of Fellows of the Royal Society* 49 (2003): 461–73.

Hull, Matthew. *Government of Paper: The Materiality of Bureaucracy in Urban Pakistan.* Berkeley: University of California Press, 2014.

Huxley, Julian S. *Evolution: The Modern Synthesis.* London: Allen & Unwin, 1942.

Huxley, Julian S. *UNESCO: Its Purpose and Its Philosophy.* Washington, DC: Public Affairs Press, 1947.

Huxley, Julian S., Alfred C. Haddon, and Alexander M. Carr-Saunders. *We Europeans: A Survey of "Racial" Problems.* London: Jonathan Cape, 1935.

Hyam, Ronald. *Britain's Declining Empire: The Road to Decolonisation, 1918–1968.* Cambridge: Cambridge University Press, 2006.

Ikin, Elizabeth W., Aileen M. Prior, Robert R. Race, and George L. Taylor. "The Distribution of the A_1A_2BO Blood Groups in England." *Annals of Eugenics* 9 (1939): 409–11.

Jacob, Marie-Andrée. *Matching Organs with Donors: Legality and Kinship in Transplants.* Philadelphia: University of Pennsylvania Press, 2012.

Jardine, Boris. "State of the Field: Paper Tools." *Studies in History and Philosophy of Science* 64 (2017): 53–63.

Jardine, Boris, and Matthew Drage. "The Total Archive: Data, Subjectivity, Universality." *History of the Human Sciences* 31, no. 5 (2018): 3–22.

Jardine, Boris, Emma Kowal, and Jenny Bangham. "How Collections End: Objects, Meaning and Loss in Laboratories and Museums." *BJHS Themes* 4 (2019): 1–27.

Jardine, Nicholas. *The Scenes of Inquiry,* 2nd ed. Oxford: Oxford University Press, 2000.

Jones, Greta. "Bell, Julia (1879–1979)." In *Oxford Dictionary of National Biography* online, https://www.oxforddnb.com/view/10.1093/ref:odnb/9780198614128 .001.0001/odnb-9780198614128-e-38514.

Kahn, Jonathan, et al. "How Not to Talk about Race and Genetics." *Buzzfeed News,* March 30, 2018, https://www.buzzfeednews.com/article/bfopinion /race-genetics-david-reich.

Kaiser, David. "Stick-Figure Realism: Conventions, Reification, and the Persistence of Feynman Diagrams, 1948–1964." *Representations* 70 (2000): 49–86.

Kakaliouras, Ann M., and Joanna Radin. "Archiving Anthropos: Tracking the Ethics of Collections across History and Anthropology." *Curator: The Museum Journal* 57 (2014): 147–51.

Kay, Lily E. "Laboratory Technology and Biological Knowledge: The Tiselius Electrophoresis Apparatus, 1930–1945." *History and Philosophy of the Life Sciences* 10, no. 1 (1988): 51–72.

Kay, Lily E. *Molecular Vision of Life: Caltech, the Rockefeller Foundation and the Rise of the New Biology*. New York: Oxford University Press, 1993.

Keating, Peter. "Holistic Bacteriology: Ludwick Hirszfeld's Doctorine of Serogenesis between the Two World Wars." In *Greater than the Parts: Holism in Biomedicine, 1920–1950*, edited by Christopher Lawrence and George Weisz, 283–302. New York: Oxford University Press, 1998.

Keating, Peter, and Alberto Cambrosio. *Biomedical Platforms: Realigning the Normal and the Pathological in Late-Twentieth Century Medicine*. Cambridge, MA: MIT Press, 2003.

Kekwick, Ralph A. "Alan Nigel Drury. 3 November 1889–2 August 1980." *Biographical Memoirs of Fellows of the Royal Society* 27 (1981): 173–98.

Kendrick, Douglas B. *Blood Program in World War II*. Washington, DC: Office of the Surgeon General, 1964.

Kendrick, T. D. *A History of the Vikings*. London: Methuen, 1930.

Keuck, Lara. "Thinking with Gatekeepers: An Essay on Psychiatric Sources." In *Invisibility and Labour in the Human Sciences*, edited by Jenny Bangham and Judith Kaplan, 107–16. Berlin: Max Planck Institute for the History of Science, 2016.

Kevles, Daniel. *In the Name of Eugenics: Genetics and the Uses of Human Heredity*. 2nd ed. Cambridge, MA: Harvard University Press, 1995 [1985].

Keynes, Geoffrey. *Blood Transfusion*. London: Henry Frowde, 1922.

Kherumian, R. "Review: The Distribution of the Human Blood Groups." *Man* 54 (1954): 156–57.

Kidd, P. "Simple Blood-Grouping Methods." *British Medical Journal* 4958 (1956), 114–15.

Killens, John Oliver. *Youngblood*. Athens: University of Georgia Press, 1982.

Kim, Jieun. "The Specter of 'Bad Blood' in Japanese Blood Banks." *New Genetics and Society* 37 (2018): 296–318.

Kirsh, Nurit. "Population Genetics in Israel in the 1950s: The Unconscious Internalization of Ideology." *Isis* 94 (2003): 631–55.

Klein, Ursula. *Experiments, Models, Paper Tools: Cultures of Organic Chemistry in the Nineteenth Century*. Stanford, CA: Stanford University Press, 2003.

Klugman, Matthew. *Blood Matters: A Social History of the Victorian Red Cross Blood Transfusion Service*. Melbourne: Australian Scholarly Publishing, 2004.

Knight, Robert L. *Dictionary of Genetics, Including Terms Used in Cytology, Animal Breeding and Evolution*. Waltham, MA: Chronica Botanica, 1948.

Kohler, Robert E. *Lords of the Fly*: Drosophila *Genetics and the Experimental Life*. Chicago: University of Chicago Press, 1994.

Kopeć, Ada C. *The Distribution of the Blood Groups in the United Kingdom*. London: Oxford University Press, 1970.

Kowal, Emma. "Orphan DNA: Indigenous Samples, Ethical Biovalue and Postcolonial Science." *Social Studies of Science* 43, no. 4 (2013): 577–97.

Kowal, Emma, Joanna Radin, and Jenny Reardon. "Indigenous Body Parts, Mutating Temporalities, and the Half-Lives of Postcolonial Technoscience." *Social Studies of Science* 43, no. 4 (2013): 465–83.

Kraus, Bertram S., and Charles B. White. "Microevolution in a Human Population: A Study of Social Endogamy and Blood Type Distributions among the Western Apache." *American Anthropologist* 58 (1956): 1017–43.

Krementsov, Nikolai. "Eugenics in Russia and the Soviet Union." In *The Oxford Handbook of The History of Eugenics*, edited by Alison Bashford and Philippa Levine, 413–29. Oxford: Oxford University Press, 2010.

Krementsov, Nikolai. *A Martian Stranded on Earth: Alexander Bogdanov, Blood Transfusions, and Proletarian Science*. Chicago: University of Chicago Press, 2011.

Kushner, Tony. *We Europeans? Mass-Observation, "Race" and British Identity in the Twentieth Century*. Aldershot: Ashgate, 2004.

Lachmann, Peter. "Robert Royston Amos (Robin) Coombs: 9 January 1921–25 January 2006." *Biographical Memoirs of Fellows of the Royal Society* 55 (2009): 45–58.

Landecker, Hannah. "Living Differently in Biological Time: Plasticity, Temporality, and Cellular Biotechnologies." In *Technologized Images, Technologized Bodies: Anthropological Approaches to a New Politics of Vision*, edited by Jeanette Edwards, Penny Harvey, and Peter Wade, 211–33. New York: Berghahn Books, 2005.

Landsteiner, Karl, and C. Philip Miller. "Serological Studies on the Blood of the Primates: I. The Differentiation of Human and Anthropoid Bloods." *Journal of Experimental Medicine* 42, no. 6 (1925): 841–52.

Landsteiner, Karl, and C. Philip Miller. "Serological Studies on the Blood of the Primates: III. Distribution of Serological Factors Related to Human Isoagglutinins in the Blood of Lower Monkeys," *Journal of Experimental Medicine* 42, no. 6 (1925): 863–77.

Landsteiner, Karl, and Alexander Wiener. "An Agglutinable Factor in Human Blood Recognised by Immune Sera for Rhesus Blood." *Proceedings of the Society for Experimental Biology and Medicine* 43 (1940): 223.

Landsteiner, Karl, and Alexander Wiener. "Studies on an Agglutinogen (Rh) in Human Blood Reacting with Anti-Rhesus Sera and with Human Isoantibodies." *Journal of Experimental Medicine* 74 (1941): 309–20.

Landsteiner, Karl, Alexander S. Wiener, and G. Albin Matson. "Distribution of the Rh Factor in American Indians." *Journal of Experimental Medicine* 76 (1942): 73–78.

Latour, Bruno. "Drawing Things Together." In *Representation in Scientific Practice*, edited by Michael Lynch and Steve Woolgar, 19–68. Cambridge, MA: MIT Press, 1990.

Latour, Bruno. *Science in Action: How to Follow Scientists and Engineers through Society*. Cambridge, MA: Harvard University Press, 1987.

Lattes, Leone. *Individuality of the Blood in Biology and in Clinical and Forensic Medicine*, translated by L. W. Howard Bertie. London: Oxford University Press, 1932.

Law, Jules. *The Social Life of Fluids: Blood, Milk, and Water in the Victorian Novel*. Ithaca, NY: Cornell University Press, 2010.

Lawler, Sylvia D. "Family Studies Showing Linkage between Elliptocytosis and the Rhesus Blood Group System." *Caryologia* 6, suppl. (1954): 26.

Lawler, Sylvia D., and M. Sandler. "Data on Linkage in Man: Elliptocytosis and Blood Groups: IV. Families 5, 6 and 7." *Annals of Eugenics* 18 (1954): 328–34.

Lederer, Susan E. "Bloodlines: Blood Types, Identity, and Association in Twentieth-Century America." *Journal of the Royal Anthropological Institute* 19 (2013): S118–29.

Lederer, Susan E. *Flesh and Blood: Organ Transplantation and Blood Transfusion in 20th Century America.* Oxford: Oxford University Press, 2008.

Lederer, Susan E., and Naomi Rogers. "Media." In *Companion to Medicine in the Twentieth Century,* edited by Roger Cooter and John Pickstone, 487–502. London: Routledge, 2013.

Lehmann, Hermann. "Blood Groups [Review of *The ABO Blood Groups,* Mourant et al.]." *Eugenics Review* 51 (1959): 108–9.

Lehmann, Hermann, and Elspeth B. Smith. "Separation of Different Haemoglobins by Paper Electrophoresis." *Transactions of the Royal Society of Tropical Medicine and Hygiene* 48 (1954): 12.

Lemov, Rebecca. *The Database of Dreams: The Lost Quest to Catalog Humanity.* Chicago: University of Chicago Press, 2015.

Levine, Philip, Peter Vogal, E. M. Katzin, and Lyman Burnham. "The Role of Isoimmunization in the Pathogenesis of Erythroblastosis Fetalis." *American Journal of Obstetrics and Gynacology* 42, no. 6 (1941): 925–37.

Lewis, D. "Cyril Dean Darlington. 19 December 1902–26 March 1981." *Biographical Memoirs of Fellows of the Royal Society* 29 (1983): 113–26.

Lindee, M. Susan. "Human Genetics after the Bomb: Archives, Clinics, Proving Grounds and Board Rooms." *Studies in History and Philosophy of Biological and Biomedical Sciences* 55 (2016): 45–53.

Lindee, M. Susan. "Scaling Up: Human Genetics as a Cold War Network." *Studies in History and Philosophy of Biological and Biomedical Sciences* 47, part A (2014): 185–90.

Lindee, M. Susan. *Moments of Truth in Genetic Medicine.* Baltimore: Johns Hopkins University Press, 2005.

Lindee, M. Susan. *Suffering Made Real: American Science and the Survivors at Hiroshima.* Chicago: University of Chicago Press, 1994.

Lindee, M. Susan, and Joanna Radin. "Patrons of the Human Experience: A History of the Wenner-Gren Foundation for Anthropological Research, 1941–2016." *Current Anthropology* 57, suppl. 14 (2016): S218–301.

Lindee, Susan, and Ricardo Ventura Santos. "The Biological Anthropology of Living Human Populations: World Histories, National Styles, and International Networks: An Introduction to Supplement 5." *Current Anthropology* 53, suppl. 5 (2012): S3–16.

Linehan, Denis. "Regional Survey and the Economic Geographies of Britain 1930–1939." *Transactions of the Institute of British Geographers* 28 (2003): 96–122.

Lipphardt, Veronika. "From 'Races' to 'Isolates' to 'Endogamous Communities': Human Genetics and the Notion of Human Diversity in the 1950s." In *Human Heredity in the Twentieth Century,* edited by Bernd Gausemeier, Staffan Müller-Wille, and Edmund Ramsden, 55–68. London: Pickering & Chatto, 2013.

Lipphardt, Veronika. "'Europeans' and 'Whites.' Biomedical Knowledge about the 'European Race' in Early Twentieth Century Colonial Contexts." *Comparativ* 25 (2015): 137–46.

Lipphardt, Veronika. "'Geographical Distribution Patterns of Various Genes': Genetic Studies of Human Variation after 1945." *Studies in History and Philosophy of Biological and Biomedical Sciences* 47, part A (2014): 50–61.

Lipphardt, Veronika. "Isolates and Crosses in Human Population Genetics; or, A Contextualization of German Race Science." *Current Anthropology* 53, Suppl. 5 (2012): S69–82.

Lipphardt, Veronika. "The Jewish Community of Rome—an Isolated Population? Sampling Procedures and Biohistorical Narratives in Genetic Analysis in the 1950s." *BioSocieties* 5 (2010): 306–29.

Lipphardt, Veronika. "Knowing Europe, Europeanizing Knowledge: The Making of 'Homo Europaeus' in the Life Sciences." In *Europeanization in the Twentieth Century: Historical Approaches*, edited by M. Conway and K. K. Patel, 64–83. London: Palgrave Macmillan, 2010.

Little, Michael A. "Human Population Biology in the Second Half of the Twentieth Century." *Current Anthropology* 53, suppl. 5 (2012): S126–38.

Little, Michael A., and Kenneth J. Collins. "Joseph S. Weiner and the Foundation of Post-WWII Human Biology in the United Kingdom." *Yearbook of Physical Anthropology* 55 (2012): 114–31.

Livingstone, F. B. *Abnormal Hemoglobins in Human Populations*. Chicago: Aldine, 1967.

Lock, Margaret. "The Alienation of Body Tissue and the Biopolitics of Immortalized Cell Lines." *Body and Society* 7, nos. 2–3 (2001): 63–91.

Loughlin, Kelly. "Networks of Mass Communication: Reporting Science, Health and Medicine in the 1950s and '60s." *Clio Medica* 75 (2005): 295–322.

Love, Spencie. *One Blood: The Death and Resurrection of Charles R. Drew*. Chapel Hill: University of North Carolina Press, 1996.

Löwy, Ilana. *Between Bench and Bedside: Science, Healing and Interleukin-2 in a Cancer Ward*. Cambridge, MA: Harvard University Press, 1996.

Löwy, Ilana. "'A River That Is Cutting Its Own Bed': The Serology of Syphilis between Laboratory, Society and Law." *Studies in History and Philosophy of Biological and Biomedical Sciences* 35, no. 3 (2004): 509–24.

Macintyre, Ben. "Opening the Book of Life." *Times*, June 27, 2000, 1.

MacKenzie, Donald A. *Statistics in Britain, 1865–1930: The Social Construction of Scientific Knowledge*. Edinburgh: Edinburgh University Press, 1981.

Magnello, Eileen, and Anne Hardy, eds. *The Road to Medical Statistics*. Amsterdam: Rodopi, 2002.

Marks, Jonathan. *Human Biodiversity: Genes, Race and History*. New York: Aldine de Gruyter, 1995.

Marks, Jonathan. "The Legacy of Serological Studies in American Physical Anthropology." *History and Philosophy of the Life Sciences* 18, no. 3 (1996): 345–62.

Marks, Jonathan. "The Origins of Anthropological Genetics," *Current Anthropology* 53 (2012): S161–72.

Marks, Jonathan. "'We're Going to Tell These People Who They Really Are': Science and Relatedness." In *Relative Values*, edited by Sarah Franklin and Susan McKinnon, 355–83. Durham, NC: Duke University Press, 2001.

Matless, David. "Regional Surveys and Local Knowledges: The Geographical Imagination in Britain, 1918–39." *Transactions of the Institute of British Geographers* 17 (1992): 464–80.

Mauss, Marcel. *The Gift: The Form and Reason for Exchange in Archaic Societies*, translated by W. D. Halls. London: Routledge, 2002.

Maxson Jones, Kathryn, Rachel A. Ankeny, and Robert Cook-Deegan. "The Bermuda Triangle: The Pragmatics, Policies and Principles for Data Sharing in the History of the Human Genome Project." *Journal for the History of Biology* 51, no. 4 (2018): 693–805.

Maxwell, Roy D. H., and Katharine N. Maxwell. "ABO Blood Groups and Hypertension." *British Medical Journal* 2, no. 4932 (1955): 179–80.

Maycock, William d'A. "Kenneth Leslie Grant Goldsmith." *Lancet* 2, no. 308 (1976): 212.

Mayer, Anna-K. "Reluctant Technocrats: Science Promotion in the Neglect-of-Science Debate of 1916–1918." *History of Science* 43, no. 2 (2005): 139–59.

Mayr, Ernst. *Systematics and the Origin of Species from the Viewpoint of a Zoologist*. New York: Columbia University Press, 1942.

Mayr, Ernst, and William B. Provine, eds. *The Evolutionary Synthesis: Perspectives on the Unification of Biology*. Cambridge, MA: Harvard University Press, 1980.

Mazumdar, Pauline. "Blood and Soil: The Serology of the Aryan Racial State." *Bulletin of the History of Medicine* 64, no. 2 (1990): 186–219.

Mazumdar, Pauline. *Eugenics, Human Genetics and Human Failings: The Eugenics Society, Its Sources and Its Critics in Britain*. London: Routledge, 1992.

Mazumdar, Pauline. "The Purpose of Immunity: Landsteiner's Interpretation of the Human Isoantibodies." *Journal of the History of Biology* 8, no. 1 (1975): 115–33.

Mazumdar, Pauline. *Species and Specificity: An Interpretation of the History of Immunology*. Cambridge, UK: Cambridge University Press, 1995.

Mazumdar, Pauline. "Two Models for Human Genetics: Blood Grouping and Psychiatry in Germany between the World Wars." *Bulletin of the History of Medicine* 70, no. 4 (1996): 609–57.

McGucken, William. "On Freedom and Planning in Science: The Society for Freedom in Science, 1940–46." *Minerva* 16, no. 1 (1978): 42–72.

M'charek, Amade. "Contrasts and Comparisons: Three Practices of Forensic Investigation." *Comparative Sociology* 7, no. 3 (2008): 387–412.

M'charek, Amade. *The Human Genome Diversity Project: An Ethnography of Scientific Practice*. Cambridge, UK: Cambridge University Press, 2005.

M'charek, Amade, Rob Hagendijk, and Wiebe de Vries. "Equal before the Law: On the Machinery of Sameness in Forensic DNA Practice." *Science, Technology & Human Values* 38, no. 4 (2013): 542–65.

M'charek, Amade, Katharina Schramm, and David Skinner. "Technologies of Belonging: The Absent Presence of Race in Europe." *Science, Technology and Human Values* 38, no. 4 (2013).

McKay, Richard A. *Patient Zero and the Making of the AIDS Epidemic*. Chicago: University of Chicago Press, 2017.

McLearn, Ida, Geoffrey M. Morant, and Karl Pearson. "On the Importance of the Type Silhouette for Racial Characterisation in Anthropology." *Biometrika* 20B, nos. 3–4 (1928): 389–400.

Medical Research Council Blood Transfusion Research Committee. "The Determination of Blood Groups," General Medical Council War Memorandum no. 9. London: His Majesty's Stationary Office, 1943. Available in the Wellcome Library, shelf mark WH420 1943M49d.

Middell, Matthias, ed. "The Invention of the European." Special issue, *Comparativ* 25, nos. 5–6 (2015): 7–228.

Miles, A. Ashley. "International Standards for Anti-A and Anti-B Blood-Grouping Sera." *Bulletin of the World Health Organization* 3 (1950): 301–8.

Miller, Clark A. "'An Effective Instrument of Peace': Scientific Cooperation as an Instrument of U.S. Foreign Policy, 1938–1950." *Osiris* 21 (2006): 133–60.

Misson, Gary P., A. Clive Bishop, and Winifred M. Watkins. "Arthur Ernest Mourant. 11 April 1904–29 August 1994." *Biographical Memoirs of Fellows of the Royal Society* 45 (1999): 331–48.

Mitchell, Robert, and Catherine Waldby. *Tissue Economies: Blood, Organs, and Cell Lines in Late Capitalism*. Durham, NC: Duke University Press, 2006.

Mol, Annemarie. *The Body Multiple: Ontology in Medical Practice*. Durham, NC: Duke University Press, 2002.

Mold, Alex, Peder Clark, Gareth Millward and Daisy Payling. *Placing the Public in Public Health in Post-War Britain, 1948–2012*. Cham, Germany: Springer, 2019.

Mollison, Patrick L. "Blood Groups." *British Medical Journal* 4697 (1951): 75.

Mollison, Patrick L. *Blood Transfusion in Clinical Medicine*. Oxford: Blackwell Scientific, 1951.

Mollison, Patrick L., Arthur E. Mourant, and Robert R. Race. "The Rh Blood Groups and Their Clinical Effects," Medical Research Council memorandum no. 19. London: His Majesty's Stationary Office, 1948. Available in the Wellcome Library, shelf mark WH420 1954M72r.

Mollison, Patrick L., and H. A. Sloviter. "Successful Transfusion of Previously Frozen Human Red Cells." *Lancet* 258, no. 6689 (1951): 862–64.

Mollison, Patrick L., and George L. Taylor. "Wanted: Anti-Rh Sera." *British Medical Journal* 4243 (1942): 561–62.

Monsalve, M. V., and E. Hagelberg. "Mitochondrial DNA Polymorphisms in Carib People of Belize." *Proceedings of the Royal Society of London. Series B: Biological Sciences* 264, no. 1385 (1997): 1217–24.

Morant, Geoffrey M. "A Contribution to Basque Craniometry." *Biometrika* 21, nos. 1–4 (1929): 67–84.

Moreau, Philippe. "The Bilineal Transmission of Blood in Ancient Rome." In *Blood and Kinship: Matter for Metaphor from Ancient Rome to the Present*, ed. Christopher

Johnson, Bernhard Jussen, David Warren Sabean, and Simon Teuscher, 40–60. New York: Berghahn Books, 2013.

Motulsky, Arno G. "Drug Reactions, Enzymes, and Biochemical Genetics," *Journal of the American Medical Association* 165, no. 7 (1957), 835–37.

Motulsky, Arno G., and Jean M. Campbell-Kraut. "Population Genetics of Glucose-6-Phosphate Dehyrogenase Deficiency of the Red Cell." In *Proceedings of the Conference on Genetic Polymorphisms and Geographic Variations in Disease*, edited by Baruch S. Blumberg, 159–91. New York: Grune and Stratton, 1961.

Motulsky, Arno G., Milton H. Paul, and E. L. Durrum. "Paper Electrophoresis of Abnormal Hemoglobins and Its Clinical Applications." *Blood* 9, no. 9 (1954): 897–910.

Motulsky, Arno G., J. Vandepitte, and G. R. Fraser, "Population Genetics in the Congo. I: Glucose-6-Phosphate Dehydrogenase Deficiency, Hemoglobin S and Malaria," *American Journal of Human Genetics* 18, no. 6 (1966): 514–37.

Mourant, Arthur E. *Blood and Stones: An Autobiography*. La Haule, Jersey: La Haule Books, 1995.

Mourant, Arthur E. "Blood Groups." In *Genetical Variation in Human Populations*, edited by G. A. Harrison, 1–15. New York: Pergamon, 1961.

Mourant, Arthur E. "Blood Groups and Anthropology." *Nature* 167 (May 5, 1951): 705–6.

Mourant, Arthur E. "The Blood Groups of the Basques." *Nature* 160 (October 11, 1947): 505–6.

Mourant, Arthur E. *The Distribution of the Human Blood Groups*. Oxford: Blackwell Scientific, 1954.

Mourant, Arthur E. "The Establishment of an International Panel of Blood Donors of Rare Types." *Vox Sanguinis* 10, no. 2 (1965): 129–32.

Mourant, Arthur E. *The Genetics of the Jews*. Oxford: Clarendon Press, 1978.

Mourant, Arthur E. "The International Biological Programme." *Eugenics Review* 55 (1964): 201–2.

Mourant, Arthur E. "A 'New' Human Blood Group Antigen of Frequent Occurrence." *Nature* 158 (August 17, 1946): 237–8.

Mourant, Arthur E. "A New Rhesus Antibody." *Nature* 155 (May 5, 1945): 542.

Mourant, Arthur E., Ada C. Kopeć, and Kazimiera Domaniewska-Sobczak. *The ABO Blood Groups: Comprehensive Tables and Maps of World Distribution*. London: Blackwell Scientific, 1958.

Mourant, Arthur E., Ada C. Kopeć, and Kazimiera Domaniewska-Sobczak. *Distribution of the Human Blood Groups and Other Polymorphisms*. London: Oxford University Press, 1976.

Mourant, Arthur E., and Donald Tills. "The International Biological Programme with Particular Reference to the Human Adaptability Section." *Institute of Biology Journal* 14 (1967): 24–27.

Mukharji, Projit Bihari. "From Serosocial to Sanguinary Identities: Caste, Transnational Race Science and the Shifting Metonymies of Blood Group B, India c. 1918–1960." *Indian Economic and Social History Review* 51, no. 2 (2014): 143–76.

Müller-Wille, Staffan. "Early Mendelism and the Subversion of Taxonomy: Episte-mological Obstacles as Institutions." *Studies in History and Philosophy of Bio-logical and Biomedical Sciences* 36, no. 3 (2005): 465–87.

Müller-Wille, Staffan. "Race et appartenance ethnique: La diversité humaine et l'UNESCO déclarations sur la race (1950 et 1951)." In *60 Ans d'histoire de l'UNESCO*, 211–20. Paris: UNESCO, 2007.

Müller-Wille, Staffan, and Hans-Jörg Rheinberger. *A Cultural History of Heredity.* Chicago: Chicago University Press, 2012.

Murphy, Michelle. *The Economization of Life.* Durham, NC: Duke University Press, 2017.

Murray, John. "A Nomenclature of Subgroups of the Rh Factor." *Nature* 154, no. 3918 (December 2, 1944): 701–2.

Mutch, J. R. "Hereditary Corneal Dystrophy." *British Journal of Ophthalmology* 28 (1944): 49–86.

Nash, Catherine. *Genetic Geographies: The Trouble with Ancestry.* Minneapolis: Uni-versity of Minnesota Press, 2015.

Nathoo, Ayesha. *Hearts Exposed: Transplants and the Media in 1960s Britain.* Lon-don: Palgrave Macmillan, 2009.

Neel, James V. "The Inheritance of Sickle Cell Anemia." *Science* 110, no. 2846 (July 15, 1949): 64–66.

Nelkin, Dorothy, and M. Susan Lindee. *The DNA Mystique: The Gene as a Cultural Icon.* Ann Arbor: University of Michigan Press, 2004.

"A New Blood-Donor Service." *Lancet* 260, no. 6736 (1952): 670.

Newton, Darrell. "Calling the West Indies: The BBC World Service and Caribbean Voices," *Historical Journal of Film, Radio and Television* 28, no. 4 (2008): 489–97.

Newton, Darrell M. *Paving the Empire Road: BBC Television and Black Britain.* Man-chester: Manchester University Press, 2012.

Nieter, H. M., dir. *Blood Transfusion Service.* 1941. Paul Rotha Productions. UK Min-istry of Information.

Nye, Robert A. "Kinship, Male Bonds, and Masculinity in Comparative Perspec-tive." *American Historical Review* 105, no. 5 (2000): 1656–66.

Oertzen, Christine von. "Hidden Helpers: Gender, Skill, and the Politics of Work-force Management for Census Compilation in Late Nineteenth-Century Prus-sia." In *Invisibility and Labour in the Human Sciences*, edited by Jenny Bangham and Judith Kaplan, 47–50. Berlin: Max Planck Institute for the History of Science, 2016.

Okroi, Mathias. "Der Blutgruppenforscher Fritz Schiff (1889–1940): Leben, Werk und Wirkung eines jüdischen Deutschen." PhD diss., University of Lübeck, 2004.

Okroi, Mathias, and Peter Voswinckel. "'Obviously Impossible'—The Application of the Inheritance of Blood Groups as a Forensic Method: The Beginning of Paternity Tests in Germany, Europe and the USA." *International Congress Series* 1239 (2003): 711–14.

Olszynko-Gryn, Jesse. *A Woman's Right to Know: Pregnancy Testing in Twentieth-Century Britain.* Berkeley: University of California Press, forthcoming.

Ottenberg, Reuben. "Studies in Isoagglutination: I. Transfusion and the Question of Intravascular Agglutination." *Journal of Experimental Medicine* 13 (1911): 425–38.

Oudshoorn, Nelly. *Beyond the Natural Body: An Archaeology of Sex Hormones*. London: Routledge, 1994.

Owen, Maureen. "Dame Janet Maria Vaughan, D. B. E. 18 October 1899–9 January 1993." *Biographical Memoirs of Fellows of the Royal Society* 41 (1995): 483–98.

Packard, Randall M. *A History of Global Health: Interventions into the Lives of Other Peoples*. Baltimore: Johns Hopkins University Press, 2016.

Palfreeman, Linda. *Spain Bleeds: The Development of Battlefield Blood Transfusion during the Civil War*. Brighton: Sussex Academic Press, 2015.

Parapia, Liakat Ali. "History of Bloodletting by Phlebotomy." *British Journal of Haematology* 143, no. 4 (2008): 490–95.

Parke, Davis & Co. *Biological Therapy: Including Vaccine Therapy, Serum Therapy, Phylacogen Therapy, Gland Therapy, Diagnostic Proteins*. London: Parke, Davis & Co., 1926.

Parolini, Giuditta. "The Emergence of Modern Statistics in Agricultural Science: Analysis of Variance, Experimental Design and the Reshaping of Research at Rothamsted Experimental Station, 1919–1933." *Journal of the History of Biology* 48, no. 2 (2014): 301–35.

Parolini, Giuditta. *"Making Sense of Figures": Statistics, Computing and Information Technologies in Agriculture and Biology in Britain, 1920s–1960s*. PhD diss., University of Bologna, 2013.

Parry, Bronwyn. "Technologies of Immortality: The Brain on Ice." *Studies in History and Philosophy of Biological and Biomedical Sciences*, 35, no. 2 (2004): 391–413.

Paul, Diane B. *The Politics of Heredity: Essays on Eugenics, Biomedicine, and the Nature-Nurture Debate*. Albany: State University of New York Press, Press, 1998.

Paul, Kathleen. *Whitewashing Britain: Race and Citizenship in the Postwar Era*. Ithaca, NY: Cornell University Press, 1997.

Pauling, Linus, Harvey A. Itano, S. J. Singer, and Ibert C. Wells. "Sickle-Cell Anemia: A Molecular Disease." *Science* 110, no. 2865 (November 25, 1949): 543.

Payne, Rose, and Mary R. Rolfs. "Fetomaternal Leukocyte Incompatibility." *Journal of Clinical Investigation* 37, no. 12 (1958): 1756–62.

Pearson, Karl, ed. *Treasury of Human Inheritance*. London: Dulau, 1912.

Pelis, Kim. "'A Band of Lunatics down Camberwell Way': Percy Oliver and Voluntary Blood Donation in Interwar Britain." In *Medicine, Madness and Social History: Essays in Honour of Roy Porter*, edited by Roberta Bivins and John V. Pickstone, 148–58. London: Palgrave Macmillan, 2007.

Pelis, Kim. "Blood Standards and Failed Fluids: Clinic, Lab and Transfusion Solutions in London, 1868–1916." *History of Science* 39, no. 124 (2001): 185–213.

Pelis, Kim. "Taking Credit: The Canadian Army Medical Corps and the British Conversion to Blood Transfusion in WWI." *Journal of the History of Medicine & Allied Sciences* 56, no. 3 (2001): 238–77.

Penrose, Lionel S. *Mental Defect*. New York: Farrar and Rinehart, 1934.

Penrose, Margaret, and Lionel S. Penrose. "The Blood Group Distribution in the Eastern Counties of England." *British Journal of Experimental Pathology* 14, no. 3 (1933): 160.

Petitjean, Patrick. "The Joint Establishment of the World Federation of Scientific Workers and of UNESCO after World War II." *Minerva* 46, no. 2 (2008): 247–70.

Pickles, M. M. "Simple Blood-Grouping Method," *British Medical Journal* 4955 (1955): 1561.

Pierce, Stephen, and Marion Reid. *Bloody Brilliant! A History of Blood Groups and Blood Groupers*. Bethesda, MD: AABB Press, 2016.

Pinkerton, J. R. H. "Simple Blood-Grouping Methods." *British Medical Journal* no. 4961 (1956): 289.

Plaut, G., M. Leitch Barrow, and J. M. Abbott. "The Results of Routine Investigation for Rh Factor at the N.W. London Depot." *British Medical Journal* 2, no. 4417 (1945): 273–81.

Polani, P. E. "John Alexander Fraser Roberts. 8 September 1899–15 January 1987." *Biographical Memoirs of Fellows of the Royal Society* 38 (1992): 307–22.

Porter, Theodore M. *Genetics in the Madhouse: The Unknown History of Human Heredity*. Princeton, NJ: Princeton University Press, 2018.

Porter, Theodore M. *Karl Pearson: The Scientific Life in a Statistical Age*. Princeton, NJ: Princeton University Press, 2004.

Porter, Theodore M. *Trust in Numbers: The Pursuit of Objectivity in Science and Public Life*. Princeton, NJ: Princeton University Press, 1995.

Posel, Deborah. "The Apartheid Project, 1948–1970." In *The Cambridge History of South Africa*, edited by Robert Ross, 319–68. Cambridge, UK: Cambridge University Press, 2011.

Proctor, Tammy M. *On My Honour: Guides and Scouts in Interwar Britain*. Philadelphia: American Philosophical Society, 2002.

Proctor, Tammy M. "(Uni)Forming Youth: Girl Guides and Boy Scouts in Britain, 1908–39." *History Workshop Journal* 45 (1998): 103–34.

Proger, L. W. "Development of the Emergency Blood Transfusion Scheme." *British Medical Journal* 4260 (1942): 252–53.

Provine, William B. *The Origins of Theoretical Population Genetics*. Chicago: University of Chicago Press, 1971.

Quirke, Viviane, and Jean-Paul Gaudillière. "The Era of Biomedicine: Science, Medicine, and Public Health in Britain and France after the Second World War." *Medical History* 52, no. 4 (2008): 441–52.

Rabinow, Paul. *French DNA: Trouble in Purgatory*. Chicago: University of Chicago Press, 1999.

Race, Robert R. "An 'Incomplete' Antibody in Human Serum." *Nature* 153 (June 24, 1944): 771–72.

Race, Robert R. "A Summary of Present Knowledge of Human Blood Groups, with Special Reference to Serological Incompatibility as a Cause of Congenital Disease." *British Medical Bulletin* 4, no. 3 (1946), 188–93.

Race, Robert R., and Arthur E. Mourant. "The Rh Chromosome Frequencies in England." *Blood* 3 (1948): 689–95.

Race, Robert R., Arthur E. Mourant, and Sheila Callender. "Rh Antigens and Antibodies in Man." *Nature* 157 (1946): 410.

Race, Robert R., and Ruth Sanger. *Blood Groups in Man*. 1st ed. Oxford: Blackwell Scientific, 1950.

Race, Robert R., and Ruth Sanger. *Blood Groups in Man*. 3rd ed. Oxford: Blackwell Scientific, 1958.

Race, Robert R., and Ruth Sanger. *Blood Groups in Man*. 4th ed. Oxford: Blackwell Scientific, 1962.

Race, Robert R., George L. Taylor, Daniel F. Cappell, and M. N. McFarlane. "The Rh Factor and Erythroblastosis Foetalis." *British Medical Journal* 4313 (1943): 289–93.

Race, Robert R., George L. Taylor, and J. Murray. "Serological Reactions Caused by the Rare Human Gene Rhz." *Nature* 155 (1945): 112–14.

Radin, Joanna. "Collecting Human Subjects: Ethics and the Archive in the History of Science and the Historical Life Sciences." *Curator: The Museum Journal* 57 (2014): 249–58.

Radin, Joanna. "Ethics in Human Biology: A Historical Perspective on Present Challenges." *Annual Review of Anthropology* 47 (2018): 263–78.

Radin, Joanna. "Latent Life: Concepts and Practices of Human Tissue Preservation in the International Biological Program." *Social Studies of Science* 43, no. 4 (2013): 484–508.

Radin, Joanna. *Life on Ice: The History of New Uses for Cold Blood*. Chicago: University of Chicago Press, 2017.

Radin, Joanna. "Unfolding Epidemiological Stories: How the WHO Made Frozen Blood into a Flexible Resource for the Future." *Studies in History and Philosophy of Biological and Biomedical Sciences* 47, part A (2014): 62–73.

Radin, Joanna, and Emma Kowal. "A Comparative Study of Indigenous Blood Samples and Ethical Regimes in the United States and Australia Since the 1960s." *American Ethnologist* 42, no. 4 (2015): 749–65.

Radin, Joanna, and Emma Kowal, eds. *Cryopolitics: Frozen Life in a Melting World*. Cambridge, MA: MIT Press, 2017.

Rajagopalan, Ramya M., and Joan H. Fujimura. "Variations on a Chip: Technologies of Difference in Human Genetics Research." *Journal of the History of Biology* 51, no. 4 (2018): 841–73

Ramsden, Edmund. "Carving up Population Science: Eugenics, Demography and the Controversy over the 'Biological Law' of Population Growth." *Social Studies of Science* 32, nos. 5–6 (2002): 857–99.

Ramsden, Edmund. "Surveying the Meritocracy: The Problems of Intelligence and Mobility in the Studies of the Population Investigation Committee." *Studies in History and Philosophy of Biological and Biomedical Sciences* 47, part A (2014): 130–41.

Rangil, Teresa Tomas. "The Politics of Neutrality: UNESCO's Social Science Department, 1946–1956." Center for the History of Political Economy at Duke University (CHOPE) working paper no. 2011-08, April 2011, https://hope.econ.duke.edu/sites/hope.econ.duke.edu/files/The%20politics%20of%20neutrality-Teresa%20Tomas.pdf.

Reardon, Jenny. *The Postgenomic Condition: Ethics, Justice, and Knowledge after the Genome.* Chicago: University of Chicago Press, 2017.

Reardon, Jenny. *Race to the Finish: Identity and Governance in an Age of Genomics.* Princeton: Princeton University Press, 2004.

Reardon, Jenny, Rachel A. Ankeny, Jenny Bangham, Katherine W. Darling, Stephen Hilgartner, Kathryn Maxson Jones, Beth Shapiro, and Hallam Stevens. "Bermuda 2.0: Reflections from Santa Cruz." *GigaScience* 5, no. 1 (2016): 1–4.

Reardon, Jenny, and Kim TallBear. "'Your DNA Is Our History': Genomics, Anthropology, and the Construction of Whiteness as Property." *Current Anthropology* 53, suppl. 5 (2012): S233–45.

Reddy, Deepa S. "Citizens in the Commons: Blood and Genetics in the Making of the Civic." *Contemporary South Asia* 21, no. 3 (2013): 275–90.

Redhead, Grace O. "Histories of Sickle Cell Anaemia in Postcolonial Britain, 1948–1997." PhD diss., University College London, 2019.

Redman, Samuel J. *Bone Rooms: From Scientific Racism to Human Prehistory in Museums.* Cambridge, MA: Harvard University Press, 2016.

Rees, Amanda. "Doing 'Deep Big History': Race, Landscape and the Humanity of H J Fleure (1877–1969)." *History of the Human Sciences* 32, no. 1 (2019): 99–120.

Reginald, Daniel G. *More Than Black? Multiracial Identity and the New Racial Order.* Philadelphia: Temple University Press, 2002.

Reid, Marion E. "Alexander S. Wiener: The Man and His Work." *Transfusion Medicine Reviews* 22, no. 4 (2008): 300–316.

Renwick, Chris. *British Sociology's Lost Biological Roots: A History of Futures Past.* London: Palgrave Macmillan, 2012.

Renwick, Chris. "Completing the Circle of the Social Sciences? William Beveridge and Social Biology at London School of Economics during the 1930s." *Philosophy of the Social Sciences* 44, no. 4 (2014): 478–96.

Research in Population Genetics of Primitive Groups: Report of a WHO Scientific Group. World Health Organization Technical Report Series no. 279. Geneva: World Health Organization, 1964.

Rheinberger, Hans-Jörg. "Scrips and Scribbles." *MLN* 118, no. 3 (2003): 622–36.

Rheinberger, Hans-Jörg. *Toward a History of Epistemic Things: Synthesizing Proteins in the Test Tube.* Stanford, CA: Stanford University Press, 1997.

Ricaut, F. X., T. Thomas, C. Arganini, J. Staughton, M. Leavesley, M. Bellatti, R. Foley, and M. Mirazon Lahr. "Mitochondrial DNA Variation in Karkar Islanders." *Annals of Human Genetics* 72, no. 3 (2008): 349–67.

Rice-Edwards, John T. "A Simple Blood-Grouping Method." *British Medical Journal* 4940 (1955): 681.

Richmond, Marsha L. "Opportunities for Women in Early Genetics." *Nature Reviews Genetics* 8 (2007): 897–902.

Richmond, Marsha L. "Women in the Early History of Genetics: William Bateson and the Newnham College Mendelians, 1900–1910." *Isis* 92, no. 1 (2001): 55–90.

Riddell, Horsley. *Blood Transfusion*. London: Oxford University Press, 1939.

Riddell, W. J. B. "A Pedigree of Blue Sclerotics, Brittle Bones, and Deafness, with Colour Blindness." *Annals of Eugenics* 10 (1940): 1–13.

Rivett, Geoffrey. *From Cradle to Grave: Fifty Years of the NHS*. London: Kings Fund, 1998.

Roberts, Derek F., Elizabeth W. Ikin, and Arthur E. Mourant. "Blood Groups of the Northern Nilotes." *Annals of Human Genetics* 20, no. 2 (1955): 135–54.

Roberts, Derek F., and Joseph S. Weiner. "Preface." In *The Scope of Physical Anthropology and its Place in Academic Studies: A Symposium Held at the Ciba Foundation, 6th November 1957*. Oxford: Wenner-Gren Foundation for Anthropological Research, 1958.

Robertson, Jennifer. "Biopower: Blood, Kinship, and Eugenic Marriage." In *A Companion to the Anthropology of Japan*, edited by Jennifer Robertson, 329–54. Oxford: Blackwell, 2005.

Robertson, Jennifer. "Blood Talks: Eugenic Modernity and the Creation of New Japanese." *History and Anthropology* 13, no. 3 (2002): 191–216.

Robertson, Jennifer. "Eugenics in Japan: Sanguinous Repair." In *The Oxford Handbook of the History of Eugenics*, edited by Alison Bashford and Philippa Levine, 431–48. Oxford: Oxford University Press, 2010.

Robertson, Jennifer. "Hemato-Nationalism: The Past, Present, and Future of 'Japanese Blood.'" *Medical Anthropology* 31, no. 2 (2012): 93–112.

Robson, Betty. "Sylvia Lawler." *British Medical Journal* 7035 (1996): 906.

Rood, J. J. van, J. G. Eernisse, and A. van Leeuwen. "Leucocyte Antibodies in Sera from Pregnant Women." *Nature* 181 (1958): 1735–36.

Roque, Ricardo. "The Blood That Remains: Card Collections from the Colonial Anthropological Missions." *BJHS Themes* 4 (2019): 29–53.

Roque, Ricardo. *Headhunting and Colonialism: Anthropology and the Circulation of Human Skulls in the Portuguese Empire, 1870–1930*. Basingstoke: Macmillan, 2010.

Rose, Nikolas. "Calculable Minds and Manageable Individuals." *History of the Human Sciences* 1, no. 2 (1988): 179–200.

Rosenfeld, L. "A Golden Age of Clinical Chemistry: 1948–1960." *Clinical Chemistry* 46, no. 10 (2000): 1705–14.

Roxby, P. M. "The Conference on Regional Survey at Newbury." *Geographical Teacher* 9, no. 2 (1917): 94–98.

Royal Anthropological Institute and Institute of Sociology. *Race and Culture*. London: Le Play House Press, 1936.

Royal Anthropological Institute of Great Britain and Ireland. *Notes and Queries on Anthropology*. 6th ed. London: Routledge, 1951.

Rucart, M. Marc. "Séance solenelle d'ouvertue." In *Bulletin de la société française de la transfusion sanguine: IIe congrès international de la transfusion sanguine*, 19–21. Paris: Baillère, 1939.

Rudavsky, Shari. "Blood Will Tell: The Role of Science and Culture in Twentieth Century Paternity Disputes." PhD diss., University of Pennsylvania, 1996.

Russell, N. C., E. M. Tansey, and P. V. Lear. "Missing Links in the History and Practice of Science: Teams, Technicians and Technical Work." *History of Science* 38, no. 2 (2000): 237–41.

Rymer, M. R. "The Editor's Page." *AABB Bulletin* 12 (1959): 484.

Saxon, R. S. "Towards Cadaver Blood Transfusions in War." *Lancet* 231, no. 5977 (1938), 693–94.

Sayers, Dorothy L. "Blood Sacrifice." In *In the Teeth of the Evidence*, 153–76. London: New English Library, 1970.

Schaffer, Gavin. "'Like a Baby with a Box of Matches': British Scientists and the Concept of 'Race' in the Inter-War Period." *British Journal for the History of Science* 38, no. 3 (2005): 307–24.

Schaffer, Gavin. *Racial Science and British Society, 1930–62*. Basingstoke: Palgrave Macmillan, 2008.

Schiff, Fritz. "The Medico-legal Significance of Blood Groups." *Lancet* 214, no. 5540 (1929): 921–22.

Schmidt, Paul J. "Blood and Disaster: Supply and Demand." *New England Journal of Medicine* 346, no. 8 (2002): 617–20.

Schmidt, Paul J. "Rh-Hr: Alexander Wiener's Last Campaign." *Transfusion* 34, no. 2 (1994): 180–82.

Schmidt, Paul J., and Paul M. Ness. "Hemotherapy: From Bloodletting Magic to Transfusion Medicine." *Transfusion* 46, no. 2 (2006): 166–68.

Schmuhl, Hans-Walter. *The Kaiser Wilhelm Institute for Anthropology, Human Heredity, and Eugenics, 1927–1945: Crossing Boundaries*. Dordrecht: Springer, 2008.

Schneider, David M. *American Kinship: A Cultural Account*. Chicago: University of Chicago Press, 1968.

Schneider, David M. "Kinship and Biology." In *Aspects of the Analysis of Family Structure*, edited by Ansley J. Coale, 83–101. Princeton: Princeton University Press, 1965.

Schneider, William H. "Blood Group Research in Great Britain, France and the United States Between the World Wars." *Yearbook of Physical Anthropology* 38 (1995): 87–114.

Schneider, William H. "Blood Transfusion Between the Wars." *Journal for the History of Medicine* 58, no. 2 (2003): 187–224.

Schneider, William H. "Chance and Social Setting in the Application of the Discovery of Blood Groups." *Bulletin of the History of Medicine* 57, no. 4 (1983): 545.

Schneider, William H. *The History of Blood Transfusion in Sub-Saharan Africa*. Athens: Ohio University Press, 2013.

Schneider, William H. "The History of Research on Blood Group Genetics: Initial Discovery and Diffusion." *History and Philosophy of the Life Sciences* 18, no. 3 (1996): 277–303.

Schneider, William H. "Introduction to 'The First Genetic Marker' Special Issue." *History and Philosophy of the Life Sciences* 18, no. 3 (1996): 273–76.

Schumaker, Lyn. *Africanizing Anthropology: Fieldwork, Networks, and the Making of Cultural Knowledge in Central Africa*. Durham: Duke University Press, 2001.

Schwartz-Marín, Ernesto and Irma Silva-Zolezzi. "'The Map of the Mexican's Genome': Overlapping National Identity, and Population Genomics." *Identity in the Information Society* 3, no. 3 (2010): 489–514.

Secord, James A. "Knowledge in Transit." *Isis* 95, no. 4 (2004): 654–72.

Selcer, Perrin. "Patterns of Science: Developing Knowledge for a World Community at Unesco." PhD diss., University of Pennsylvania, 2011.

Selcer, Perrin. "The View from Everywhere: Disciplining Diversity in Post-World War II International Social Science." *Journal of the History of the Behavioral Sciences* 45, no. 4 (2009): 309–29.

Sellen, Abigail J., and Richard H. R. Harper. *The Myth of the Paperless Office*. Cambridge, MA: MIT Press, 2001.

Shaw, Jennifer. "Documenting Genomics: Applying Archival Theory to Preserving the Records of the Human Genome Project." *Studies in History and Philosophy of Biological and Biomedical Sciences* 55 (2016): 61–69.

"Shortage of Rh Testing Sera." *British Medical Journal* 1, no. 4645 (1950): 108–9.

Silverman, Rachel. "The Blood Group 'Fad' in Post-War Racial Anthropology." In *Racial Anthropology: Retrospective on Carleton Coon's The Origin of Races (1962)*, edited by Jonathan Marks, 11–27. Berkeley: Kroeber Anthropological Society, 2000.

Silverstein, Arthur M. *A History of Immunology*. 2nd ed. Cambridge, MA: Academic Press, 2009.

Simon, Jonathan. "Emil Behring's Medical Culture: From Disinfection to Serotherapy." *Medical History* 51, no. 2 (2007): 201–18.

Simpson, George Gaylord. *Tempo and Mode in Evolution*. New York: Columbia University Press, 1944.

Sluga, Glenda. "UNESCO and the (One) World of Julian Huxley." *Journal of World History* 21, no. 3 (2010): 393–418.

Smith, Roger. "Biology and Values in Interwar Britain: C. S. Sherrington, Julian Huxley and the Vision of Progress." *Past & Present* 178, no. 1 (2003), 210–42.

Smithies, Oliver. "Grouped Variations in the Occurrence of New Protein Components in Normal Human Serum." *Nature* 175, no. 4450 (February 12, 1955): 307–8.

Smocovitis, Vassiliki Betty. "Unifying Biology: The Evolutionary Synthesis and Evolutionary Biology." *Journal of the History of Biology* 25, no. 1 (1992): 1–65.

Smocovitis, Vassiliki Betty. "Humanizing Evolution: Anthropology, the Evolutionary Synthesis, and the Prehistory of Biological Anthropology, 1927–1962." *Current Anthropology* 53, suppl. 5 (2012): S108–25.

Snape, Robert. *Leisure, Voluntary Action and Social Change in Britain, 1880–1939*. London: Bloomsbury Academic, 2018.

Snyder, Laurence H. *Blood Grouping in Relation to Clinical and Legal Medicine*. Baltimore: Williams and Wilkins, 1929.

Snyder, Laurence H. "The 'Laws' of Serologic Race-Classification Studies in Human Inheritance IV." *Human Biology* 2 (1930): 128–33.

Solomon, Susan Gross, Lion Murard, and Patrick Zylberman, eds. *Shifting Boundaries of Public Health: Europe in the Twentieth Century*. Rochester: University of Rochester Press, 2008.

Sommer, Marianne. "Biology as a Technology of Social Justice in Interwar Britain: Arguments from Evolutionary History, Heredity, and Human Diversity." *Science, Technology, & Human Values* 39, no. 4 (2014): 561–86.

Sommer, Marianne. "DNA and Cultures of Remembrance: Anthropological Genetics, Biohistories and Biosocialities." *BioSocieties* 5, no. 3 (2010): 366–90.

Sommer, Marianne. "History in the Gene: Negotiations Between Molecular and Organismal Anthropology." *Journal of the History of Biology* 41, no. 3 (2008): 473–528.

Sommer, Marianne. *History Within: The Science, Culture and Politics of Bones, Organisms, and Molecules*. Chicago: University of Chicago Press, 2016.

Spörri, Myriam. *Reines und Gemischtes Blut: Zur Kulturgeschichte der Blutgruppenforschung, 1900–1933*. University of Bielefeld: Verlag Bielefeld, 2013.

Stanton, Jenny. "Blood Brotherhood: Techniques, Expertise and Sharing in Hepatitis B Research in the 1970s." In *Technologies of Modern Medicine*, edited by Ghislaine Lawrence. London: Science Museum, 1994.

Star, Susan Leigh, and Karen Ruhleder. "Steps toward an Ecology of Infrastructure: Design and Access for Large Information Spaces." *Information Systems Research* 7, no. 1 (1996): 111–34.

Stark, Laura. *Behind Closed Doors: IRBs and the Making of Ethical Research*. Chicago: University of Chicago Press, 2012.

Stark, Laura. "The Bureaucratic Ethic and the Spirit of Bio-Capitalism." In *Invisibility and Labour in the Human Sciences*, edited by Jenny Bangham and Judith Kaplan, 13–24. Berlin: Max Planck Institute for the History of Science, 2016.

Starr, Douglas. *Blood: An Epic History of Medicine and Commerce*. New York: Harper Perennial, 2009. Originally published in 1998.

Steinberg, Arthur G., and Charles E. Cook. *The Distribution of the Human Immunoglobulin Allotypes*. Oxford: Oxford University Press, 1981.

Stepan, Nancy. *The Idea of Race in Science: Great Britain, 1800–1960*. London: Macmillan, 1982.

Stocking, George W. Jr., ed. *Bones, Bodies, Behavior: Essays in Biological Anthropology*. Madison: University of Wisconsin Press, 1988.

Stone, Dan. "Race in British Eugenics." *European History Quarterly* 31, no. 3 (2001): 397–425.

Strandskov, Herluf H. "Blood Group Nomenclature." *Journal of Heredity* 39, no. 4 (1948): 108–12.

Strandskov, Herluf H., and Sherwood L. Washburn. "Genetics and Physical Anthropology." *American Journal of Physical Anthropology* 9, no. 3 (1951): 261–64.

Strasser, Bruno J. "Collecting, Comparing, and Computing Sequences: The Making of Margaret O. Dayhoff's Atlas of Protein Sequence and Structure, 1954–1965." *Journal for the History of Biology* 43, no. 4 (2010): 623–60.

Strasser, Bruno J. "Genbank: Natural History in the 21st Century?" *Science* 322, no. 5901 (October 24, 2008): 537–38.

Strasser, Bruno J. "Laboratories, Museums, and the Comparative Perspective: Alan A. Boyden's Quest for Objectivity in Serological Taxonomy, 1924–1962." *Historical Studies in the Natural Sciences* 40, no. 2 (2010): 149–82.

Stratton, Frederick J. M. "International Council of Scientific Unions." *Notes and Records of the Royal Society* 4 (1946): 168–73.

Stratton, F., F. A. Langley, and U. Lister. "Haemolytic Disease of the Newborn in One of Dizygotic Twins." *British Medical Journal* 4387 (1945): 151–52.

Sturdy, Steve. "Reflections: Molecularization, Standardization and the History of Science." In *Molecularizing Biology and Medicine: New Practices and Alliances, 1910s—1970s*, edited by Soraya de Chadarevian and Harmka Kamminga, 254–71. Amsterdam: Harwood, 1998.

Sturdy, Steve, and Roger Cooter. "Science, Scientific Management and the Transformation of Medicine in Britain c. 1870–1950." *History of Science* 36, no. 114 (1998): 421–66.

Suárez, Edna. "Models and Diagrams as Thinking Tools: The Case of Satellite-DNA." *History and Philosophy of the Life Sciences* 29, no. 2 (2007): 177–92.

Sunseri, Thaddeus. "Blood Trials: Transfusions, Injections, and Experiments in Africa, 1890–1920." *Journal of the History of Medicine and Allied Sciences* 71, no. 3 (2016): 293–321.

Swanson, Kara W. *Banking on the Body: The Market in Blood, Milk and Sperm in Modern America*. Cambridge, MA: Harvard University Press, 2014.

Szreter, Simon. "The Population Health Approach in Historical Perspective." *American Journal of Public Health* 93, no. 3 (2003), 421–31.

TallBear, Kim. *Native American DNA: Tribal Belonging and the False Promise of Genetic Science*. Minneapolis: University of Minnesota Press, 2013.

Tapper, Melbourne. *In the Blood: Sickle Cell Anemia and the Politics of Race*. Philadelphia: University of Pennsylvania Press, 1999.

Taussig, Karen-Sue. *Ordinary Genomes: Science, Citizenship, and Genetic Identities*. Durham: Duke University Press, 2009.

Taylor, George L., and Aileen M. Prior. "Blood Groups in England, I: Examination of Family and Unrelated Material." *Annals of Eugenics* 8 (1938): 343–55.

Taylor, George L., and Aileen M. Prior. "Blood Groups in England, II: Distribution in the Population." *Annals of Eugenics* 8 (1938): 358–61.

Taylor, George L., and Aileen M. Prior. "Blood Groups in England, III: Discussion of the Family Material." *Annals of Eugenics* 9 (1939): 18–44.

Taylor, Griffith. "The Evolution and Distribution of Race, Culture, and Language." *Geographical Review* 11, no. 1 (1921): 54–119.

Teslow, Tracy. *Constructing Race: The Science of Bodies and Cultures in American Anthropology*. Cambridge, UK: Cambridge University Press, 2014.

Thackeray, Anne I. "Leakey, Louis Seymour Bazett (1903–1972)." In *Oxford Dictionary of National Biography*. Oxford University Press, 2004, https://doi.org/10.1093/ref:odnb/31343. Accessed June 8, 2020.

Thomson, Arthur Landsborough. *Half a Century of Medical Research*. London: Medical Research Council, 1987.

Thorsby, E. "A Short History of HLA." *Tissue Antigens* 74 (2009): 101–16.

Tilley, Helen. *Africa as a Living Laboratory*. Chicago: University of Chicago Press, 2011.

Tilley, Helen. "Racial Science, Geopolitics, and Empires: Paradoxes of Power." *Isis* 105 (2014): 773–81.

Titmuss, Richard M. *The Gift Relationship: From Human Blood to Social Policy*. London: Allen & Unwin, 1970.

Turda, Marius. "From Craniology to Serology: Racial Anthropology in Interwar Hungary and Romania." *Journal of the History of Behavioural Sciences* 43, no. 4 (2007): 361–77.

Turda, Marius. "The Nation as Object: Race, Blood, and Biopolitics in Interwar Romania." *Slavic Review* 66, no. 3 (2007): 413–41.

Turda, Marius, and Paul Weindling, eds. *"Blood and Homeland": Eugenics and Racial Nationalism in Central and Southeast Europe, 1900–1940*. Budapest: Central European University Press, 2007.

Turda, Marius, and Paul Weindling. "Eugenics, Race and Nation in Central and Southeast Europe, 1900–1940: A Historiographic Overview." In *"Blood and Homeland": Eugenics and Racial Nationalism in Central and Southeast Europe, 1900–1940*, edited by Marius Turda and Paul Weindling, 1–20. Budapest: Central European University Press, 2006.

Turton, Keren. "Films and Blood Donation Publicity in Mid-Twentieth Century Britain." MPhil diss., University of Cambridge, 2019.

UNESCO. *Creation of UNESCO*. Video. London: UNESCO TV, 1945, http://www.unesco.org/archives/multimedia/index.php?s=films_details&pg=33&id=15.

UNESCO. "Four Statements on the Race Question." Paris: UNESCO, 1969, https://unesdoc.unesco.org/ark:/48223/pf0000122962.

UNESCO. *The Race Concept: Results of an Inquiry*. Paris: UNESCO, 1952, https://unesdoc.unesco.org/ark:/48223/pf0000073351.

UNESCO. *What Is Race? Evidence from Scientists*. Paris: UNESCO, 1952.

UNESCO. "UNESCO Constitution," November 16, 1945, http://portal.unesco.org/en/ev.php-URL_ID=15244&URL_DO=DO_TOPIC&URL_SECTION=201.html.

Vaughan, Janet M., and Philip N. Panton. "The Civilian Blood Transfusion Service." In *The Emergency Medical Services*, vol. 1, edited by C. L. Dunn, 334–55. London: Her Majesty's Stationary Office, 1952.

Vavilov, Nikolai I. "The Problem of the Origin of the World's Agriculture in the Light of the Latest Investigations." In *Science at the Crossroads: Papers Presented to the International Congress of the History of Science and Technology 1931*, edited by N. I. Bukharin, 97–106. London: Kniga, 1931.

von Dungern, Emil, and Ludwik Hirszfeld. "Über Vererbung gruppenspezifischer Strukturen des Blutes, II." *Zeitschrift für Immunitätsforschung und experimentelle Therapie* 6 (1910): 284–92.

Wailoo, Keith. *Drawing Blood: Technology and Disease Identity in Twentieth-Century America.* Baltimore: Johns Hopkins University Press, 1999.

Wailoo, Keith. *Dying in the City of the Blues: Sickle Cell Anemia and the Politics of Race and Health.* Chapel Hill: University of North Carolina Press, 2001.

Wailoo, Keith, Alondra Nelson, and Catherine Lee, eds. *Genetics and the Unsettled Past: The Collision of DNA, Race, and History.* New Brunswick, NJ: Rutgers University Press, 2012.

Wailoo, Keith, and Stephen Pemberton. *The Troubled Dream of Genetic Medicine: Ethnicity and Innovation in Tay-Sachs, Cystic Fibrosis, and Sickle Cell Disease.* Baltimore: Johns Hopkins University Press, 2006.

Waldby, Catherine, and Melinda Cooper. *Clinical Labor: Tissue Donors and Research Subjects in the Global Bioeconomy.* Durham, NC: Duke University Press, 2014.

Walker, C. B. V., and H. G. Dennis. "Anti-A Haemolysin in Group O Blood Donors: An East Anglian Survey." *British Medical Journal* 2, no. 5162 (1959): 1303–5.

Walker, William. "Refresher Course for General Practitioners: Haemolytic Disease of the Newborn." *British Medical Journal* 2, no. 4740 (1951): 1142–46.

Washburn, Sherwood L. "The New Physical Anthropology." *Transactions of the New York Academy of Sciences* 13, no. 7 (1951): 298–304.

Washburn, Sherwood L. "Physical Anthropology . . . an Appraisal." *American Scientist* 45, no. 1 (1957): 79–87.

Waters, Chris. "'Dark Strangers' in Our Midst: Discourses of Race and Nation in Britain, 1947–1963." *Journal of British Studies* 36, no. 2 (1997): 207–38.

Watkin, I. Morgan. "Blood Groups in Wales and the Marches." *Man* 52, no. 6 (1952), 83–86.

Webster, Charles. *The National Health Service: A Political History.* 2nd ed. Oxford: Oxford University Press, 2002 [1998].

Webster, Charles. *Problems of Health Care: The National Health Service before 1957.* London: Her Majesty's Stationary Office, 1988.

Weiner, Joseph S. "Physical Anthropology . . . an Appraisal." *American Scientist* 45, no. 1 (1957): 79–87.

Weiner, Joseph S., and E. M. Lourie. *Human Biology: A Guide to Field Methods.* Great Britain: International Biological Programme, 1969.

Weiss, Sheila Faith. "After the Fall: Political Whitewashing, Professional Posturing, and Personal Refashioning in the Postwar Career of Otmar Freiherr von Verschuer." *Isis* 101, no. 4 (2010): 722–58.

Werskey, Gary. *The Visible College.* London: Allen Lane, 1978.

Weston, Kath. "Kinship, Controversy, and the Sharing of Substance: The Race/Class Politics of Blood Transfusion." In *Relative Values: Reconfiguring Kinship Studies,* edited by Sarah Franklin and Susan McKinnon, 147–74. Durham, NC: Duke University Press, 2001.

Whitby, L. E. H. "The Hazards of Transfusion." *Lancet* 239 (1942): 581–85.

White, Luise. "Cars out of Place: Vampires, Technology, and Labor in East and Central Africa." *Representations* 43 (1993): 27–50.

White, Luise. *Speaking with Vampires: Rumor and History in Colonial Africa*. Berkeley: University of California Press, 2000.

White, Luise. "Tsetse Visions: Narratives of Blood and Bugs in Colonial Northern Rhodesia." *Journal of African History* 36, no. 2 (1995): 219–45.

Whitfield, Nicholas. "A Genealogy of the Gift: Blood Donation in London, 1921–1946." PhD diss., University of Cambridge, 2011.

Whitfield, Nicholas. "Who Is My Donor? The Local Propaganda Techniques of London's Emergency Blood Transfusion Service, 1939–45." *Twentieth Century British History* 24, no. 4 (2013): 542–72.

Whitfield, Nicholas. "Who Is My Stranger? Origins of the Gift in Wartime London, 1939–45." *Journal of the Royal Anthropological Institute* 19, (2013): S95–117.

Widmer, Alexandra. "Making Blood 'Melanesian': Fieldwork and Isolating Techniques in Genetic Epidemiology (1963–1976)." *Studies in History and Philosophy of Biological and Biomedical Sciences* 47, part A (2014): 118–29.

Widmer, Alexandra, and Veronika Lipphardt, eds. *Health and Difference: Rendering Human Variation in Colonial Engagements*. New York: Berghahn Books, 2016.

Wiener, Alexander S. "George Lees Taylor." *Science* 102, no. 2638 (July 20, 1945): 55.

Wiener, Alexander S. "The Rh Factor and Racial Origins." *Science* 96, no. 2496 (October 30, 1942): 407–8.

Wiener, Alexander S. "Theory and Nomenclature of the Hr Blood Factors." *Science* 102, no. 2654 (November 9, 1945): 479–82.

Wilson, Duncan. *The Making of British Bioethics*. Manchester: Manchester University Press, 2014.

Wilson, G. S. "The Public Health Laboratory Service: Origin and Development of Public Health Laboratories." *British Medical Journal* 4553 (1948): 677–82.

Winlow, Heather. "Anthropometric Cartography: Constructing Scottish Racial Identity in the Early Twentieth Century." *Journal of Historical Geography* 27 (2001), 507–28

Winlow, Heather. "Cartographic Representations of Race: c.1850–1930." PhD diss., Queen's University of Belfast, 1999.

Wolfe, Audra J. "The Cold War Context of the Golden Jubilee, or, Why We Think of Mendel as the Father of Genetics." *Journal of the History of Biology* 45, no. 3 (2012): 389–414.

World Health Organization. *The First Ten Years of the World Health Organization*. Geneva: World Health Organization, 1958.

World Health Organization Department of Vaccines and Biologicals. *WHO Global Action Plan for Laboratory Containment of Wild Polioviruses*, 2nd ed. World Health Organization, 2003, https://apps.who.int/iris/bitstream/handle/10665/68205/WHO_V-B_03.11_eng.pdf.

Wright, S. "Evolution in Mendelian Populations." *Genetics* 16, no. 2 (1931): 97–159.

Wyman, Leland C., and William C. Boyd. "Human Blood Groups and Anthropology." *American Anthropologist* 37, no. 2 (1935): 181–200.

Yates, F., and K. Mather. "Ronald Aylmer Fisher: 1890–1962." *Biographical Memoirs of Fellows of the Royal Society* 9 (1963): 91–129.

Zallen, Doris T., D. A. Christie, and E. M. Tansey, eds. "The Rhesus Factor and Disease Prevention." London: Wellcome Trust, 2004.

Zallen, Doris T. "From Butterflies to Blood: Human Genetics in the United Kingdom." In *The Practices of Human Genetics*, edited by Michael Fortun and Everett Mendelsohn, 197–216. Kluwer Academic, 1997.

Zeitlin, R. A. "A Simple Blood-Grouping Method." *British Medical Journal* 4945 (1955): 970–71.

Zimmerman, Andrew. *Anthropology and Antihumanism in Imperial Germany*. Chicago: University of Chicago Press, 2001.

Zoutendyk, A., "The Blood Groups of the Hottentots." *American Journal of Physical Anthropology* 13, no. 4 (1955): 691–97.

Zoutendyk, A., Ada C. Kopeć, and Arthur E. Mourant. "The Blood Groups of the Bushmen." *American Journal of Physical Anthropology* 11, no. 3 (1953): 361–68.

Index

Page numbers in italics refer to figures.